P9-BXY-175

Five Easy Lessons

Strategies for Successful Physics Teaching

Randall D. Knight

California Polytechnic State University, San Luis Obispo

Addison
Wesley

San Francisco Boston New York
Cape Town Hong Kong London Madrid Mexico City
Montreal Munich Paris Singapore Sydney Tokyo Toronto

Acquisitions Editor: Adam Black
Marketing Manager: Christy Lawrence
Production Coordinator: Thompson Steele
Cover Designer: Blakeley Kim
Manufacturing Buyer: Vivian McDougal

ISBN 0-8053-8702-1

Copyright ©2002 Pearson Education, Inc., publishing as Addison Wesley, 1301 San-
some St., San Francisco, CA 94111. All rights reserved. Manufactured in the United
States of America. This publication is protected by Copyright and permission should
be obtained from the publisher prior to any prohibited reproduction, storage in a
retrieval system, or transmission in any form or by any means, electronic, mechanical, photocopying, recording, or likewise. To obtain permission(s) to use material
from this work, please submit a written request to Pearson Education, Inc., Permissions Department, 1900 E. Lake Ave., Glenview, IL 60025. For information regarding permissions, call 847/486/2635.

Many of the designations used by manufacturers and sellers to distinguish their
products are claimed as trademarks. Where those designations appear in this book,
and the publisher was aware of a trademark claim, the designations have been
printed in initial caps or all caps.

45678910—CRS—05 04 03 02
www.aw.com/bc

Preface

If I don't lecture, what else can I do?

You know the story. You meet someone new at a party who, upon learning that you're a physicist, replies, "Physics! That was my *worst* subject. You must be *so* smart to understand it!" Perhaps an ego booster the first couple of dozen times you hear this, but it slowly dawned on me that this remark was saying something significant—and disagreeable!—about the subject I love. These people—intelligent, educated, articulate—*all* had a frustrating, intimidating experience with physics. This is made all the more puzzling when you find that most of these folks really are interested in science—in black holes, Schrödinger's cat, superconductors, what have you.

Must physics be like this?

I don't believe it has to. I'm convinced that we can teach physics so that people look back on it as an empowering experience, as a course where they really learned to understand the world around them, as a mind-expanding intellectual journey rather than as a forced march through 1200 pages of facts and formulas. To succeed in this endeavor, we—as teachers—need to do a much better job of understanding our primary subject: the students themselves. Who are they? What are they like? What conditions are most conducive for learning, understanding, and appreciating physics?

I've been teaching introductory physics for over 20 years. I began, as do so many young faculty, by thinking that I would be a superior teacher by virtue of my exceptionally lucid lectures, by clearly explicating the subtle points of physics to an eager audience. Alas, it soon became clear that my abilities in the classroom had little or no beneficial effect on the students. Their performance on exams showed only a rudimentary ability to manipulate equations, and they clearly had not experienced any form of intellectual awakening. I was creating the very students who, in a few years, would reply, "Physics? That was my *worst* subject!"

My first reaction was to blame the students. They were unprepared for the course, unmotivated, unwilling to do the hard but necessary work. Talking with students, however, revealed a different picture. Many of the students were bright, their mathematical skills, if not polished, were certainly adequate for the course, they were willing to work hard, and they really wanted to succeed. But they just weren't "getting it." Could it be that my instructional method was grossly mismatched to my students' instructional needs?

About this time, in the mid-1980s, I became aware of the young but growing field of physics education research. As I read, the pieces began to fit together. Research findings gave me an intellectual framework for understanding *why* my students found physics to be so difficult. And research also began to offer hints and suggestions for instructional approaches that would help students work through their difficulties.

Over the next several years, using research as a guide, I completely changed my teaching style. Instead of talking *about* physics to a passive audience, I began to use class time—a precious commodity—to have students actively engaged in *doing* physics. This is the essence of what is called *active learning* or *interactive engagement.*

The active-learning process, at first sight, appears to be woefully inefficient. And to a physicist, who already knows the material, it would be. But beginning students aren't physicists. They don't think like physicists, and they don't respond to material as physicists would. Despite the apparent inefficiency, classroom testing finds that actively engaged students are learning *more* even while they appear to be "covering" less. The students are more satisfied, and, as an added benefit, I find this way of teaching to be a more rewarding experience.

I've now spent over a decade developing a research-based curriculum and an active-learning teaching style. Along the way, I've written a calculus-based introductory textbook, *Physics: A Contemporary Perspective*, Preliminary Edition (Addison-Wesley, 1997; forthcoming in 2004 as *Physics for Scientists and Engineers: A Strategic Approach,* First Edition), that is based on physics education research and is designed to support an active-learning classroom. While I encourage you to take a look at it, *Five Easy Lessons: Strategies for Successful Physics Teaching* is quite generic, independent of the textbook you use. It is written specifically for the instructor who would like to bring fresh ideas to the teaching of introductory physics but isn't sure where to start.

Note that *Five Easy Lessons* is an instructor's guidebook, not an instructor's manual. This is not a how-to book that will prescribe how you should teach your class. Instead, this guidebook is a *resource.* It contains background information about physics education research, an extensive bibliography, and a vast number of ideas and suggestions, organized by topic, for how to bring these research findings into the classroom.

The chapter *Physics Education Research* provides an overview of 25 years of research. It won't make you an expert, but it will provide useful background information and supply the rationale for my pedagogical approach. An extensive bibliography will let you find the original papers if you want more details.

The chapter *An Active-Learning Classroom* is intended to answer the fundamental question, "If I don't lecture, what else can I do?" It describes a dozen classroom-tested methods for getting your students to be active participants rather than passive listeners. Some will suit your circumstances and interests, others won't. The point is that there are *lots* of things you can do in place of lecture, regardless of whether you teach to an intimate group of 20 or a less-than-intimate group of 500. If one of these approaches strikes your fancy, the *References and Resources* chapter will point you to the literature for more information.

Nineteen chapters then provide an extensive collection of specific suggestions—demonstrations, exercises, discussion questions, sample problems, and sample exam questions—for teaching most of the major topics of introductory physics in an active-learning style. These suggestions are research based and classroom tested.

This guidebook is based on the research findings of many individuals. I'm indebted, in particular, to Arnold Arons, Uri Ganiel, Richard Hake, Ibrahim Halloun, David Hestenes, Leonard Jossem, Priscilla Laws, John Mallinckrodt, Lillian McDermott, Edward "Joe" Redish, John Rigden, Bruce Sherwood, David Sokoloff, Ronald Thornton, Shelia Tobias, and Alan Van Heuvelen. Many early users of my textbook have provided useful suggestions. I am grateful to my colleague, Matt Moelter, for a careful review of the manuscript. And finally, I would like to thank my editors at Addison Wesley for encouraging me to publish this guidebook in a form that would be useful to all physics instructors, not just users of my textbook.

Randy Knight
California Polytechnic State University
San Luis Obispo, California

Contents

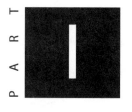

PART I

Teaching Introductory Physics

1

Introduction

*To some extent science is hard because it simply **is** hard. That is to say, the material to be learned involves a great many concepts, some of which are very counterintuitive. . . . This fact is well understood by the students, the professor, and the general public. What is not as well understood are the various ways in which this already hard subject matter is made even harder and more frustrating by the pedagogy itself.*

Eric
A physics student in Sheila Tobias' study of science education, 1990

Sheila Tobias (1990) has spent many years as an astute observer of science education. In an interesting experiment, she recruited graduate students from other disciplines to enroll in introductory physics at a large state university. They all had the academic skills needed to succeed in the course, and they offered a perspective on the course more articulate and somewhat more sophisticated than the typical freshman. Uniformly, they found the standard introductory physics course to be boring, crammed with too much material, narrowly focused on numerical manipulation and computation, and biased against any attention to the "big picture" of what physics is all about. Many of our students likely see the course in the same way as Eric.

But introductory physics need not be this way. The last 20 years have seen the advent of new knowledge from physics education research, new approaches to physics teaching, and new thoughts about the appropriate content of introductory physics. Little of this information is reflected in most of today's textbooks, which trace their lineage to the appearance of the first edition of Resnick and Halliday in 1960.

This guidebook is intended to help bring these new ideas and new techniques into practice. My primary goals are

- to move the results of physics education research into the classroom,
- to foster an active-learning classroom environment,
- to balance quantitative reasoning and problem solving with qualitative reasoning and conceptual understanding,

- to teach students better problem-solving skills, and
- to help students come away from introductory physics with improved outcomes *and* with better memories of the experience.

All of these goals can be met if you're willing to make some modest changes in the use of class time, in your use of recitations and laboratories, and in the tasks you assign to students. Textbooks and other materials that are based on physics education research will produce even larger improvements. But even if that is not an option, which it may not be if you're teaching at a large school where decisions are made by committee, there are still plenty of ideas in this guidebook for how to improve the introductory physics course—not only for your students, but also for *you*.

Desired Student Outcomes

What do we expect a student who completes introductory physics to have learned and accomplished? Are our teaching materials and methods designed to achieve these outcomes? Are our expectations reflected in what we ask of students on exams?

We all expect students to become adept at problem solving. That has long been a paramount goal of introductory physics, and there is no reduction of commitment to that goal in this guidebook. In fact, there is good evidence that the approaches suggested here—which, superficially, appear to be *less* focused on quantitative problem solving—actually *improve* students' ability to attack and solve the complex, ill-framed problems of real science and engineering, problems that differ substantially from the contrived, exactly determined problems of typical physics texts.

In addition, I will argue that we should expect students to learn to *reason* qualitatively and logically about physics phenomena, to express their knowledge in multiple forms (verbal, pictorial, and graphical as well as mathematical), to achieve a modest understanding of the conceptual structure of physics (modest, because real perfection cannot be attained in one year), and to have a decent understanding of science as a process.

Most instructors will agree that few, if any, of these goals are met in a typical course. We are all painfully aware of the many students who, when we reach charged-particle motion in electricity, seem utterly oblivious to having ever seen Newton's laws or energy conservation before. The problem-solving skills of only a handful of students ever seem to rise above random equation hunting.

Alan Van Heuvelen (1991a) has characterized the education process as an "educational transformer." As physicists, we know that maximum power is delivered from a source to a load only if their impedances are matched by a transformer. The physics instructor is a vast source of factual, conceptual, and procedural knowledge about physics, but this source is totally mismatched to the student "load." If the impedances are mismatched, knowledge is either *transmitted* (well characterized by "In one ear, out the other") or *reflected* (regurgitating what the professor wants to hear, then forgetting it immediately after the exam), but little is *absorbed*. To achieve

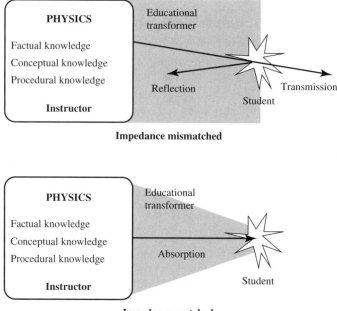

Impedance mismatched

Impedance matched

the desired student outcomes, we must use the correct educational transformer—that is, teaching materials and methods that are properly matched to an accurate assessment of the knowledge state of the students.

We must also bear in mind that there is a *maximum* possible power, or knowledge, transfer. Even an optimum transformer will not turn freshmen into physicists in a year's time. As instructors who love their subject, we all want to package insights that we've gained over many years into little nuggets that we can insert directly into our students' minds. Our good intentions not withstanding, students cannot reach a sophisticated level of understanding until they, too, wrestle with these ideas for many years. It is important to have realistic expectations of what even talented, motivated students can learn in a first course.

My goal in writing this guidebook has been to make design improvements in the educational transformer so as to improve the chances of achieving the desired student outcomes.

Five Lessons for Teachers

Physics education research does not provide a "formula" for optimal teaching. Nonetheless, research does provide general guidance as to instructional methods that are *likely* to be effective. Over the years, I've distilled this information into five basic lessons that I've tried to incorporate into my teaching. These five lessons are discussed at more length in the next chapter, but I list them here as an introduction to the route ahead.

Lesson One—Keep students actively engaged and provide rapid feedback.

Lesson Two—Focus on phenomena rather than abstractions.

Lesson Three—Deal explicitly with students' alternative conceptions.

Lesson Four—Teach and use explicit problem-solving skills and strategies.

Lesson Five—Write homework and exam problems that go beyond symbol manipulation to engage students in the qualitative and conceptual analysis of physical phenomena.

Just how to put these ideas into practice is what this guidebook is all about.

Toward an Active-Learning Environment

I've written this guidebook primarily to support an active-learning classroom environment. Active learning is described more fully in later chapters. Suffice it to say, for the moment, that much research has shown the standard lecture-format class to be ineffective as a vehicle for learning. Effective learning requires the students to be active participants in the process, not passive listeners. The chapter *An Active-Learning Classroom* will describe a number of techniques that can be used in large classes as well as small.

In an active-learning environment, the textbook is to be seen as a *resource*—a source for factual knowledge, for explication of new and difficult ideas, and for worked examples. Class time is then used to *discuss* ideas (rather than to present them), to answer questions, to clarify points of confusion, to demonstrate physical process, and to practice doing physics while the instructor is present to give feedback.

The downside, of course, is that students must *read* the text. Unfortunately, a situation has evolved in which students in science courses expect the instructor to *tell* them everything that is important. To the extent that we comply, this expectation is reinforced. Most of us, having been educated in a system where "teaching by telling" was the norm, have also come to see class as a time when the instructor primarily repeats information in the text, giving students no incentive to read it for themselves. Indeed, surveys show that many students read no more than the chapter summaries because experience has shown them that this is sufficient to pass the typical exam.

Lest we think this is the only option, contrast a typical introductory physics class with an introductory English class or an introductory history class. An English professor doesn't spend class time giving plot summaries or reading the literary works. Students are expected to have read the works, and class time is spent *discussing* them, allowing students to gain understanding of how the novel is constructed or how it is related to other literary and historical events. The history professor doesn't use class time listing names, places, and dates. Students are expected to have read the text, and class time is used to fit this information into a larger historical framework. Admittedly, some students do not read the assignment, and most of them end up doing poorly in the class. But they know what's *expected* of them, and they know they have only themselves to blame for not doing it. Why is the expectation different in science classes?

Whatever the reason, it is an expectation that is difficult to overcome. Yet it must be overcome to establish a more effective learning environment. I have found three ingredients essential to success. First, lay out your expectations *and the reasons for them* on the first day of class. In particular, be clear that you are not going to repeat the factual knowledge in the text but that they *will* be required to know it on the exams. Make students feel they are partners in a learning process, not just passive recipients of knowledge. But once is not enough—you'll have to repeat these expectations several times as the term progresses. Second, hand out a day-by-day schedule, then stick to it. Students do have busy lives, and you can't expect them to do the reading if they don't have a clear schedule of when you'll start each chapter. Third, *don't relent!* There will be classes early on where it is clear that hardly anyone has done the reading. Press on anyway with whatever demonstrations or discussion activities you had in mind, even though most of the students won't be able to answer questions. This is, admittedly, hard to do with an unprepared and unresponsive audience. But if you give in and start lecturing, then the game's up! They'll know you were just bluffing about not telling them what they need to know. Better just to dismiss class early and tell them to spend the time reading than for you to start a formal lecture!

If *you* can last through this painful transition period, it doesn't take most of the class long to begin completing the reading assignments. One device that can speed the process is to give short *reading quizzes* at the beginning of class. This can be every class, at the beginning of each new chapter, or on a random, unannounced basis. Multiple choice questions work well if grading time is a concern. A sample multiple-choice reading quiz question is:

In an inelastic collision
a. energy is conserved.
b. momentum is conserved.
c. both energy and momentum are conserved.
d. neither energy nor momentum are conserved.
e. Inelastic collisions were not covered in this reading assignment.

Obviously, you do occasionally have to give questions where e is the right answer for students to take this as a serious option. Sample reading quiz questions are given in the subject descriptions later in this guidebook.

Another option, and one I often use, is to give a *vocabulary quiz* at the beginning of class on days when a new chapter is started. We tend to overlook terminology in physics, as we focus on concepts and mathematics, but many students simply don't know the most basic definitions of the terms you will be using in class. What will they gain from class discussion without knowing the terms? This is information that should be acquired on an initial reading of each chapter. All textbooks highlight terms as they are introduced, and the quiz should be based on these. Two or three terms in a five-minute quiz is sufficient to get most students to read the chapter.

When grading a reading quiz, you should not look for mastery of the material but simply for evidence that the student has read the chapter. On vocabulary quizzes, I can judge this by skimming the answers, allowing me to grade a set of 40 quizzes in less than 20 minutes.

Does It Work?

Do active learning and the many activities I suggest in this guidebook make any difference? Does a stronger focus on conceptual understanding and quantitative reasoning come at the expense of problem-solving skills? To find out, we conducted a detailed comparison of active learning with lecture-mode instruction in my own department at Cal Poly.

In 1998 we opened a "studio classroom" as an alternative to our regular calculus-based introductory physics. The goals of the studio have been to

- Eliminate the boundary between lecture and laboratory, and
- Promote an active-learning environment.

Class meets in the studio three days a week for two hours. The use of class time is based on many of the ideas and suggestions in this guidebook. Students carry out experiments and activities taken from Thornton and Sokoloff's *RealTime Physics* and Law's *Workshop Physics.* They engage in think/pair/share discussions, and they practice problem solving using Van Heuvelen–like worksheets and a structured approach. All of these are described more thoroughly in the chapter "An Active-Learning Classroom." There is little or no formal lecture. The studio class uses my textbook, *Physics: A Contemporary Perspective* (Addison Wesley, 1997), which is based on physics education research and is designed to support an active-learning environment.

In fall 1998, we carried out a direct comparison of three studio sections with three of our regular lecture sections of the calculus-based intro course. To eliminate any "instructor factor," the studio sections and the lecture sections were taught by the same three instructors. They ranged from a relatively new Ph.D. with only one year of teaching experience to a senior faculty member who retired at the end of the year. None of the three had any particular expertise with physics education research or innovative pedagogies, but all were willing to try something new. (I was not one of the three!)

We collected data about the students from the registrar, including class standing, SAT scores, high school GPA, and previous physics classes. We administered the Force Concept Inventory (FCI) of Hestenes, Wells, and Swackhamer (1992a) to all students as a pretest on the first day of class. The FCI tests students' conceptual understanding of Newtonian mechanics. We found no pretest differences between the students in the lecture classes and those in the studio.

Each class then went its own way until the end of the quarter, when we gave all students a common final exam and the FCI posttest. We also tabulated student responses to the "Overall instructor rating" question from our standard course evaluation. The final exam—typical, quantitative exam problems—was written by a faculty member who was not teaching the course. To assure fairness, students were identified on the exam only by a code number, the exams from all students in all six sections were randomly mixed, and a single instructor then graded one problem in its entirety.

Here's what we found:

	Lecture	**Studio**	**Difference significant at**
FCI posttest score (%)	68.0 ± 2.1	79.0 ± 1.4	>99.9% level
Final exam score (out of 150)	86.4 ± 2.0	93.9 ± 2.7	98% level
Overall instructor rating (1-to-5 scale)	3.51 ± 0.09	3.92 ± 0.07	>99.9% level

Not only did the studio classes do significantly better on both qualitative (FCI) and quantitative (final exam) measures, but the same three instructors received significantly higher student evaluations in the studio than in their lecture classes.

A closer look at the results finds that students' performance on the standard physics problems of the final exam was highly correlated with their conceptual understanding, as measured by the FCI posttest. The figure shows the results for all 225 students in the study. (The FCI scores are out of 30 points.) The correlation coefficient is 0.59.

David Hestenes, co-author of the FCI, has defined the *Newtonian threshold* as an FCI score >60%. Students below the Newtonian threshold continue to hold significant misconceptions about mechanics. *Newtonian mastery* is a score ≥90%. A very

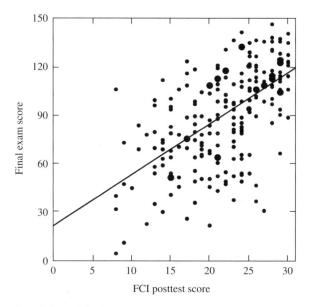

Correlation of final exam scores (quantitative) with FCI posttest scores (qualitative).

interesting pattern emerges if we look at students' final exam scores (out of 150) based on their FCI classification:

	Lecture		Studio	
	% of students	**final exam (out of 150)**	**% of students**	**final exam (out of 150)**
FCI posttest <19 out of 30	43	71 ± 4	14	66 ± 5
FCI posttest 19–26 (over Newtonian threshold)	40	89 ± 4	57	91 ± 3
FCI posttest 27–30 (Newtonian mastery)	17	117 ± 4	29	112 ± 4

Within each range of FCI posttest scores, there are *no differences* between the lecture students and the studio students. The studio's higher average on the final exam comes about entirely by having a much higher fraction of students crossing the Newtonian threshold (86% above threshold in studio as compared to 57% in the lecture sections).

So the answer is "yes," active learning *does* make a difference. Not only do students acquire better conceptual and qualitative understanding, which was to be expected, but their quantitative problem-solving skills also improve. The strong correlation between FCI scores and final exam scores carries an important implication: *Problem-solving skills are best improved by giving students a stronger conceptual foundation from which to work.*

This conclusion is bolstered by a large-scale study carried out by Richard Hake (1998). Hake collected data from high schools and colleges across the country on how students perform on the force concept inventory. He measures the effectiveness of instruction by the *gain G* on the FCI, defined as

$$G = \frac{\text{posttest average \% } - \text{ pretest average \%}}{100 \; - \text{ pretest average \%}}$$

$$= \text{fraction of the maximum possible gain.}$$

Gains range from 0 (posttest average = pretest average, no learning) to 1 (posttest average = 100, perfect learning). Our studio sections had an average gain $G = 0.60$.

With over 6000 students in his sample, Hake finds that conventional lecture classes have gains $G = 0.22 \pm 0.05$—that is, the class average goes up only 22% of what is possible. This result appears to be true regardless of the pretest score and independent of the instructor.

By contrast, Hake finds that active-learning classes have $G = 0.52 \pm 0.10$, more than twice the gain of conventional instruction. This result has high statistical significance. It is matched by data showing that the quantitative problem-solving skills of students in these classes increase by a comparable amount.

Both Hake's large-scale study and our detailed comparison lead to the same conclusion: Conceptual understanding, as measured by the FCI, seems to be a *prerequisite* to good problem-solving ability.

2

Teaching
Introductory Physics

It is the basic premise of the vast majority of introductory physics courses taught at the present time that, if one takes a huge breadth of subject matter and passes it before the students at sufficiently high velocity, the Lorentz contraction will shorten it to the point at which it drops into the hole which is the student mind. In such courses, final grades, which are invariably adjusted to allow "passing" of a reasonable fraction of students, cannot possibly be an index of the kind of intellectual development and achievement to which most of us render lip service.

Arnold Arons, 1979

The introductory physics experience can range from a pre-planned package tour, with only limited scope for variation by individual instructors, to a fully independent trek. At large universities with multiple sections, instructors usually have to follow a predetermined syllabus, use a text selected by committee, and perhaps even use exams written by others. Instructors at smaller schools have much more freedom to tailor the course to the interests of their students. Fortunately, even the most structured course leaves a fair bit of room for individual initiative.

Things to Consider

Audience: It would be ideal to teach introductory physics just to science majors, or even just to physics majors. But few of us have such a luxury. For the calculus-based intro course, the audience most of us teach is overwhelmingly engineering students. The algebra-based introductory course has a more varied audience, although usually dominated by life science and premedical students. The needs of our audience place some constraints on the course.

Nearly all engineering students will take three big courses that introduce them to the fundamentals of engineering. These are

- engineering mechanics,
- thermodynamics, and
- electrical circuits.

An issue at some schools is an assumption that the engineering college wants us to teach a very specific list of topics to prepare their students for these courses. I no longer believe that this is the case. In talking with engineering faculty at several schools about their "expectations" of physics, their initial response is, indeed, usually a long list of topics. But if pressed, they'll soon admit that a knowledge of specific topics isn't their real concern. These fundamental engineering courses start over from the basics, and they presume very little prior knowledge. What the engineering college *really* wants is students who have developed some reasonable problem-solving skills, who can think logically and coherently about technical issues, and who can apply *basic concepts* of physics to engineering situations.

We'll meet the needs and expectations of our engineering audience if they leave introductory physics with a *basic* working knowledge of

- force and motion,
- energy, energy transformations, and conservation laws, and
- charge, field, and potential.

Not surprisingly, this doesn't really differ from the expectations we hold for physics majors. If you haven't had a heart-to-heart talk with your engineering college in a while, you might find it quite an enlightening experience.

The audience needs in the algebra-based course are not so clearly defined. However, life science students are generally going to find energy, heat, and thermal physics to be more relevant than Newtonian mechanics. Unfortunately, algebra-based textbooks have tended to mimic the calculus-based textbooks for engineers, putting an undue emphasis on mechanics and rarely giving a coherent presentation of thermal physics. You can compensate for this to some extent with a judiciously designed syllabus that shortens and omits topics in Newtonian mechanics in order to give more emphasis to energy and thermodynamics.

Content: The typical introductory physics textbook has, over the years, become encyclopedic in scope. It's simply not possible to do justice to all the topics in a typical text unless you have *at least* three semesters, not counting any follow-up modern physics course. An attempt to "cover" the book will have students attempting to drink from a fire hose. It's simply a torrent of information rushing past. Is this the memory we want students to have of physics?

There's no reason not to omit or significantly shorten a number of topics to produce a more focused, coherent course. More and more instructors are doing so, opting for "depth over breadth," not to mention a pace that allows a bit of breathing room. Even instructors who have to teach from a common syllabus can get away with dropping or shortening some topics that are never used in subsequent terms. But what to omit?

Chapter-length subjects that are candidates for omission are

- statics,
- solid body motion and rotational dynamics,
- Newton's law of gravity,
- fluid dynamics,
- geometric optics,
- special relativity, and
- AC circuits.

Although I like these subjects as well as anyone, they are not directly relevant to the main thrust of introductory physics and no subsequent topics depend on these. The omission of some or all of these allows students to focus better on more essential topics.

If students have acquired a functional understanding of Newtonian physics as applied to particles, a later extension of that knowledge to solid bodies should be straightforward. Students do learn about center of mass and moments of inertia in calculus, so they are not entirely bereft of this knowledge if we don't include it in physics. It is important to include an introduction to angular momentum in your coverage of circular motion, but such an introduction need not invoke the full vector cross product definition.

Special relativity is a beautiful subject, but the conceptual difficulties with it are profound. The cursory treatment in standard texts just adds more formulas to the students' repertoire without producing any significant awareness of what the subject is about. Although the ideas of relativity are accessible at the freshman level, the few innovative curricula that have taught relativity successfully devote at least three weeks to it. This appears to be the minimum time needed to establish the conceptual framework and to allow students to wrestle with the difficult and counterintuitive ideas.

Examples of more specific topics that could be omitted from within chapters include

- macroscopic phenomena: Thermal expansion, heat transfer, phase changes, and many of the details about heat engines,
- waves: The Doppler effect, decibels, musical acoustics, intensities of interference and diffraction patterns, and
- electricity and magnetism: Gauss's and Ampere's laws, Maxwell's equations, circuits other than single loops, dielectrics, magnetic materials, and instrumentation.

I'm not suggesting that you omit *all* of these, since many of these topics are fun and add spice to the course, but that you should not feel a moral duty to cover them all. It does the students little good to "see" these if they go by so fast as to prevent any real learning.

Electricity and magnetism have come to dominate the second half of traditional textbooks, even though most engineering students are in majors other than electrical engineering. Any attempt to reduce the encyclopedic scope of introductory physics has to find some way to separate the essential from the nonessential topics in electricity and magnetism. Part of the decision can be made on mathematical grounds, as discussed in the next section, and on the basis of physics education research, as described later in this guidebook. On purely pragmatic grounds, the large majority of engineers and scientists only need to know about *static* fields. Even the majority of electrical engineers, as circuit designers, will deal with time dependence only on a phenomenological, circuit-level basis. Only physicists themselves and the minority of electrical engineers who work with antennas or waveguides will ever need to use time-dependent fields at the level of Maxwell's equations, and these students will all take an intensive fields-and-waves course at a later time. A thorough grounding in the *fundamentals* of electricity and magnetism—in particular, acquiring a working understanding of the abstract and difficult concepts of *field* and *potential*—will be more than adequate to prepare students for more advanced, upper-division courses.

The goal of reducing the encyclopedic scope of introductory physics has led to an unfortunate trend to reduce or even eliminate the treatment of thermodynamics. While the motive may be sincere, the solution is not particularly cognizant of the needs of our primary audience. The rationale for skipping on thermodynamics is that "this is covered in chemistry." Well, not quite. It's true that chemistry teaches temperature scales, gas laws, and a bit of heat and calorimetry, but chemistry provides no conceptual framework for thinking about thermodynamics as the science of energy and energy transformations. Of the students who have taken chemistry before physics, I've seen very few who can work a basic calorimetry problem, and even then it's a matter of manipulating memorized formulas rather than any realization that calorimetry is about energy and energy conservation.

It's true that the thermodynamics chapters in many textbooks have a collection of miscellaneous topics that could be omitted, such as thermal expansion. But the basic *concepts* of thermodynamics—knowing what energy is and how it is changed from one form to another—are essential ingredients of physics and important knowledge for our audience.

Use of Mathematics: Schools differ as to whether students in the calculus-based physics course are starting calculus concurrently or have already completed a term of calculus. If calculus is concurrent, students will—at best—be learning of derivatives in calculus just as they arrive at the idea in kinematics. However, any integration used in kinematics will be *prior* to students having seen integrals in calculus. It will be essential to allow adequate time to teach the mathematics along with the physics. Although we use only "simple" derivatives and integrals early in the course, they hardly appear simple to students facing these ideas for the first time.

Even when students have some prior experience with calculus, many physics textbooks soon outpace the students' mathematical knowledge. Partial derivatives enter with waves, and vector integrals rear their heads with Gauss's law, if not sooner. An examination of calculus texts and discussions with mathematics faculty have convinced me that most students will *not* have seen these topics in math before we typically use them in physics. This is certainly true for students starting calculus concurrently with physics. It is likely true in the case of vector integrals even for students with an early start, because this topic is often not reached until late in the third semester of calculus.

Although students are slowly becoming adept at the *mechanics* of taking a derivative or doing an integral, few yet have any functional understanding of when or where it is appropriate to do so. Most have a long way to go before calculus becomes a useful tool of analysis. Physics can help them gain that skill, but we have to provide instruction in how to "think with calculus" rather than assume they know just because they're taking calculus. Rates and summations are a natural part of physics, so simple calculus fits nicely in an introductory course. But for students struggling to grasp many new and difficult concepts, too high a level of mathematics detracts from, rather than aids, the *physics* we want them to learn. There is ample time in upper division courses for a more formal and rigorous treatment. It's counterproductive to burden students with unfamiliar and frightening mathematical baggage during their first exposure to the subject. Sections of the textbook that utilize partial derivatives or vector integrals are good candidates for omission.

A few words about vectors: Physics instructors tend to assume that students are familiar with vectors from high school or from math courses. Surveys (Knight, 1995) have found that this is not the case. In a state university with a somewhat above average student body, only one-third of the students entering physics had a working knowledge of vector arithmetic (e.g., vector addition and subtraction done graphically and by components). Another one-third had a partial knowledge of vectors but still made significant errors on even simple vector-addition questions, and the final one-third had no usable knowledge of vectors at all. Only a tiny fraction of students were familiar with dot products, and not a single student in the sample could calculate a cross product. Further, calculus courses don't introduce vectors until typically the third semester, so you cannot assume that students are learning about vectors elsewhere.

The lesson for physics instructors is clear: Students need *explicit* instruction and practice with vector components and vector arithmetic. These portions of the text cannot be passed over quickly as a "review." Careful examination of students' work shows that the failure to solve problems using Newton's laws, even if the conceptual barriers are overcome and the forces identified correctly, often stems from a failure with the vector quantities and the vector mathematics. That many students have failed to understand vectors becomes clear months later when they are asked to calculate electric forces and fields via superposition when charges are, for example, arranged on the corners of a rectangle. Few students succeed at this task without another review, at that time, of vector addition.

Quantitative/Qualitative Balance: Most traditional texts are heavily weighted toward a quantitative and analytical approach to physics. There is a tacit assumption that students will "pick up" the necessary conceptual ideas through extensive practice with problem solving, and students are rarely or never asked to form a qualitative argument to *explain* a physical process. In recent years, physics education research has shown convincingly that students can excel at finding and manipulating formulas while still holding vast misconceptions about physics concepts and principles. (The background for this statement is given in the chapter on physics education research.) Physical understanding and physical intuition are *not* an automatic by-product of working lots of quantitative problems.

To a physicist, reasoning with and about principles *is* "doing the physics." Mathematics allows precision, but detailed mathematics is a latecomer to the analysis process—simply one tool among many needed to understand a situation. To students, however, physics is synonymous with mathematics. They see no distinction between the two. This attitude stems in part from the large number of "physics problems" that appear in calculus textbooks, problems that ignore any conceptual ideas and focus purely on the mathematics. But an even larger factor is the way we usually teach science in high school and college, with textbooks, homework problems, and exams that continuously reinforce a message that physics is done by selecting and manipulating equations.

As a consequence, the majority of students harbor serious misconceptions and misunderstandings of physical principles. They do not see how the information they are learning in physics class has any bearing on understanding the physical phenomena they see and experience around them every day. This situation is immediately brought to light if you ask students on an exam to give a qualitative *explanation* of some physical process. A significant number of students don't know what it means "to explain," and simply restate the question in different terms. Others try to give an "explanation" by stating an equation in words, such as "Because the kinetic energy equals one-half the mass times the velocity squared." (After all, they were told to be qualitative!) Of those that even attempt an explanation, very few can state a relevant physical principle, then reason logically to a correct conclusion. Most see only superficial aspects of a situation, such as "Because it's moving in a circle," and are satisfied with that as an explanation.

Without conceptual understanding and the ability to *reason* with it, students cannot pass beyond the rote memorization stage and cannot proceed to develop the skills needed to solve open-ended, indeterminate, real-world problems. To help students develop more sophisticated problem-solving skills, the introductory course needs to establish a better quantitative/qualitative balance than is usually found in the textbook. That is, quantitative analysis and problem solving need to be balanced by an emphasis on conceptual understanding and qualitative reasoning. In particular, the class needs to

- devote extra attention to discussing, justifying, and clarifying the concepts and principles of physics, with less emphasis on mathematical derivations. Many ideas for doing this are contained in later chapters of this guidebook.
- give extended examples of *qualitative* reasoning and explaining.
- develop explicit problem-solving strategies that emphasize *qualitative* analysis steps to describe and clarify the situation, with mathematics entering only at the end.
- assign homework problems that explicitly ask for analysis and interpretation.

Needless to say, an emphasis on conceptual understanding and qualitative reasoning does take time. Some instructors fear that time spent on conceptual issues and class discussions, at the expense of practice on quantitative analysis, will hurt students' problem-solving abilities. The evidence, however, is just the opposite. Students in classes that emphasize active learning and conceptual understanding score *higher* on the Mechanics Baseline Test, a test of basic *quantitative* problem solving, than do students in traditional classes (Hake, 1998). The conclusion seems to be that students often do poorly on quantitative problems not because their math skills are weak, but because they don't understand the situation and don't know *what* to compute.

The evidence is now strong that the route to learning better problem-solving skills, for any problems more significant than single-equation plug-and-chug, is through emphasis on conceptual understanding and qualitative analysis of the situation. If we want our students to acquire these broader skills, however, we must devote class time to qualitative discussions and demonstrations, assign homework that involves qualitative reasoning, and place qualitative questions on exams. The bottom line for students, as always, is "Will it be on the test?"

From the Concrete to the Abstract: Arnold Arons, a pioneer in the physics education research community, has long emphasized the need for students to be able to answer the questions "How do we know . . . ?" and "Why do we believe . . . ?" The ability to answer such questions shows true learning and understanding. Physics is about understanding and explaining physical phenomena, yet you would hardly know that from an examination of most introductory physics textbooks. Such texts tend to enunciate theories and principles, then derive consequences—working from the abstract to the concrete.

Physics education research has found that students learn best when they can connect a theory and the underlying phenomena/experiments. To help establish this connection, start with *observations* about the real world (concrete experience), often observations that are puzzling or at odds with "common sense," then gradually consider the concepts and principles needed by a *theory* (abstract formalism) that will make sense of the observations. Not only does this "from the concrete to the abstract" approach better illustrate how science operates, it better matches how most students learn. Increasingly abstract and subtle concepts are firmly grounded on real-world observations, rather than presented as accomplished facts.

Then, once the pieces of a theory are in place, you—or, in some cases, the students themselves through homework problems—need to go back to the original phenomena and observations to ask, "Does this theory explain the observations?" Not only does this procedure provide closure, it is an excellent opportunity to practice conceptual and qualitative reasoning.

More specific suggestions will be given later, but it is worth mentioning the role of "from the concrete to the abstract" in teaching electricity and magnetism. Unlike mechanics, where students have spent nearly twenty years observing and participating in dynamics, most students have essentially *no* familiarity with even the most basic phenomena of electricity and magnetism. Their experience is limited to not much more than turning on lights, changing batteries, and using refrigerator magnets—and even with these, as will be shown, students have major misconceptions. Traditional texts make a few casual remarks about charges and magnets, then proceed to build a highly abstract and mathematical theory of fields and potentials. To students, this theory is a mathematical exercise with no connection to reality.

All texts introduce magnetism by reminding students that they are familiar with magnets sticking to metal surfaces—permanent magnets. After this "motivation," the text develops a theory of *electro*magnetism—a classic case of bait-and-switch. Little or no effort is made to connect electromagnetism to permanent magnets—it's assumed obvious that they are the same force—and few textbooks ever allow students to finally answer the most basic question: "*How* does the magnet stick to the refrigerator?" That is, the concrete phenomena are not successfully explained by the theory that has been developed.

Because "teaching by telling" has limited effectiveness, this guidebook urges instructors to make extensive use of demonstrations and laboratories to illustrate the phenomena. Then, as the theory is developed, you need to refer back to the underlying phenomena and to ask if the theory is helping to provide an explanation. At the end, the students *are* then in a position to answer questions such as "Why do we believe that permanent magnetism and electromagnetism are really the same thing?" and "Just how is it that the refrigerator magnet sticks to the door?"

Physics in Context: Physics is not simply a collection of facts and formulas, although students want to see it that way. Science, including physics, is a mode of human thought and a human activity. As such, it has a history and a context. One goal of an introductory course should be for students to learn something about what science is and how it functions. This goal is met, in part, by the frequent reference to the experimental basis of our knowledge. In addition, short digressions and stories about the history of the ideas can help bring the subject to life. It's occasionally useful even to discuss "wrong" theories as examples of how ideas that were perfectly reasonable *at the time,* maybe even "obvious," were later found to be inadequate. This is not a history of science course, and instructors should not spend too much time on historical topics, but you do want students to recognize that science evolves because of real people struggling with real issues, and that our present understanding wasn't handed down on stone tablets.

From a pedagogical perspective, historical issues provide a good entry point into a class discussion about difficult conceptual issues. Many of the common-sense alternative conceptions held by students are similar to ancient or medieval concepts— such as an impetus theory of motion or a caloric theory of heat. A discussion of how "obvious" these ideas are can be followed by a lecture demonstration that clearly contradicts the prediction of the early theory. Now you've set the stage for true learning to occur as students struggle to reconcile this paradox.

Problem-Solving Strategies and Homework Problems: We want students to become good problem solvers, but the typical approach is well characterized by David Hestenes (1987):

Problem solving is traditionally taught by providing examples for the students to emulate. The drawback is the students tend to emulate what they see. What they see, typically, is that after a little talk some formulas are written down from which a numerical solution is obtained by manipulation and substitution. The teacher or textbook may say that it's important to do such things as "draw a figure," but they seldom say why, and the student can see that the answer comes from a formula, so why bother with a diagram? Little wonder that students come to see selection of the correct formula as the key to problem solving. . . . This strategy is especially effective for homework problems when the necessary formulas can be found in the chapter from which the problems are assigned. Dedicated students learn this strategy well by working a lot of assigned problems, for they know that "practice makes perfect!"

Hestenes goes on to say that "Students are not easily weaned from a formula-centered problem-solving strategy that has been successful in the past."

If we want students to learn better and more effective problem-solving skills, several ingredients are necessary. First, they must receive explicit instruction in a better strategy than mere formula seeking. Second, the instructor (and, ideally, the textbook) have to *use* the strategy—including all the small steps that we, as instructors, tend to skip over as being obvious. Third, students must have an opportunity to practice the steps of the strategy in isolation, not just during full-blown problem solving. Fourth, homework—and exam!—problems must be such as to discourage a simple formula-seeking approach.

Many of the steps in a problem-solving strategy are taken for granted by experienced instructors but are especially difficult for students. Much of the difficulty arises in not knowing how to "translate" a word problem into an appropriate equation, or set of equations, that are based on relevant principles. Without a systematic approach, students resort to searching for a formula that seems to have the right quantities—hopefully with only one unknown that they can then solve for!

I am a strong advocate of the worksheet approach to problem solving developed by Alan Van Heuvelen (1991a, 1991b) as part of his Overview, Case Study teaching method. This is described more in the chapter on active learning. The worksheets provide a template for students, making sure that they follow the steps of the strategy. My experience is that many students chafe at being required to use the worksheets, with comments that it's "too rigid" or it "stifles my creativity." But what they are really saying, as becomes clear if you make the worksheets optional, is that it prevents them from simply searching for a formula to plug numbers into. If you model a problem-solving strategy in class, carefully showing and explaining *all* the steps, and insist that students follow this strategy and solve problems on worksheets, most students eventually begin to realize that they're learning an important skill. My end-of-term evaluations often contain statements such as, "I never knew there is a *method* to solving problems! Why didn't they teach me this in my other classes?"

Problem-solving strategies break the words-to-symbols translation problem into discrete steps that are easily followed. Even so, students need to practice the steps of the translation *independently* of doing calculations. You can give students a "problem" but, rather than solving it, have them merely draw an appropriate figure and coordinate system, or simply identify the relevant forces, or simply draw the free body diagram. After working on the individual steps of the procedure in isolation— and these exercises make great classroom activities—students are in a much better position to approach full problems at the end of the chapter.

Instructors often assign large numbers of problems because, as Hestenes noted, "practice makes perfect." Unfortunately, students who lack good problem-solving skills are mostly practicing bad habits and not becoming better problem solvers. A few well-written problems studied carefully—with feedback and, sometimes, in-class discussion—are a far better learning experience than many problems practiced blindly.

Exams: Ultimately, we do have to give our students exams to assess how they're doing and what they've learned. As students know all too well, you're not serious about a topic unless it's on the exam. If your exams are all straightforward numerical computations, then no amount of class time spent on conceptual understanding, problem-solving strategies, or other issues will dissuade students from an equation-memorization, formula-seeking strategy.

If your goal is for students to *reason,* and to have conceptual understanding as well as calculational skills, then exams must test for this—and students must be told up front and repeatedly what kinds of questions will be on exams. My exams are a roughly 50-50 mix of qualitative and quantitative problems. Qualitative questions range from simple exercises (draw a free-body diagram or interpret a wave function) to graphical interpretation (draw an acceleration graph that matches a given velocity graph) to short-answer questions to more extended qualitative analyses ("Why does such and such happen? Give a an *explanation* based on physical concepts and principles.").

My quantitative exam problems often start with a simple calculation, so that all students get off to a good start, but then focus more on questions that don't tell the student what to calculate. These can't be too extended or difficult in an exam setting, but even a simple problem that can't be solved by immediate application of an equation will quickly reveal how much a student has learned.

In this day of hand calculators with vast alphanumeric memory, and even built-in equation libraries, there's little point in completely closed-book exams. For one thing, it discriminates against students who can't afford fancy calculators. More important, it conveys a hidden message that success will go to the students who memorize the most formulas. If we want students to believe we're testing them on their ability to do physics, not on their ability to find and use equations, then there's no reason not to relieve the memory pressure by making equations available. I tend to oscillate between two approaches, without a clear sense of either being superior. In some terms I allow students to prepare and bring a single 3×5 card with whatever information they want. This is easily sufficient to write down all pertinent equations they might need—and many others!—and it does force them to organize their knowledge, which is a good learning opportunity. In other terms, I prepare a short page of equations—with no labels or names—and guarantee that all exam problems can be worked from these. The students get a copy of the page a few days before the exam, so they study using it, but they can't bring that copy to the exam. A fresh copy is handed out with the exam.

Sample exam questions are given later with the discussions of specific topics.

Laboratories and Recitations

Physics instructors hold a wide variety of views about the purposes of the introductory physics laboratory. Some see it as an opportunity to "verify" the theory learned in class, others see it as an opportunity to learn about measurements and data analysis, and yet others look upon it as an opportunity for students to "discover" principles of physics on their own. There is a fairly extensive literature on the usefulness and effectiveness of laboratories. Sadly, the overall conclusion is that *standard* laboratory experiences produce little or no measurable benefit!

Part of the reason seems to stem from the fact that standard do-a-measurement-and-analyze-it experiments carry with them a tacit assumption that students think more or less like physicists, that they understand the role of measurements, and that they realize how conclusions are drawn from data. Because few students really share these insights, most are going through the motions of the laboratory but gaining little from the experience.

I noted above how little experience most students have with the basic phenomena of physics. If students are unfamiliar with the properties of charges, if they think—as many do—that current is "used up" in a circuit and that a battery is a constant current source rather than a voltage source, then a standard experiment to "verify" Ohm's law by using voltmeters and ammeters is conceptually over their heads.

However, laboratories that explore at length the properties of charged objects and the differences between insulators and conductors can make a real contribution to a student's familiarity with the phenomena of electricity. Similarly, the use of simple battery-and-bulb circuits to explore the basic properties of currents and batteries quickly exposes misconceptions that most students hold about electricity and gives them an opportunity to develop scientifically correct conceptions. Simple experiments with magnets reveal that permanent magnets don't stick to all kinds of metal, and that if one end of a bar magnet attracts a steel sphere—then so does the other end! This is a surprising discovery to many students, who confidently predict that the opposite end of the bar magnet will *repel* the steel sphere.

I recommend—at least early in each part of the course—simple *experiential* labs in which students predict, observe, and explain rather than measure. The purpose of such labs is to acquire basic familiarity with the phenomena that the theories of the textbook/lecture are intended to explain. Such labs also are very effective at confronting the alternative conceptions many students hold and nudging them toward acceptance of scientifically correct conceptions. Only after they have acquired basic familiarity and roughly correct concepts do more traditional measurement-oriented labs begin to make sense.

Some instructors think that such labs are too low level—high school, or even junior high school science. I agree that such experiences could be, and probably should be, provided at a much earlier age. Nonetheless, our students do come to us without these experiences, with little awareness of physical phenomena, and holding many misconceptions. Until we deal with this situation—and the laboratory is an ideal place to do so—students will not make much progress toward understanding the formal theories of physics.

The *RealTime Physics* labs developed by Sokoloff, Thornton, and Laws (1999) are highly effective at allowing students to learn concepts of force and motion. These are computer-based laboratories that use a sonic ranger and force probes to acquire data in real time. The experiments themselves are focused on establishing *connections* between force and motion rather than on detailed measurements or analysis. Effective use of these labs requires the instructor to interact closely with the students, moving through the room to see that they are getting decent data (some students will happily and obliviously record garbage if the probes aren't working right), stopping to ask questions to make sure they're interpreting the data correctly, and making suggestions about techniques. These are not sit-back-while-the-students-do-measurements labs!

Later in the course there are fewer commercially available labs that have proven to be effective. I have developed a semester's worth of experiential labs on charges and circuits that I use with my own classes. Two weeks are spent exploring the properties of charges, following a procedure similar to that of the first chapter of Chabay and Sherwood's *Electric and Magnetic Interactions* (1999). These experiments are started in lab a few weeks *before* starting electricity in the textbook / lecture so that students already have some familiarity with charging phenomena. This is followed by ten weeks of circuits labs, starting with simple battery and bulbs, developing operational ideas of current, voltage, and resistance through a series of experiments modeled on the description of Evans (1978), then learning more traditional resistance and capacitance circuits and culminating with learning to use an oscilloscope to measure RC circuits. These labs have proven to be popular with students, they give students a strong sense of self confidence that they can learn useful information, and they significantly improve students' conceptual understanding of electricity and circuits. Steinberg's *CASTLE Project* (available from PASCO) is somewhat similar and accomplishes many of the same goals.

Regardless of whether you use commercially supplied equipment or experiments designed and built-in house, it is definitely worth modifying the student activities to place more emphasis on simply *observing,* on *explaining* observations, and on *predicting* the outcome of an experiment. For example, if students are measuring the speed of a cart at a point on an air track, ask "If the cart's mass is doubled, will the speed at that point increase, decrease, or remain the same?" Have them make a prediction, *justify* their prediction in writing, and only then do the experiment to find the outcome. If the outcome differs from their prediction, then they have to reconcile the difference and write up a reason. Extensive discussion with lab partners and neighbors is encouraged during these activities.

Recitation sections, if your course has them, are a golden opportunity for active-learning experiences. Recitations tend to be dreary affairs, either with instructors working away at the board in front of a restless audience or with students called up to write **the equation** on the board and show how they manipulated it to get the answer. Attempts to question the students about their reasoning meet with such dismal results that most instructors give up after a few tries—aversive conditioning at its finest!

It need not be this way. Recitation sections are an excellent opportunity to have students work in pairs or small groups on either homework problems, special questions that you bring in, or prewritten exercises. Especially for large lecture classes, where group activities are difficult, recitations offer the best location for active-learning activities. As with lecture, though, you must establish up front how the recitations will function and not yield to demands that *you* work the problems while the students just watch.

McDermott's tutorials (1998) and Heller's problem solving groups (1992a, 1992b), both discussed in the chapter on active learning, are specifically designed for use in recitations, and both have proven to be extremely effective. Other good recitation activities are the exercises in the *Student Workbook* that goes with my textbook.

Regardless of the task at hand, it is important that the class—not you or the TA—work collectively on an answer. Because many questions about homework problems are of the form "I don't know what equation to use," it's best to direct the class back to the issue of "What principles are relevant?" and let them work through the steps of a problem-solving strategy. When using tutorials or workbook exercises, which are designed to elicit answers based on common alternative conceptions, let groups report their answers *to each other* and then reconcile any differences. With rare exceptions, they'll be able to judge the correct answer without your intervention other than in your role as a facilitator.

An issue at many schools is that lab and recitation sections are taught by graduate teaching assistants. For these approaches to work, the graduate TAs must go through an initial training period about the goals of the lab/recitation and what their role is. Several large universities that have switched to active-learning recitations and experiential laboratories report that their graduate students are hesitant at first, but quickly become enthusiastic supporters—more so, in many cases, than the faculty! The first year is the most difficult, when it is new for everyone, but by the second year you'll have experienced TAs who can help train new TAs.

3

Physics Education Research

It's not what you don't know that hurts you. It's what you know that ain't so!
Mark Twain

Just who are the students who inhabit our classrooms? What knowledge do they bring with them? What motivates them? How do they learn?

It's a natural human tendency to think back to our own undergraduate experiences, the difficulties we faced and the methods we followed, and to ascribe similar characteristics to our students. In other words, to think of students as younger versions of ourselves. Unfortunately, demographics suggest that very few students are much like us, that we—today's physics instructors—are anomalies and not the norm.

Although the statistics vary widely among colleges and universities, the national average is that only about 1 student in 30 in a calculus-based introductory physics class will major in physics. Perhaps 1 in 100 will pursue some form of graduate education in physics, and less than 1 in 300 will eventually acquire a Ph.D. in physics.

The fraction of students who are "like us" is vanishingly small. To extrapolate from our experiences to college students as a whole is a risky business. Fortunately, there's little need to make such a leap. Physics education research provides a detailed picture of what "real" students are like and how they learn. This chapter will provide an overview of what we know from physics education research and what the results imply for physics teaching.

What Is Physics Education Research?

Over the last twenty-five years, a growing number of physicists and psychologists have been studying just how it is that students learn physics. This research into physics education is an interdisciplinary blend of physics, psychology, and cognitive science. This field of research was pioneered in the United States by Arnold Arons (see especially Arons, 1997) and Lillian McDermott (see especially McDermott, 1984 and 1991) at the University of Washington and by Fred Reif (see Reif and Heller, 1982) at the University of California, Berkeley. A recent resource letter by McDermott and Redish (1999) gives a detailed bibliography of physics education research.

Physics education research has had two major thrusts:

- studying the *concepts* that students hold about the physical world and how those concepts are altered as a result of various methods of instruction, and
- studying the *problem-solving* techniques and strategies of students.

This section of the guidebook will summarize what has been learned by physics education research and the implications of these findings for physics instruction. Further details will be given in the chapters on specific topics.

It is important to note at the outset that physics education research does not give us a formula for the "best" teaching method or methods. Good teaching will still rely on the accurate judgment of an instructor as to the difficulties his or her students are facing and how they are responding to specific situations. But physics education research does offer *general* guidance as to why some teaching methods are likely to be more effective than others. Perhaps more important, physics education research provides the instructor with a powerful set of tools for recognizing the *knowledge state* of his or her students and how that knowledge state is likely to respond to different instructional strategies.

Most physics education research has followed a two-step methodology. First, a relatively small number of students are interviewed in detail about their understanding of a particular situation. In a typical interview, the student is presented with a piece of apparatus, such as a cart on an air track or a group of magnets, then asked to *predict* how the apparatus will respond to a specified set of circumstances. The interviewer tries to draw out the *conceptual beliefs* by which the student makes his or her prediction. The transcripts of these interviews are extremely revealing about what students really think. In many circumstances, researchers have found that student beliefs fall into a small number of fairly distinct categories.

The second stage of the research is to devise a multiple choice test with answers that correspond to the categories that were discovered in the interviews. (The tests are conceptual—"What will happen if . . . "—rather than computational.) The multiple choice tests can then be administered to large numbers of students to learn what fraction of students hold each type of belief. In some research projects, a test is given before any instruction to measure the *initial state* of the students. A similar test is given after instruction to measure the effectiveness of the instruction at changing student beliefs and concepts.

The multiple choice tests appear so simple, even trivial, that many physics instructors have confidently predicted that *their* students would all score near 100%. It has been a shock to many—including professors at highly selective universities—to discover just how low the scores are *after* instruction. Class averages of 50% or less are typical on many such questions. The discovery of how little conceptual knowledge students gain from ordinary physics instruction has been a major finding of physics education research.

Summary of Findings from Physics Education Research

This section will look at some of the most general findings of physics education research. This is intended as an overview, not as a full review article, and only a few general citations are given. The next section will illustrate these findings with a few select examples. The citations and the bibliography provide entry points into the literature for those wanting to know more.

In brief summary, physics education research has revealed that

- Students enter our classroom not as "blank slates," *tabula rasa,* but filled with many prior concepts. These are called, by various researchers, misconceptions, preconceptions, alternative conceptions, or common-sense conceptions. Students' concepts are rather muddled, not well differentiated, and contain unrecognized inconsistencies. By the standards of physics, their concepts are mostly wrong. Even so, they are the concepts by which students make decisions about physical processes.
- Students' prior concepts are remarkably resistant to change. Conventional instruction—lecture classes, homework, and exams that are predominately or exclusively quantitative—makes almost no change in a student's conceptual beliefs.
- Students' knowledge is not organized in any coherent framework. At the end of instruction, their knowledge of physics consists of many discrete facts and formulas only loosely connected to each other. This is in contrast to a physicist's knowledge, which is organized in terms of physical principles. Whereas a physicist sees "a Newton's second law situation," then retrieves specific knowledge as needed, most students see "a falling body problem" or "an inclined plane problem" or "a pulley problem," with little or no recognition of the similarities. Their organization of knowledge (or lack thereof) is largely responsible for their formula-seeking problem-solving strategies. Our typical admonition that "Newton's laws are all you need to remember" is meaningless to students who lack the knowledge organization that we have.

As a result, most students don't develop a functional understanding of physics, they can't apply their knowledge to problems or situations not previously encountered, and they can't reason correctly about physical processes. Now let's flesh these ideas out a bit.

It's useful to distinguish three general categories of physics knowledge:

Factual Knowledge: Knowledge of specific events and situations. Some of our factual knowledge is through experience (objects fall when you release them, two similarly charged objects repel) and some is accepted on the basis of authority (the earth is round, electrons are negative). Increasing experience, such as in the laboratory, can move individual pieces of factual knowledge from the *authority* category to the *direct experience* category, but much of scientific knowledge is accepted on the basis of authority because none of us can independently reconfirm all the experiences of others. Needless to say, some factual knowledge held by students is partially or entirely wrong!

Conceptual Knowledge: Knowledge of physical principles, knowledge that provides a unified understanding of many pieces of factual knowledge. We generally think of conceptual knowledge as having explanatory and predictive power. "Net force causes acceleration" is conceptual knowledge. So is "motion requires a force," although this particular example of commonly held conceptual knowledge turns out to be wrong.

Procedural Knowledge: Knowledge of how to apply factual and conceptual knowledge to specific problem-solving situations; knowing how to *use* what you know.

The boundaries between these categories are rather fuzzy, but this will do for our purposes.

Students certainly enter physics classes with many mistaken beliefs and incorrect pieces of factual information (e.g., there is no gravity in outer space). The main focus of physics education research, however, is on their conceptual knowledge and their procedural knowledge. Students, as they enter our classes, have had 18 or more years' experience as "experimental physicists," exploring their physical environment and making hypotheses about how it works. Although not systematic in a scientific sense, students have developed common-sense theories of the physical world that have proven time and again to be satisfactory for their day-to-day existence. Furthermore, these hypotheses are not trivial. Many student beliefs have been advocated by eminent scientists and philosophers, from the time of Aristotle to the Middle Ages and on into the nineteenth century.

These student beliefs are sometimes called *misconceptions,* but that is a rather pejorative term with the implication that students are at fault for holding such views, that they should have known better. The terms *preconceptions* or *alternative conceptions* are preferable. We can say the following about the alternative conceptions of students:

- These are strongly held beliefs about how the world works. They are used by students to explain and predict physical processes. These views are highly stable and very resistant to change, which is not surprising because they have been developed and tested—successfully, for the most part!—over many years.
- Student beliefs differ from, and are generally incompatible with, the concepts of physicists. Because concepts are "filters" through which we see and interpret the world, students often interpret what the instructor says or demonstrates, or what the textbook says, in ways quite different from what the instructor or the text intended.
- These alternative conceptions must be altered or eliminated if students are to achieve a satisfactory understanding of physics.

Examples of alternative conceptions are

■ Gravity gets stronger closer to the ground, which is why objects speed up as they fall.
■ Motion of an object implies a force acting on the object.
■ If half of a lens is covered by a piece of opaque paper, half of the image on a screen will disappear.
■ Electric current is "used up" in a circuit.

These are not small, minority views. Some of these beliefs are held by well over half the students in a typical university physics class. These beliefs generally don't prevent students from finding the right formula and getting the "right" answer on a typical exam problem, so many instructors remain unaware that students continue to hold these beliefs *after* instruction.

The traditional lecture mode of instruction is built on an unexamined assumption that students are lacking information and that knowledge is gained when the instructor supplies the information in a clear, intelligible fashion. But from the alternative conceptions perspective, students already have a complete, even if inconsistent and erroneous, set of concepts. Their mental slots are already full! New information is not plugged into empty and waiting slots, as conventional instruction assumes, but instead is filtered and reinterpreted by the existing concepts.

Halloun and Hestenes (1985b), in their interviews with students, found that students held firm in their mistaken beliefs even when shown a demonstration that directly contradicted their predictions. Rather than question their own beliefs, the students tried to argue that the demonstration was not relevant to the question they had been asked! So perhaps it is little wonder that students' alternative conceptions are little changed by conventional instruction.

As noted, it's easy for instructors to be unaware of what students *really* think. Many students, even at the beginning of the course, can cite Newton's laws or Galileo's law of falling objects. Nearly all can at the end of the course. Given a sufficiently straightforward problem, most can select the right equation and arrive at the right answer. But a little probing beneath the surface reveals that this is simply "book learning," that instruction has made little or no impact on students' beliefs about the world or on how they *reason* about physical processes. In effect, students are now living in two parallel universes—the universe of physics class, where they dutifully learn the algorithms for producing answers their instructor likes to see, and the *real* (to them) universe governed by their unaltered alternative conceptions.

Alan Van Heuvelen (1991a), in his research on instruction, found that 20% of the students in a typical university class entered the class as Newtonian thinkers—that is, with a set of beliefs generally in accord with the Newtonian concepts of physicists. The net impact of a semester of conventional instruction was to raise this figure to 25%. Others have reported similar figures.

Halloun and Hestenes (1985a) were the first to show (since confirmed at many other schools) that the conceptual gain of students is *independent of the instructor* for those instructors giving traditional lecture courses. We all pride ourselves on thinking that logical, carefully crafted lectures, with good demonstrations and a touch of wit, will clarify difficulties for students and lift them to higher levels of understanding and achievement. Unfortunately, studies have found that students of teachers perceived by their students as dull and boring learn just as much (or just as little) as students of dynamic, award-winning lecturers. Now good lecturers can certainly be more inspirational and can make the class more enjoyable (I'm not advocating poor, ill-prepared lectures!), but the evidence is strong that good lecturers are no more effective than any other instructor in terms of increasing their students' understanding of physics.

So we can say that

- Conventional, lecture-based instruction has little impact on students' conceptual understanding of physics. Most leave with the same erroneous concepts that they brought with them to the class.
- This result is independent of the instructor.

Lest this conclusion sound too dismal and hopeless, it is important to note the recurring qualification "conventional, lecture-based instruction." Effective instruction requires more than dedication and a firm knowledge of physics. It also requires a knowledge of how students think and learn, and it requires teaching methods and materials that are cognizant of the knowledge state of students. The good news is that knowledgeable instructors, using appropriate methods and materials, *can* help students to make significant conceptual gains.

Examples of Students' Alternative Conceptions

The preceding ideas will be illustrated with two examples: students' conceptions about force and motion, and students' conceptions about electrical circuits. These are two of the more heavily researched topics, with an extensive literature.

Students' Conceptions about Force and Motion

Most student beliefs have, in the past, been advocated by eminent pre-Newtonian scientists and philosophers, from the time of Aristotle to the Middle Ages and even beyond. Aristotle was an astute observer, but he lacked the mathematical tools to make his ideas precise. The essential ideas of Aristotelian dynamics are

- Every motion has a cause—namely, a force. In the absence of any force, an object immediately comes to rest.
- Forces come in two types—contact forces (pushes and pulls) and inherent forces, which are the tendencies of objects to seek their *natural place*. Note that, in our modern terminology, an inherent force would be a property *of* the object rather than an action *on* the object.
- Gravity is the tendency of a heavy object (predominantly earth or water) to fall. Thus gravity is an inherent force.
- Heavy bodies fall faster than lighter bodies in proportion to their weights. Note that Aristotle had a concept of speed, but not of acceleration.

■ A medium, such as air, has a *motive power* to propel objects through it. Because all motions must have causes, this was how Aristotle understood the continued motion of an arrow shot horizontally.

It's often said that beginning students are Aristotelian thinkers. Many students do tend to believe the first, third, and fourth statements, but few think of objects as having natural places or the air as having a motive power. Only a small percentage of students are accurately characterized as Aristotelian. Far more would be described as accepting the basic tenets of the medieval *impetus theory.*

Medieval scientists were particularly critical of Aristotle's idea about the motive power of media such as air or water. They offered an alternative explanation—that when an object is thrown, the active agent (the person throwing) imparts an *impetus,* or motive power, to the object. Students, of course, don't call it by the word *impetus,* but they do make frequent reference to concepts such as "the force of the throw." The impetus is an *intrinsic force* that sticks with the object (thus is a property *of* the object) and is the *cause* that keeps the object moving. In most cases, the impetus weakens with time and distance—hence objects slow down and eventually stop. Impetus is a common-sense idea that is accepted, to one degree or another, by over two-thirds of the students beginning university physics.

These are not naive ideas. Much of day-to-day experience does, indeed, suggest that motion needs a cause and that an object's impetus keeps it going. Newton's ideas, on the other hand, are not at all obvious. They are very difficult ideas, and their acceptance depends on making precisely clear what is meant by *acceleration* and *force.* Scientists of the stature of Galileo and Descartes failed to solve the "problem of motion," and even Newton struggled for much of his life to get the ideas pinned down. We shouldn't be surprised if students find their common-sense ideas more appealing than Newton's ideas, and that it will take more than a little effort for them to accept and use a Newtonian perspective.

Fortunately for physics instruction, we can put students' beliefs to the experimental test. Only after their ideas are repeatedly found to be in conflict with experience can we look toward better alternatives. So just what are students' ideas about force and motion?

First, most students do not have clear ideas about just what force and motion actually are. They do not discriminate between *velocity* and *acceleration.* To many students, it's simply motion. Similarly with force: students readily interchange the concepts that we would call *force, inertia, momentum, energy,* and *power.* They are all simply the "cause" of motion. Transcripts of student interviews show them using phrases like "The force of inertia" or "The energy or force you shot at it" or "The power also has a force." Most of us hear our own students using such phrases, but we tend to overlook them as minor inaccuracies. In fact, they reveal major misunderstandings about what a force is. Without a clear and functional understanding of the basic concepts of force and acceleration, students can't possibly reach a Newtonian view of the connection between force and motion.

Many forces recognized in physics are *not* considered forces by students. Friction, for example, is "what makes it stop," but many students don't consider friction to be a force. Likewise, a significant fraction of students don't recognize gravity as a force—it's simply the tendency of objects to fall (a fairly Aristotelian concept). Objects such as tables and walls are often not considered as exerting forces. They are simply "obstacles" that "get in the way" of motion.

There are two common themes in this list of what is and is not recognized as a force. First, many students think that only *animate* objects can exert force. If you give a student one end of a rope and tug on the other end, then ask him what kind of force he's experiencing, he'll likely answer, "You're pulling on me." The rope's tension is not recognized as the force that he experiences. Second, student's tend to view the issue of force from *their* perspective as the *applier* of force, rather than from the object's perspective as a system *experiencing* forces. If a student pushes a box across the floor, her analysis is "I have to apply a force to keep it moving. If I remove the force, the box stops 'because of friction.'" She's seeing the situation from *her* perspective, rather than the box's perspective, and not recognizing friction as a second force on the box. From this perspective, the conclusion "motion requires a force" certainly seems valid. We can say that students tend to focus on *applied force* rather than on the Newtonian idea of *net force*. An important instructional task is to get students to shift to the object's perspective.

With that said, we can summarize the most common student beliefs about the connection between force and motion. These, in essence, are the student versions of Newton's laws!

1. If there's no force on an object, the object is at rest or will *immediately* come to rest. But the converse is *not* true. An object being at rest does *not* imply no net force.
2. Motion requires a force or, alternatively, force causes motion. In most cases, speed is proportional to force. The force may have to "overcome" an obstacle or resistance, such as friction, before the motion begins.
3. Active agents impart an *impetus* to an object that cause the object to continue to move after the active agent is removed (the *impetus principle*).
4. When two objects interact, the larger object "wins" by exerting a larger force on the smaller object (the *dominance principle*). Depending on the context, "larger" may mean more massive, faster, more powerful, or some other measure.

This list is based on Halloun and Hestenes (1985b), who give a more complete description and discussion. Note that this is a summary of the *most common* student beliefs. Not every student holds every belief. Equally important, students are not at all consistent in the application of these beliefs; they will apply an idea in one situation but not in another. Needless to say, students don't describe their beliefs in these terms and probably wouldn't recognize them stated as such. But this list is an important tool that allows *instructors* to know how and why their students think about different situations. Let's look at some examples.

A student pushing on a heavy crate that refuses to move sees the situation as one in which there *is* a force, but the object is at rest. Thus the comment that the converse of Statement 1 isn't true. The difficulty here is one of the wrong perspective, where the student is considering only applied forces and not recognizing friction as a force.

Because students don't equate "no motion" with "no force," they don't accept the standard argument for the existence of the normal force. As instructors, we may point to a book on the table and say, "It's at rest, so there's no net force on it, so there must be a normal force to balance the gravitational force." Students don't accept the premise of this argument—and, in fact, most students don't believe the table exerts a force on the book. They may *tell you* there's a normal force, because they learned so in high school, but they don't *believe* there is. To these students, gravity pulls down on the book while the table just "gets in the way" and prevents the book from falling. There's no conflict in their mind between an apparent net force and no motion. More direct tactics, as described in Chapter 7, must be followed to convince them of the reality of the normal force.

Statement 2, that motion requires a force and that force determines speed, seems perfectly obvious. A horse pulling a cart has to continue exerting a force. If he stops, the cart stops. The harder the horse pulls, the faster the cart goes. Similar reasoning applies to riding a bicycle: the harder you pedal, the faster you go. Who could doubt Statement 2? As already noted, the fallacy is using the wrong perspective, the perspective of the animate *applier* of force rather than the perspective of the object experiencing the force.

Statement 2 begins to get especially interesting when coupled with the impetus principle of Statement 3. Shove a block on the table. It moves a distance after you release it, then stops. Because motion requires a force, you apparently supplied the block with a "force of motion" or "force of the push" or "force of your hand" that keeps it going. This force gradually weakens until eventually friction "overcomes" the force of motion, then the block stops. What could be simpler?

In the vertical dimension, this situation is the "coin toss problem." Clement (1982), in an early study of students' concepts in mechanics, gave students the following problem:

> A coin is tossed from point A straight up into the air and caught at point E. On the dot to the left of the drawing, draw one or more arrows showing the direction of each force acting on the coin when it is at point B. Draw longer arrows for larger forces.

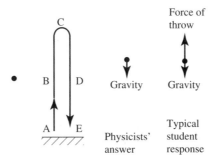

Before instruction, 88% of the students (engineering students) answered incorrectly, with virtually all errors being the inclusion of an *upward* force vector at B. Students who were later interviewed referred to this vector as "the force of the throw" or "the force I'm giving it." This is clearly an impetus that is seen to stick to the coin and is necessary to carry the coin in an upward direction against gravity. Eventually the impetus weakens and is "overcome" by gravity, and at that point the coin reverses direction and begins to fall. Most students answer correctly that gravity is the only force at point D.

This is such a simple situation that we would expect significant improvement in the responses following instruction. Yet the post-instruction error rate was still 75%. Instruction made only marginal improvement in students' ability to answer this question correctly. This illustrates how resistant students' alternative conceptions are to change! Situations where the motion is opposite in direction to the net force, as at point B, seem to be especially difficult for students to understand from a Newtonian perspective.

Van Heuvelen (1991a) reports scores from 152 engineering students on a conceptual multiple-choice exam given at the *end* of instruction in mechanics. Two problems are shown below. In each, answers d and e represent impetus thinking, with a force pushing the object in the direction of motion. These and other studies show that about 60% of the students are still impetus thinkers *at the end of instruction.* They have not—despite an ability to use Newton's laws mathematically—grasped the primary concepts of Newtonian mechanics.

(For questions 1 and 2, * = correct response.)

1. A ball rolls down an incline and off the horizontal ramp. Ignoring air resistance, what force(s) act on the ball as it moves through the air after leaving the horizontal ramp?

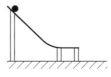

		Percent after instruction
*a.	The weight of the ball vertically down.	38
b.	A horizontal force that maintains the motion.	1
c.	A force whose direction changes as the direction of motion changes.	2
d.	The weight of the ball and a horizontal force.	29
e.	The weight of the ball and a force in the direction of motion.	30

2. A ball swings at the end of a string in a circular path in a vertical plane. Which free-body diagram shown below best represents the forces acting on the ball when at the bottom of the circle and moving toward the right? The ball moves at constant speed.

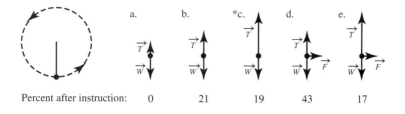

Percent after instruction: 0   21   19   43   17

The dominance principle is evident in second law situations, where one force is seen as "overcoming" another, but it is especially important for interacting objects. If a truck and a mosquito collide head-on, students overwhelming believe that the truck exerts a larger force on the mosquito than the mosquito does on the truck. This belief is little changed by conventional instruction.

Students' conceptual understanding of mechanics has been measured for thousands of students by the Force Concept Inventory, or FCI, of Hestenes, Wells, and Swackhamer (1992a). Many of the results reported here were either learned from or confirmed with the FCI. Hake (1998) has compiled data from many different colleges and universities. His results show convincingly that the average "gain" in conceptual knowledge following conventional instruction is not only low, it's fairly predictable once the pre-instruction score of the class is determined. The good news, however, is that his data also show convincingly that *interactive engagement* teaching methods produce substantial increases in the average gain. This will be discussed further in the section on Implications for Instruction.

Students' Conceptions about Circuits

It seems fairly clear why students have alternative conceptions about mechanics. It's not so clear why they should have alternative conceptions about electrical circuits. Very few students have ever played with a circuit. Their direct experience is pretty much limited to turning on light switches and changing batteries. Nonetheless, researchers have found that a large fraction of students hold an identifiable set of alternative conceptions about circuits and, as was the case in mechanics, that these conceptions are little changed by conventional instruction.

Students' lack of familiarity with basic ideas about circuits is easily shown by giving each student a battery, a light bulb, and a piece of copper wire, then asking them to make the bulb light. Only about half of a typical class can do this without hesitation, and a sizable minority may struggle with trial-and-error attempts for ten minutes or more until getting it. Many of their incorrect attempts reveal that

- Many students are not aware of the need for a complete circuit.
- Many students are not aware that a light bulb is a two-terminal device and that current must flow *through* the bulb to make it light.

Systematic studies have presented students with the two situations shown below and asked them to rank order the light bulbs from the brightest to the least bright. A common response to Example 1 is A = B > C (A and B equally bright, C dimmer), and a common response to Example 2 is D > E = F. McDermott and Schaffer (1992) found in a class of predominantly engineering students that only 15% could answer both correctly prior to instruction, and that there was no significant improvement after instruction. (Different versions of the problems, as well as interviews, show that students' incorrect answers are not due to thinking of "real" batteries, with internal resistance, rather than the assumed "ideal" batteries.)

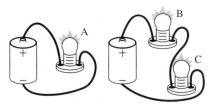 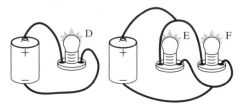

Rank order the brightness of bulbs A, B, and C.

Example 1

Rank order the brightness of bulbs D, E, and F.

Example 2

What can account for these responses? Two more battery-and-bulb circuits, shown below, can clarify the nature of student thinking.

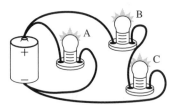 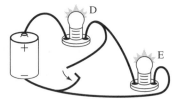

What happens to the brightness of bulbs A and B if bulb C is removed from its socket?

Example 3

What happens to the brightness of bulbs D and E if the switch is closed?

Example 4

In Example 3, students are asked what will happen to the brightness of bulbs A and B when bulb C is removed. Most predict that B will go out (correct) and that A will get brighter (incorrect). In the circuit of Example 4, they are asked what will happen to the brightness of bulbs D and E when the switch is closed. This question generates a wide array of responses, almost none of which are correct, but the most common response is that E will get dimmer, but not go out, and that D will be unchanged—both incorrect.

The most common student beliefs about electrical circuits are found to be

- Current is "used up" as it moves through a circuit.
- A battery is a source of constant and unchanging current.
- Current divides equally when it reaches a junction.

Before discussing these, we have to note that students don't have a well-defined concept of *current*. We saw in the previous section that students have a single concept of *motion* that is not differentiated into concepts of velocity and acceleration. Similarly, most students seem to have a vague concept of *electricity,* which is used inter-changeably to mean either *current* or *energy.* (Note that the questions don't ask directly about current, but student beliefs about current and energy can be inferred from their predictions about brightness.)

The student prediction that B > C in the series circuit of Example 1 reflects a belief that the current is partially "used up" by B, so there's less for C. (A few clever students think this is a trick question, and so answer C > B. They know that current is "really" the flow of electrons. Because electrons come from the negative end of the battery, which is closer to C, it will be C that's brightest!) This belief stems from not distinguishing between *energy,* which is "used up" (transformed, to be precise), and *current,* which isn't. This conception does seem fairly easily changed, and most post-instruction students will correctly use conservation of current in a single wire to answer B = C.

The most persistent student belief is that batteries are constant-current sources. This thinking is seen in their response to removing bulb C in Example 3: The "unchangeable" current of the battery is now all shunted through A, making it brighter. Similarly, closing the switch in the circuit of Example 4 doesn't change the brightness of D because the battery's current remains fixed. It is constant-current thinking that leads to the prediction that A = B in the circuits of Example 1 (same current leaving the battery in both cases) and that D > E = F in Example 2 (same current leaving the battery in both cases, but on the right it gets split between E and F). Very few students change their belief that a battery is a constant-current source as a result of conventional instruction.

It seems fairly obvious in the right circuit of Example 2 that the current will split equally between the two bulbs, but the response to shorting a bulb with a switch is surprising. The common response in Example 4, that E dims but doesn't go out, reflects a belief that the current at the junction splits *evenly*—50% through each side—regardless of what lies along the path. In other words, current decisions are made by thinking locally (one wire splits into two at this point) rather than globally (looking ahead at the resistance of the two paths).

The student model of circuit behavior is purely a *current model.* The concept of potential difference does not enter into their thinking, and even with prompting they find it difficult to reason using potential differences (Cohen et al., 1983). This stems from perhaps two causes. First, potential is a very abstract concept, whereas current is a concrete idea. Second, instruction doesn't make a convincing link between the definition of potential and the use of potential differences in circuits.

The problem on the next page asks a qualitative question about potential differ-ences. This question is used only after instruction, so at least the terms, if not the con-cepts, are familiar. A large fraction of the students answer e—the potential difference is zero. The most common reasoning pattern is, "There is no resistance between 1 and 2, so according to Ohm's law $V = IR$ is zero." There are multiple errors in this reason-ing—mistaking "no resistor" for "no resistance," trying to apply Ohm's law inappro-priately, and not recognizing the meaning of potential difference between two points. Almost no students spontaneously use Kirchhoff's voltage law to reason that the potential difference is 3.0 V, and most have difficulty even being led through this line of reasoning.

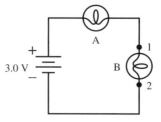

Bulb B is removed from the circuit. After it is removed, the potential difference between points 1 and 2 is:

a. 3.0 V.
b. Between 1.5 V and 3.0 V.
c. 1.5 V.
d. Between 0.0 V and 1.5 V.
e. 0.0 V.

These results suggest that physics instruction needs to give much more attention to the mental models by which students think about electric (and magnetic) phenomena. Later chapters will make specific suggestions.

Knowledge Structures and Problem Solving

The preceding discussion has been of students' conceptual knowledge. Another area of physics education research has been concerned with *procedural* knowledge—how students organize and use their knowledge within the context of solving a problem. Problem solving in various domains has long been studied by psychologists, and there is a vast literature. Reif and Larkin pioneered the study of problem solving in physics starting in the 1970s. They did so by careful studies that compared and contrasted the problem-solving abilities of "novices" (students in introductory physics classes) and "experts" (physics professors).

The methodology of these studies was to have subjects solve, or attempt to solve, an elementary physics problem while continuously talking out loud about what they were thinking and doing. After completing the problem, or reaching a dead end, the interviewer would ask more questions about why the subject pursued some avenues but not others. Verbatim transcripts and the subject's work were then analyzed to determine the underlying reasoning patterns.

The results of this research show that novices are *not* simply slower, more error-prone versions of experts. The expert does, indeed, have more factual knowledge and is able to work more quickly, but the major difference between experts and novices is due to their procedural knowledge. As stated by Reif and Heller (1982),

> Observations by Larkin and Reif and ourselves indicate that experts rapidly redescribe
> the problems presented to them, often use qualitative arguments to plan solutions

before elaborating them in greater mathematical detail, and make many decisions by first exploring their consequences. Furthermore, the underlying knowledge of such experts appears to be tightly structured in hierarchical fashion.

By contrast, novice students commonly encounter difficulties because they fail to describe problems adequately. They usually do little prior planning or qualitative description. Instead of proceeding by successive refinements, they try to assemble solutions by stringing together miscellaneous mathematical formulas from their repertoire. Furthermore, their underlying knowledge consists largely of a loosely connected collection of such formulas.

In other words, experts have i) a *structured* knowledge that allows them to recognize the "deep structure" of a problem, and ii) a *strategy* that exploits the structured knowledge by describing and analyzing the problem in steps of increasing refinement and sophistication. Mathematical details are added only at the end. Novices, by contrast, have i) an unstructured knowledge of loosely connected facts and formulas, so they see only the "surface structure" of a problem, and ii) no strategy other than equation searching. We all regularly see the lack of a coherent strategy when we talk with students or observe their work, but we don't normally associate the lack of a strategy with the student's underlying knowledge structure. Research implies that students have little choice but to follow an equation-seeking strategy, and our exhortations will not change this until we deal with the larger issue of their knowledge structure.

Knowledge structures are studied two ways. One is by analyzing how knowledge is accessed and used during a problem-solving interview. Another way, more direct, is to give a subject a group of problems written on cards with the instructions to sort the problems into meaningful categories. Experts sort the problems into relatively few categories, such as "Problems that can be solved by using Newton's second law" or "Problems that can be solved using conservation of energy." Novices, on the other hand, make a much larger number of categories, such as "inclined plane problems" and "pulley problems" and "collision problems." That is, novices see primarily surface features of a problem, not the underlying physical principles.

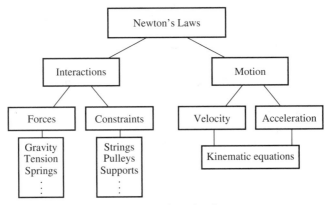

Expert's knowledge structure of mechanics.

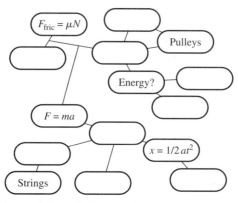

Novice's knowledge structure.

The preceding figure shows examples of an expert's knowledge structure and a novice's knowledge structure in the realm of single-particle dynamics. (These are meant to be illustrative. Experts may disagree over whether this adequately portrays the nuances of the expert's knowledge structure.) The main point is that the expert's knowledge structure is hierarchical, allowing him or her to pursue a chain of reasoning and to evaluate the consequences of different possible decisions. The hierarchical structure also proceeds from the general to the specific, avoiding unnecessary details during the initial analysis. The novice's loosely connected, non-hierarchical knowledge structure discourages logical reasoning and analysis. In addition, all knowledge is on an equal footing so that $x = \frac{1}{2}at^2$ seems just as significant as the law of conservation of energy.

The problem-solving efficiency of experts is related to the fact that an organized, hierarchical knowledge structure allows the expert to recognize and work with larger *patterns* in the problem. This idea is well illustrated in work with chess players. Twenty or twenty-five chess pieces are placed on the board at their positions in the middle of a real game, and a subject is allowed to study the board for just a few seconds. He or she is then asked to reproduce the positions from memory. A novice player does well to place four or five pieces correctly, but a chess master can usually reproduce the entire board. If, however, the pieces are placed on the board at random, the chess master does no better than a novice.

It's well known that short-term memory can store only few—typically 4 to 6—"chunks" of information. To a novice, the position of one chess piece is a chunk of information. The expert, however, has his knowledge organized in such a way that whole groups of pieces forming a pattern—"castled on the queen's side," for example—are represented as a single chunk of information.

A typical physics problem has far more than six pieces of information. Students who see each piece of information as a distinct chunk are already starting to forget the beginning of a problem by the time they finish reading it! Experts, however, will recognize patterns in the information—perhaps noting that all the initial conditions and forces are described—that allow them to "grasp" the entire problem in one reading and to begin planning a solution. Even so, for any problem other than the most simple an expert almost always draws some sort of sketch. A novice rarely does. A sketch, in effect, is an *extended memory* that allows the problem solver to track and work with information on paper rather than in short-term memory.

An expert proceeds by

- Describing the physical systems—particles, interactions, charges, fields, or whatever else is relevant.
- Clarifying geometrical relationships, usually with the aid of a coordinate system.
- Identifying relevant physical principles, such as Newton's laws or conservation laws.
- Identifying simplifying assumptions, such as whether or not to use a particle or a continuum model, whether or not to include air resistance, and so on.
- Adding details as needed, such as drawing a free-body diagram.
- Inserting the relevant quantities into the relevant equations and solving for the answer.
- Finally, checking the units and *interpreting* the result.

To physicists, the first five steps *are* the physics. If they are done correctly, the equations should be perfectly obvious and the mathematical manipulations are of little interest. But to students, who gloss over or omit the first five steps, physics is synonymous with mathematics.

Unfortunately, conventional instruction does little to improve the situation. In working sample problems in class, we say a few words about the problem and then start writing equations. Most of our thought processes and decision-making processes are tacit, so the performance students *see* is that we read the problem and went immediately to an equation.

Much of the pattern recognition and initial analysis is so "obvious" to an expert that we do it unconsciously—*second nature,* or *physical intuition.* Students are completely unaware that this aspect of problem solving even exists, so it seems equally "obvious" to them that success in physics equates with successfully finding the right equation.

Better problem-solving skills have to be explicitly taught; very few students acquire them on their own. Three instructional strategies can be effective. First, students need explicit instruction in a step-by-step analysis procedure that parallels the expert's problem-solving strategy. Second, the instructor—and, ideally, the textbook—have to "think out loud" as they work examples, explicitly stating assumptions and demonstrating the steps in the strategy, no matter how trivial they seem. Third, the instructor and the textbook need to provide opportunities that help students organize their knowledge in a more coherent structure. Students will not acquire the knowledge or expertise of a physicist in one course, but several studies (Reif and Heller, 1982; Mestre and Touger, 1989; Van Heuvelen, 1991b) have shown these instructional strategies can significantly improve students' problem-solving skills.

A few students—through persistence and above average self-imposed feedback—do manage to make the novice-to-expert transition on their own. Nearly all physics instructors are in this group. Even so, most of us have said, "I never really understood the subject well until I taught it." We, as instructors, need to try placing ourselves in the student mind and to become acutely aware that students need to have all the reasoning steps spelled out in detail.

(Author aside: On my Ph.D. oral exams at Berkeley, an eminent solid-state theorist asked me to estimate how long it would take a pencil standing on its point to fall over. I confidently began by saying, "Well, if we write down the Lagrangian . . . " He immediately stopped me and said, "No, just use a straightforward Newtonian analysis." I was stumped! I hadn't tried to think of such problems using "elementary concepts" since I was a freshman. And, clearly, I hadn't carried much lasting knowledge away from my freshman experience. I must have eventually put together a partially coherent explanation, but it was a humbling experience. It's also a good experience for me to keep in mind when I see students stumbling through the steps of solving a problem. What seems "obvious" to me appears so only from many long years of experience.)

Many of the pedagogical tactics described in this book are designed to help students acquire a better knowledge structure and better problem-solving strategies. These include multiple representations of knowledge (essential for the description and qualitative analysis of problems), explicit problem-solving strategies that are modeled in detail, explicit teaching of many reasoning skills (such as identifying forces and drawing free-body diagrams), problem-solving worksheets that guide students through the steps, and knowledge structure tables.

Implications for Instruction—The Five Lessons

Several investigators have tried to develop a theoretical framework for understanding student learning and the student mind. Interested readers are referred to Dykstra et al. (1992), Hammer (1996), and the references therein for more information. Our goals are more practical—what are the implications of physics education research for actual instruction? Three individuals who have looked at this issue are David Hestenes ("Toward a Modeling Theory of Physics Instruction," 1987), Lillian McDermott ("What We Teach and What Is Learned—Closing the Gap," 1991), and Edward Redish ("Implications of Cognitive Studies for Teaching Physics," 1994).

These three, and others, emphasize that physics education research does not provide a "formula" for optimal teaching. Individuals are too, well, *individual* for a one-size-fits-all teaching strategy. The previous sections have outlined some of the most prevalent modes of student thinking, but probably no particular student fits this model exactly, and certainly no student would consistently and faithfully think this way in every instance. Nonetheless, research does provide general guidance as to instructional methods that are *likely* to be effective—as well as methods that, for most students, will be ineffective. Perhaps more significantly, physics education research has provided important new tools for thinking about and understanding our students, allowing us to adapt our instruction to their ever-changing conditions and needs.

Conventional instruction, provided by lectures and most textbooks, is often referred to as a *transmissionist model.* We hold the knowledge, the students lack it, so we will simply tell it to them—often known as "teaching by telling." This is highly efficient from the instructor's perspective, allowing the instructor to pass along large quantities of information to many students in a limited period of time. Unfortunately, one of the clear and repeatedly confirmed results of education research is that the transmissionist model is grossly ineffective from the students' viewpoint. The information may be broadcast flawlessly, but little is received by passive students just listening to it.

This is not to say that lectures are never effective. The lecture mode appears to work best where instructor and students share a common set of beliefs and assumptions, such as in graduate classes. Even in the introductory class, short periods of instructor-centered discourse can clarify difficult issues or provide background information. But extended lectures, particularly formal lectures of deriving results, appear to be the *least* effective mode of instruction.

Physics education research has brought out the need for instruction to be student-centered, explicitly recognizing the knowledge state of the students and the activities that will transform them to the desired state. In this view of instruction, the teacher is less an authoritative source of knowledge and more a *facilitator,* providing opportunities and feedback through which the students can develop correct knowledge structures and mental models. Implicit in this view is that students *construct* their mental models from their experience; we cannot hand them an already built, fully operational mental model. This model of instruction is called *scientific constructivism.*

To quote McDermott (1991):

> We can briefly summarize the constructivist view of how scientific knowledge is acquired as follows: All individuals must construct their own concepts, and the knowledge they already have (or think they have) significantly affects what they can learn. The student is viewed not as a passive recipient of knowledge but rather as an active participant in its creation. Meaningful learning, which connotes the ability to interpret and use knowledge in situations not identical to those in which it was initially acquired, requires deep mental engagement by the learner. The student mind is not a blank slate on which new information can be written without regard to what is already there. If the instructor does not make a conscious effort to guide the student into making the modifications needed to incorporate new information correctly, the student may do the rearranging. In that case, the message on the slate may not be the one the instructor intended to deliver.

Redish (1994) has this to say about constructivism:

> People must build their *own* mental models. This is the cornerstone of the educational philosophy known as constructivism. . . . An extreme statement of constructivism is: *You cannot teach anybody anything. All you can do as a teacher is make it easier for your students to learn.* . . . Constructivism should not be seen as disparaging teaching, but as demanding that we get feedback and evaluations from our students to see what works and what does not. It asks us to focus less on what we are teaching, and more on what our students are learning.

Scientific constructivism is the underlying teaching philosophy of this guidebook. But a word of caution is in order. The term *constructivism* is used in other disciplines, usually in the sense of *social constructivism,* with quite different meaning and intent. In particular, some people see constructivism as a license to make up any rules you want, as long as everyone agrees to adhere to them (a social contract), and it has been fashionable in recent years for sociologists to critique science through this filter. By contrast, I use the term *scientific* constructivism as being nearly synonymous with *scientific method.* It is shorthand for a teaching philosophy in which students actively build their knowledge and concepts by constantly testing them against the harsh judge of physical reality. There's nothing arbitrary about the outcome we wish them to achieve.

The results of physics education research can be sliced and rearranged in many patterns, but I see Five Lessons for teachers:

Lesson One—Keep Students Actively Engaged and Provide Rapid Feedback. Active engagement is the essence of the constructivist approach, because students must *build* their mental models rather than receive them from the instructor. Active learning is the subject of the next chapter of this guidebook, where various techniques are discussed in more detail, but a short list of active engagement methods includes

- Interactive lecture demonstrations.
- Nearest neighbor discussion activities.
- Collaborative group activities.
- Microcomputer-based laboratories or other guided-discovery laboratories.
- Take-home experiments.

The common theme is that students are engaged in *doing* or *talking about* physics, rather than listening to physics. Many of these activities will directly address students' misconceptions, thus also contributing to another of the Five Lessons.

An important corollary is that students must receive prompt feedback if active engagement is to be effective. It does little good to confront a student's misconceptions unless the student gets the real-time feedback needed to recognize the conflict. Many active-learning activities ask students to predict the outcome of an experiment. Then, within a few seconds, they discover if their prediction is right or wrong. This provides rapid feedback on the prediction, but that isn't enough. This activity needs immediate follow-up to discuss the implications and, where necessary, to have the student "try on" different conceptual models.

Lesson Two—Focus on Phenomena Rather than Abstractions. The goal of physics is to understand physical phenomena. Mathematics and other abstractions are important and useful tools, but they're just that—tools. If we want students to reason correctly about physical processes and to develop physical intuition, instruction must remain focused on the phenomena. In particular,

- Use experiential labs, where possible, to provide familiarity with basic phenomena.
- Work inductively, from the concrete to the abstract. This keeps theory grounded in reality rather than, as happens now for too many students, becoming "just math."
- Ask students the questions "How do we know . . . ?" and "Why do we believe . . . ?"
- Ask students to *explain* the outcome of an experiment by using qualitative reasoning but no equations.

Certainly we want our students to be able to solve problems, but we want them to solve *physics* problems, not just math problems in physics clothing. As noted, several studies have shown that problem-solving ability *increases* when instruction is shifted away from derivations and theory and toward building a coherent knowledge structure. There is ample time in upper division courses to establish more formal theories.

Lesson Three—Deal Explicitly with Students' Alternative Conceptions. McDermott, 1991 noted, as have many others, that the student's slate is already full when they enter our class—but likely with much false and misleading information. Nonetheless, the student can't write new information on the slate until *the student* erases information that's already there. One of our most important tasks as teachers is to persuade them to erase the incorrect information, then to provide them with reasons to build better mental models. As you no doubt recognize by now, simply *telling* them what's wrong with their conceptions and *telling* them the "right" conceptions will have little or no effect.

As many researchers have found, the learning cycle that appears to be most effective is

- Confront student misconceptions directly. This is most often done through experiments or lecture demonstrations known to elicit common misconceptions. Students are asked to make a prediction, and the instructor or assignment usually asks them to be explicit about their reasoning (this forces them to *use* their mental model, rather than just guess). Then the experiment or demonstration is performed.
- Explore the fact that many predictions were wrong. This can't be glossed over quickly. Students have to recognize and accept that there really is a conflict between their prediction and reality. Left to themselves, many students will brush the conflict aside as of no relevance.
- Consider alternative models. This must include not only the hypotheses of the model (such as $F = ma$), but clarifying and differentiating the terms of the model (such as distinguishing velocity and acceleration, rather than the students' undifferentiated idea of *motion*). Be explicit about the reasoning steps from the hypotheses of the model to the prediction of a specific experimental outcome.
- Reiterate. Students' alternative conceptions are highly resistant to change, and one example of a conflict is unlikely to have much effect. They need to see repeatedly that their conceptual model fails, when put to the test, but that an alternative model succeeds.

There's a delicate balance here for the instructor. You need to challenge students' misconceptions, but you don't want to put students down or make them feel dumb for holding such views. Emphasizing two items can help. First, nearly everyone holds these misconceptions, including many very smart people in other disciplines. They're not alone. Second, the concepts of physics *are* difficult and aren't obvious. Galileo couldn't figure them all out, and even Newton struggled with them for many years. But everything's easier in hindsight than it was to discover, so they *can* learn these ideas if they keep practicing.

Lesson Four—Teach and Use Explicit Problem-Solving Skills and Strategies. As Hestenes (1987) noted, "Students are not easily weaned away from a formula-centered problem-solving strategy that has been successful in the past." Many students have breezed through high school science classes by being skilled equation hunters. Although such a simple strategy fails when facing the increasingly complex problems of college courses, students have no other alternative strategy at their disposal. Exhortations to "just remember a few general principles" are meaningless to students because they don't know—unless they're taught—how to reason this way.

A major goal of the introductory physics course is for students to learn more sophisticated problem-solving skills. But such an outcome does not happen automatically for many students. To succeed, we must

- Teach students the specific skills needed to solve complex problems. These include interpretation skills, pictorial skills, graphical skills, and reasoning skills.
- Show students how those skills are assembled into a powerful problem-solving strategy and demonstrate their use, in detail, in the example problems we work in class.
- Make explicit the assumptions, decisions, and reasoning that are part of an expert's problem-solving strategy but which usually go unsaid.
- Help students organize their knowledge in a more coherent, hierarchical, easily searched structure.

A coherent knowledge structure is essential if students are to follow a more sophisticated problem-solving strategy, but how is such a knowledge structure built? This is a bit of a chicken-and-egg problem. One way is to ask students for significant qualitative reasoning and explanations, activities that promote learning the logical connections between ideas rather than the memorization of formulas. Another is to *require* the students to follow the specific steps of a problem-solving strategy— either instructor provided or given in the text. (The worksheets described later are for this purpose.) This forces students to consider other issues besides "find the right formula," and with practice this technique aids them in building a coherent knowledge structure.

Lesson Five—Write Homework and Exam Problems that Go Beyond Symbol Manipulation to Engage Students in the Qualitative and Conceptual Analysis of Physical Phenomena. Students certainly must practice to build problem-solving skills. But they must engage in the *right kind* of practice, using the right kind of problems. As Reif (1995) commented, "When students working on homework problems spend hours floundering, what they really practice is floundering. And if they spend most of their time haphazardly grabbing miscellaneous equations, they certainly do not practice valuable problem-solving skills." It does students little good to work "lots" of problems, as we frequently urge them to do, if the problems are poorly chosen and the students are simply reinforcing an equation-seeking strategy.

Appropriate homework and example problems need to

- Balance qualitative and quantitative reasoning.
- Emphasize reasoning, de-emphasize formulas and equations.
- Deal directly with phenomena and observations. Derivations and "Show that . . . " problems have little efficacy for students at this level.

This is not to say that students shouldn't or won't learn to solve quantitative problems. The point is that problems should focus students' attention on interpretation, on analysis, on reasoning—that is, on doing *physics*—rather than just on getting a number.

It is important that exam problems follow the same guidelines. As we all know only too well, students give only lip service to issues or procedures that aren't on the exam. There's no better measure of what we *really* want them to know than the types of questions on our exams. It's true that more sophisticated problems, asking for qualitative analysis and explanation, are longer than the typical equation-oriented exam problems, especially if you require students to follow all the steps in a problem-solving strategy. An hour exam might have only two such problems. Students should have adequate time to think and not have to race through the exam.

Most of us, as instructors, feel we need an exam problem for each of the significant topics that has been covered, but this way of thinking leads to exams with many short equation-oriented problems. What is this testing other than a student's ability to memorize and manipulate equations under time pressure? Two or three deeper problems, with adequate time to think about them, do a better job of testing what we *really* want students to know about physics.

To achieve these Five Lessons takes a somewhat slower pace than a conventional course, giving students more time to assimilate ideas and practice using the new information. Lecture mode is highly efficient *for the instructor,* because it's not hard to package all the major ideas of a typical chapter into a single lecture. The student, however, needs to see the ideas repeated from several perspectives, needs to see examples and demonstrations, and needs to confront and resolve conceptual issues—all time-consuming affairs.

These Five Lessons don't dictate the exact mode of instruction. Active engagement can be carried out in large lecture halls, in small classes, in a *workshop physics* setting such as that of Laws (1991), or in a computer-based *studio physics* setting such as that of Wilson (1994). Some activities work better than others in each of these settings, as will be discussed more in the next chapter, but significant student learning is possible in each of these classroom environments.

An Active-Learning Classroom

I understand the material. I just can't solve the problems.
Attributed to many thousands of anonymous
students who have taken introductory physics
throughout the years.

How many times have you heard students say the above? They've followed your crystal-clear lectures and the textbook's lucid explanations and derivations ("It all makes sense."), but homework and exam problems seem to be written in an incomprehensible foreign language. The difficulty, of course, is that students have a very different definition of *understand* than we do. We expect understanding to be demonstrated by an ability to work problems, answer questions, conduct experiments, and so on, not by memorization of facts and formulas. This level of understanding is especially difficult when it involves conceptual change and an ability to reason with and about ideas.

But how else might we help students gain a higher level of understanding? The fundamental question seems to be, "If I don't lecture, what else could I do?" This chapter provides an overview of a variety of alternatives.

So What's Wrong with Lectures?

The evidence is now in: The lecture mode of instruction is simply not an effective vehicle to help most students reach a satisfactory level of understanding.

Alan Van Heuvelen (1991a) sums up the situation:

> Historically, we have relied on expository lectures—telling students the physical rules that seem to guide the universe and demonstrating how to use the rules to solve problems. . . . This is a very efficient method to transmit information in terms of the

time interval needed. We know the concepts and techniques, and students do not. Why not just tell them? Study after study indicates that this expository method is very ineffective—the transmission is efficient but the reception is almost negligible.

The great majority of instructors were educated entirely in lecture mode—and were likely in the tiny minority for whom lecture-based instruction works well (although it's impossible to rerun the experiment to find out how we would have fared with more active learning). It's hard for us to see the difficulties, much less to recognize alternatives. A few trouble spots are

- Listeners to serious information have an attention span—at best—of 10 to 15 minutes.
- Lecture information passes by too quickly for critical or contemplative thought. But without such thought, most of the information is lost to the vagaries of short-term memory. How much do you recall, other than the broadest outline, from the physics seminar or colloquium you attended last week? Would you want to be tested on it?
- Most students don't know *how* to listen to a lecture. Note taking can help commit important points to memory, but have you ever looked at student notes to one of your lectures? They're frightening to behold—a few random equations that you wrote on the board, but without connecting thoughts and, in most cases, without even specifying the context or defining the symbols.
- Most lectures simply reiterate material already in the textbook.
- Most lectures focus on formal issues—derivations and deductions—rather than on physical phenomena or concepts.

Richard Hake (1998) has collected data from high schools and colleges across the country on student performance on the Force Concept Inventory of Hestenes, Wells, and Swackhamer (1992a). The FCI tests students' conceptual understanding of Newtonian mechanics, and it is given as a pretest (first day of class) and a posttest (end of term). Several studies have shown that FCI scores correlate well with problem-solving ability: Students who score the highest on the FCI posttest usually score the highest on quantitative problems. More significantly, conceptual understanding as measured by the FCI seems to be a *prerequisite* to good problem-solving ability.

Hake measures the effectiveness of instruction by the *gain G* on the FCI, defined as

$$G = \frac{\text{posttest average } \% \; - \; \text{pretest average } \%}{100 \; - \; \text{pretest average } \%}$$
$$= \text{fraction of the maximum possible gain},$$

where "average" means the class average. Possible gain ranges from 0 (posttest average = pretest average, no learning) to 1 (posttest average = 100, perfect learning). With over 6000 students in his sample, Hake finds that conventional lecture-mode instruction classes have gains $G = 0.22 \pm 0.05$ — that is, the class average goes up only 22% of what is possible. This result appears to be true regardless of the pretest score and independent of the instructor.

By contrast, Hake finds that interactive-engagement classes have $G = 0.52 \pm 0.10$, more than twice the gain of conventional instruction. This result has high statistical significance. It is matched by data showing that the problem-solving skills of students in these classes increase by a comparable amount.

Active Learning

But what is interactive engagement, or *active learning* as it's also called? The cornerstone of a constructivist teaching philosophy is that students must *construct* their knowledge, through interaction with the ideas and materials, rather than simply receive knowledge. This basic tenet has been implemented in a variety of different formats, but all successful implementations have the following characteristics:

- Students spend much of class time actively engaged in doing/thinking/talking physics—not listening to someone else talk about physics.
- Students interact with their peers.
- Students receive immediate feedback on their work.
- The instructor is more a facilitator, less a conveyor of knowledge. "A guide on the side, not a sage on the stage" is a simpleminded but memorable cliché that makes the point.
- Students take responsibility for their knowledge. This includes participating in activities, studying the text, and completing the assignments. "You didn't talk about this in class" is not an acceptable excuse for missing an exam question— assuming that the question was based on material that was well described in the textbook.

Before getting into specific examples, it is worth saying what active learning is *not.* Lecture classes that encourage student questions and discussion are—with some exceptions—not active learning environments. Only a small fraction of students ask questions or participate in open discussions. The majority of the class are still passive watchers and listeners. In fact, most students probably listen to their peers even less closely than to the instructor. Questions and discussion can be important *components* of an active learning environment, particularly in smaller classes, but they don't constitute active learning in and of themselves.

"Active learning" does not connote a single type of teaching. There are a variety of approaches to active learning, many described in the next section, and an instructor needs to select what will work best in his or her local situation. Some approaches work well in large lecture halls, others really are dependent on smaller groups. Some require training TAs (although several large universities have reported that this is not as high a barrier as many would assume), others can be carried out by the instructor alone. Some can be effective with small changes in teaching style, others require a significant overhaul. The point to make, as Hake has documented, is that *they all work!* Interactive engagement of *any* form is more effective than conventional lecture instruction.

A Dozen Things You Can Do to Change Physics Education

This section summarizes 12 different approaches to active learning. They range from relatively small changes in lecture (Mazur's peer instruction) to the total elimination of lecture (Laws' *Workshop Physics*). The following table indicates, roughly, the suitability of each of the approaches to large classes (>40 or 50), small classes (<40 or 50), recitations, and laboratories.

The key to the table is:

+ Works very well.
? Might work well if local adaptations are made.
− Generally not suited to this format.

Activity	Large class	Small class	Recitation	Laboratory
OCS (Van Heuvelen)	+	+	+	?
Cooperative groups (Heller)	?	+	+	?
SDI labs (Hake)	−	?	?	+
MBL (Thornton and Sokoloff)	−	−	?	+
Interactive demos	+	+	−	?
Peer instruction (Mazur)	+	+	?	−
Think/pair/share	+	+	+	?
Ranking tasks	+	+	+	?
Tutorials (McDermott)	−	?	+	?
Workshop Physics (Laws)	−	← + integrated environment + →		
Studio physics (Wilson)	−	← + integrated environment + →		
Research-based textbooks	+	+	+	−

Laboratory activities indicated as ? are not traditional laboratory activities. However, some schools might consider supplementing, or even replacing, traditional labs with alternative active-learning activities if there are few active-learning opportunities in lectures or recitations.

The descriptions that follow are not exhaustive, but they should give you the flavor of a variety of alternative approaches to physics instruction. Check out the references for more details on methods that might work for you and your situation.

Overview, Case Study Physics (Van Heuvelen, 1991a, 1991b): *Overview, Case Study Physics* is both an approach to the course format and a set of supplemental materials intended for use in class. These techniques were designed with large classes in mind, but they work equally well in smaller classes. Alan Van Heuvelen divides the material for the semester into three large blocks. Each block (Newtonian mechanics, for example) begins with a qualitative *overview* that establishes conceptual ideas and confronts students' alternative conceptions. The material is then developed quantitatively, and students learn expert-like problem-solving techniques. Finally, each block concludes with more sophisticated *case studies* that require integrating several concepts and techniques. Case studies help students to see how the different concepts are related and to build a coherent knowledge structure.

The instructor may spend some class time showing how different techniques are applied, but there is no formal lecture. Much of class time is spent having students work on various exercises and problems presented to them on Active Learning Problem Sheets (available as an ALPS Kit (Van Heuvelen, 1994)). The problems are conceptual and qualitative during the overview phase, gradually becoming more quantitative during the development and case study phases. Students first work individually, then discuss their results with a neighbor. The instructor may poll the class and, if there are differences of opinion, ask students to support their position or engage them in a class discussion. Finally, the instructor summarizes the correct response, giving students immediate feedback. An instructor's answer or solution is an important factor, even when nearly all students report the correct answer. Many students are not confident in their results and need reinforcement on the reasoning steps.

During the development phase of each block, where quantitative information is learned, the ALPS worksheets reinforce an expert-like problem-solving strategy. Students are required to begin with a problem description, proceed through intermediate analysis stages (adding coordinate systems, free-body diagrams, and so on), and reach the equation stage only at the end.

Van Heuvelen has compared engineering students in a conventional calculus-based physics class to similar students in an OCS class:

	Conventional	OCS
Qualitative questions on Newtonian physics - pretest	44%	48%
Qualitative questions on Newtonian physics - posttest	53	86
Quantitative problems from the Advanced Placement Test	36	55

The 4% pretest difference is not statistically significant, but the qualitative and quantitative posttest differences are highly significant. OCS students excel on qualitative problems, as might be expected because of the strong emphasis on conceptual understanding. But particularly striking is the equally strong improvement in their quantitative problem-solving skills.

Cooperative Groups (Heller, Keith, et al., (1992) and Heller and Hollabaugh (1992)): Patricia Heller and her colleagues at the University of Minnesota have used somewhat more formal cooperative groups to teach good problem-solving skills. Their activities were carried out in recitation sections of approximately 18 students, but they could easily be classroom activities in smaller classes. Implementation in large lecture halls would be difficult, but not impossible. On the other hand, large lectures are typically accompanied by recitations. The use of cooperative problem-solving groups in recitation sections could supplement more modest active-learning activities during lecture. The authors report no significant difficulties training graduate student TAs to run the cooperative groups.

First, lecture is used to teach—through demonstrations and supplemental handouts—an explicit five-step expert-like problem-solving strategy. Then in recitations, students are organized into groups of three, or occasionally four, to practice solving *context-rich problems.* Unlike most standard textbook problems, context-rich problems are set within a short story that establishes a reason for the problem, generally do not specify the unknown quantity ("Will this design work?"), often have irrelevant information, and sometimes are missing information that students need to estimate. Groups report their results to the instructor or, sometimes, to other groups.

Several factors are found important to make this a valuable learning experience. First, the *instructor* must form the groups. At Minnesota, the initial groups are formed at random. Once the first test scores are known, mixed ability groups are formed with one student each from the top, middle, and bottom thirds of the class. Students rotate to a different group every three weeks. Group sizes of three are optimal, two are below critical mass, and five or more generally finds some members not engaged. As for gender balance, groups should be homogeneous or have two women, one man. Groups with a single woman tend to not function well, even if the woman is the "top third" member. Members need to face each other, so a room with moveable seats is preferable. If seating is fixed, students need to sit two in one row with the other behind, rather than three in a row.

In cooperative groups, as opposed to more informal groups, students are each assigned a *role* to play. In groups of three, one is the Manager (takes charge, keeps on task), one is the Skeptic (purposefully raises as many questions and objections as possible), and one is the Recorder (keeps notes, writes up any solution to be turned in). The roles rotate every week, so each member has each role once during the three-week life of a group. The goal is to try to prevent one student from dominating a group. Students do need some initial instruction about the roles, and one task of the instructor is to ensure that students stay in their roles.

An essential feature for designing group activities is referred to as "positive interdependence but individual accountability." There has to be a tangible reward for functioning well as a group, but weaker students can't "coast" by letting the group do all their work. At Minnesota, the basic method for accomplishing this is to divide exams into group problems and individual problems. On the group problem, each group submits a single answer and all members receive the same score. This is typically 20–25% of the total exam points. The remainder of the exam is worked individually, but the problems are the same type of context-rich problems students have been learning to do in groups. There's a strong motivation for each group to function well as a group, and individuals have a strong motivation for learning problem-solving skills from their peers.

Results from several years of cooperative groups at Minnesota show the following:

- Group solutions are better than the individual solutions of the best student. The group is not just being "carried" by the best student but is able to go beyond what any individual in the group can do.

- Problem-solving abilities increase for *all* students in the group. The weaker students are not just coasting; they're definitely learning from the experience.
- Students' problem-solving abilities are significantly better than those learned from conventional instruction, as measured by identical final exam problems in classes with and without cooperative group activities. The cooperative group students averaged approximately 20 percentage points higher than the conventional students.
- The improvement in problem-solving skills was most pronounced in the qualitative analysis and conceptual understanding of the problem, least in the mathematical aspects. This suggests that students have benefited by sharing conceptual and procedural knowledge, the portions of problem solving that are generally less explicit.

The essential features of these activities seem to be: 1) The explicit problem-solving strategy, and 2) students interacting with and learning from each other. Many students don't feel comfortable talking with the instructor, but they can interact with and receive feedback from their peers. This is much more beneficial than studying alone and apparently more beneficial than the unstructured group activity of "studying with friends."

Socratic Dialog Inducing Labs (Hake, 1987, 1992): Richard Hake has developed a series of laboratories at Indiana University that he calls *Socratic Dialog Inducing Labs,* or SDI labs. These are a replacement of more traditional measurement-oriented labs. In an SDI lab, a group of students carries out fairly simple physical activities (such as pushing blocks under various conditions or swinging a bucket of water over their heads), then works through a series of questions asking them to analyze and explain their observations. The questions are designed to elicit students' well-known alternative conceptions and to provoke discussion among group members. At any point where they are stuck, students call over a *Socratic dialogist,* as the instructor is called. The instructor's role is not to answer questions, but to *ask* questions in a Socratic fashion. The dialogist hopes to guide students to a correct interpretation through leading questions about their observations and their conflicts.

In conjunction with these lab activities, lectures are devoted to problem solving based on qualitative analysis, contrasting different conceptions, and demonstrations. There is no effort to give derivations or to paraphrase the text, although students are responsible for text material.

Students from SDI labs score significantly higher on the FCI than students taking regular labs. Their success seems to be due to two features: 1) The activities are closely focused on primary physical phenomena and the conceptual understanding of those phenomena, and 2) students interact with their peers, although in a less formal structure than the cooperative groups at Minnesota.

Because many of the materials are simple (masses, spring gauges, etc.), some schools may have enough of them to do SDI lab activities in recitation sections or small classes. Large lecture classes that are unable to use much or any interactive engagement in the classroom may find these, or similar, activities to be an excellent, low-cost substitute for traditional labs.

Microcomputer-Based Labs (Thornton and Sokoloff, 1990, 1998; Sokoloff et al. 1999): Ron Thornton and David Sokoloff, in collaboration with Priscilla Laws, have developed a series of widely used laboratory experiments based on the real-time acquisition and display of data. These activities share many pedagogical features with Hake's SDI labs, but the microcomputer-based labs (MBLs) emphasize data collection and are closer in spirit to a traditional lab. The mechanics experiments are available as the *RealTime Physics* lab manual published by Wiley. Other experiments and the associated computer interfaces are available from Vernier Software and PASCO.

MBLs have two basic premises: 1) That computers can take the drudgery out of collecting and graphing data, allowing students in a limited time period to focus on the physics rather than on the mechanics of data collection, and 2) that *real time* display of graphs, as the data is being collected, allows students to associate the shape of the graph with the behavior of the object. No prior computer experience is necessary. Experience shows that students often adapt to this form of laboratory quicker and easier than the instructor!

The mechanics laboratories are the most highly developed. The basic tools are an ultrasonic motion detector and a force probe. Students begin by measuring the motion of their own bodies, then the motion of a cart on a track. Numerical differentiation produces real-time velocity and acceleration graphs, in addition to the position graphs, and structured questions lead students to a solid understanding of velocity and acceleration. The force probe shows clearly that force is directly related to acceleration, and the collision of two force probes gives by far the clearest and most convincing demonstration of Newton's third law that I have ever seen.

Many activities require students first to predict what will happen, later to write an explanation. Much of the success of these labs comes from peer interactions as students argue with each other about these. Because most groups have divided opinions, it's quite an experience for the instructor to be in a room where students are actively—sometimes heatedly—arguing *physics,* rather than arguing about which equation to use. The arguments are often rough and prone to errors, but this is real learning taking place.

The experiments, similar to SDI labs, are focused on basic physical phenomena rather than on more elaborate measurements that require detailed analysis. These activities would be a good precursor to more traditional measurement-oriented labs because they acquaint students with the basic concepts of observing and measuring. Like SDI labs, these would be an effective use of labs at schools where large lecture classes make other interactive-engagement activities difficult.

Comparison with conventional instruction shows that MBLs are highly effective for teaching a conceptual understanding of motion and force. However, the developers point out that it is not the computer tools themselves that are the key element, rather the curriculum in which the tools are used. Students must be guided to study the appropriate phenomena and asked the right questions about what they've measured with the computer.

Interactive Demonstrations: We all do lecture demonstrations in class, but in many cases they're not as effective as we would like to think. We tend to rush through them, because we have too much else to "cover," and we tend to do them after-the-fact as "proofs" of the theory we've been deriving. We rarely spend enough time describing the apparatus itself for students to understand what they're seeing (another issue that's obvious to us, from long familiarity, but not to students), and we hardly ever explore the implications of the outcome. Students do find them entertaining, but their pedagogical usefulness is questionable.

Demonstrations go right to the heart of our subject—we're observing phenomena, rather than writing equations on the board—and they *can* be highly effective with a little extra planning and time. The key is to have students engaged in the demonstration along with you rather than being passive observers. There are several ways to do this.

First, take the time to explain the apparatus and how it will be used. Not much explanation is needed for a ball rolling down an inclined plane, but later in the year—thermodynamics or electricity, for example—explanations may need to be fairly detailed. It's also good to note what some of the *conceivable* outcomes might be, being careful not to reveal what's actually going to happen.

Ask each student to write down a prediction of the outcome, then have them discuss it with a neighbor. Now they have a stake in the outcome, so it's no longer merely entertainment. Get a sample of the predictions from students before actually beginning the demonstration, perhaps with their reasons. Next, conduct the demonstration— maybe more than once if many students have given wrong predictions. Finally, discuss the implications. In what ways does this contradict students' alternative conceptions, and how is this seen as evidence of the physics concepts you want them to accept as correct? Consider doing variations that have slightly different outcomes, or let students suggest a variation for you.

Demonstrations with a rather unexpected outcome can generate lots of interest, but it's also good to include a reasonable number that most students will predict correctly so as to reward their learning. The main points are to make the demonstration a significant part of the class and to spend enough time on it to explore it thoroughly. The subject matter chapters later in this guidebook will suggest specific demonstrations.

Another way to make a demonstration interactive is in situations where the nature of the outcome is clear but the size of the effect is not. After the initial explanation, have students—working in pairs or groups—calculate the expected result as a problem-solving exercise. For example, "I'm going to roll the ball down the ramp, and at the bottom it will strike the pendulum. To what angle will the pendulum swing?" Once again, they have a vested interest in the outcome and are no longer passive observers. After the demonstration, you should review the solution. Of course, it's good to know in advance that the outcome will be acceptably close to the calculation!

Many of the MBL laboratory activities can be very effective as lecture demonstrations if you have the ability to project the computer display (Sokoloff and Thornton, 1997). This is especially true for collisions, where the force probes are excellent for revealing interactions and impulses during the collision of carts on a track. You really can enact the situation "If a mosquito collides head on with a truck, how does the force of the truck on the mosquito compare with the force of the mosquito on the truck?" An entire class spent on variations of this demonstration, whose outcomes are always surprising to students, is time well spent.

Vernier Software offers a sample of MBL mechanics labs in a package called *Interactive Lecture Demonstrations.* These come with response sheets for student use and with helpful hints about using them effectively. Alternatively, you can make your own MBL interactive lecture demonstrations by simple modifications of *Real-Time Physics* experiments.

These activities can be performed in large classes as easily as small ones. The keys are

- To focus attention on the demonstration as an important issue, rather than a quick piece of entertainment.
- To engage the students in predicting and discussing the outcome. Computer-controlled experiments with graphical displays are nice, but they're certainly not essential for carrying out very effective interactive demonstrations.

Peer Instruction (Mazur, 1997): Eric Mazur has developed active learning techniques that work well for instructors teaching large classes. These are described in his book *Peer Instruction: A User's Manual.* He advocates "minimally invasive" techniques that can be used with any textbook, don't require significant changes in content coverage, and can be used by an individual instructor without regard to what happens in other sections or in recitations and laboratories (although the effectiveness will be maximized if recitations and laboratories have a similar orientation). They do require the instructor to change how class time is used and how exams are written.

Mazur's approach is based on the recognized difficulties with lectures and on the positive influence students can have on each other. He divides each hour lecture into three or four well-defined parts, each with a specific point to make. During each part, he gives a 7–10 minute mini-lecture followed by a 5–8 minute *ConcepTest.* The mini-lectures contain no definitions or derivations but, instead, focus on concepts, reasoning, and intuition. The *ConcepTest* is a short multiple-choice *conceptual* question tied to the mini-lecture. An example he gives, which would follow a mini-lecture on the simple pendulum, is

> A person swings on a swing. When the person sits still, the swing oscillates back and forth at its natural frequency. If, instead, two people sit on the swing, the natural frequency of the swing will be
> a. greater.   b. the same.   c. smaller.

The *ConcepTests* are on transparencies (or could be presented on a computer display) so that they can be presented quickly. Once the question is posed, students get 1 minute to think about it individually and *write down* an answer. Next—this is the peer instruction—students spend 1–2 minutes discussing their answer with a neighbor, at the end of which they can, if they wish, revise their answer. The instructor polls the class to discern the distribution of answers, goes over the correct answer, and takes questions. One mini-lecture/*ConcepTest* cycle is intended to last 12–18 minutes.

Polling the class is important because students need to know that they're not alone when they make mistakes. A show of hands for each response is a possible procedure, but many students—especially those unsure of their answer—are hesitant to raise their hand or may switch choices just to be in the majority. It's better to have a simultaneous response by the entire class. Some schools are experimenting with an electronic keypad system called *Classtalk*. Students enter a response on a hand-held keypad during a time window of a few seconds, then a computer tallies and displays the results. This system has the advantage that results can be saved for later analysis and comparison.

A simpler version used successfully at Southeastern Louisiana University and elsewhere (Meltzer and Manivannan, 1996) gives students five large flashcards, labeled A to F, that they keep in their notebook. At the instructor's request, all students simultaneously raise a flashcard to show their answer. The instructor scans the room for a rough percentage distribution and writes the result on the board. Students can't easily see each others cards, so they remain fairly anonymous to each other.

ConcepTests give immediate feedback to the instructor. Depending on the class response to a *ConcepTest,* the instructor can proceed to the next topic or can reiterate the previous topic, followed by a second *ConcepTest* on the same topic. Because of the need for flexibility, the instructor must be ready to shorten or omit the last planned mini-lecture for the day. Nonetheless, Mazur feels it is better to cover less and have understanding than to rush ahead without adequate understanding.

The optimum *ConcepTests* have an initial student response rate in the range 40–60% correct. Peer instruction is not successful for lower response rates because too few students know the right answer to help their neighbor. Mazur has collected response sheets from students showing their initial answer to a *ConcepTest* and their answer after peer interactions. He finds that about half the initially wrong responses change to a correct response after peer interaction, but very few initially right responses are changed to wrong. Thus peer interaction appears to be highly effective.

Because the mini-lectures occupy only one-third to one-half the time that a conventional lecture would, the peer instruction approach (as is common with most of the activities described here) places a much larger responsibility on the student to learn material by reading the textbook. Mazur enforces this by a short *reading quiz* at the beginning of every class period. He continues to cover the same amount of material as before beginning to teach this way, but achieves this in part by not working example problems in class.

Many instructors who would like to use his techniques will probably want to slow the pace, show more examples, and cover less material. However, instructors locked into a common syllabus with parallel sections may not have that choice.

Think/Pair/Share: Think/pair/share is a less structured but more flexible version of Mazur's peer instruction. It is an active learning technique that can be used in classes of any size in a wide variety of circumstances. The essence of think/pair/share is that the instructor poses a question to the class, students think about it individually (for a time period ranging from a few seconds to a couple of minutes, depending on the question), students *write down* their response, then students share and discuss their answer with their nearest neighbor. The instructor usually joins in with one or two pairs during the discussion to hear what they're saying. Once discussions reach some closure, the instructor can poll the class by a show of hands or can simply ask a few students what they think. By finding students with differing responses, the instructor can ask one student to explain his answer to another, then ask the second if she agrees or disagrees.

Many students need initial prompting and reassurance to engage in these discussions, but most soon come to find them enjoyable experiences. (I find that engineering students in the calculus-based introductory class are initially less willing to take part in this activity than students in the algebra-based intro course. Students in the algebra-based course are always eager to talk, but I sometimes need to direct engineering students to *discuss* their answer with their neighbor rather than just saying "I think it's option b" and stopping!) It is especially important that the instructor give serious consideration to all answers and not in the slightest way put down incorrect answers. Students are highly sensitive to this issue. Students need assurance that errors during the learning process are never penalized and can even be healthy, that the goal is to be error-free by exam time.

Think/pair/share is especially effective for interactive demonstrations, where you ask students to predict the outcome of a demonstration and then to share with their neighbor. It also works very well with conceptual questions, which might be pre-arranged questions, similar to Mazur's *ConcepTests,* or simply impromptu questions that come up during class.

I often use the exercises from my *Student Workbook* as think/pair/share questions in class. Some of these exercises are conceptual and graphical questions. Others, such as drawing free-body diagrams, are focused on developing specific aspects of a problem-solving strategy. A few well chosen questions for use and discussion in class can help clarify conceptual difficulties and give students a head start on the end-of-chapter problems.

Another good idea for think/pair/share questions are the "Physics Jeopardy" questions developed by Alan Van Heuvelen and David Maloney (Van Heuvelen and Maloney, 1999). In these, as in the TV game show, the student is given the answer and asked to describe the situation or even to write the problem! As an example,

An electric circuit is described by

$$12 \text{ V} = \frac{I}{[1/(5 \, \Omega + 6 \, \Omega) + 1/(8 \, \Omega)] + 14 \, \Omega}.$$

Draw the circuit, then determine the current.

Jeopardy questions are popular with students, and they provide you with an opportunity to focus on physical principles rather than on how to select and use a formula. Later sections of this guidebook will provide some sample Jeopardy questions.

Think/pair/share also works very well for having students solve sample problems using a structured problem-solving strategy, such as Van Heuvelen's OCS method. An entire problem is too big to take in one bite, so students are asked to do *just the first step* in the strategy (such as draw an appropriate coordinate system or draw a motion diagram). Then they share with a neighbor, get feedback from the instructor, and only when everyone has the first step correct do you move them on to the second step. This is time consuming (even a fairly straightforward problem may take 20 minutes of class time to work through this way), but it's *highly* effective instruction in problem solving because each student is being guided through the steps with peer interactions and instructor feedback at each stage. One or two such example problems a week gives students confidence that they can apply the problem-solving procedures on their own for homework. Needless to say, students must have read the chapter and you must have first spent class time working through some of the conceptual issues before getting to the problem-solving aspects of a topic.

For think/pair/share to be effective, it's important that students *write down* their response from their initial thinking. This forces them to commit to a position that they then may have to defend. Without writing down their own response, it's too easy to turn to their neighbor and ask, "What do you think?"

Ranking Tasks (O'Kuma, Maloney, and Hieggelke, 2000): Ranking tasks are pencil-and-paper exercises that ask students to make a comparative judgment about several similar physical situations. The possible responses are chosen to elicit common student misconceptions.

Ranking tasks are quick and easy to administer, and they lend themselves nicely to a Peer Instruction or think/pair/share approach. They can be used in class, assigned as homework, or used on exams. As an example, the ranking task shown below asks students to compare the forces on six charges in an electric field.

> A large region of space has a uniform electric field, as shown by the arrows. At the origin, the electric field is $30 \, \hat{i}$ V/m and the electric potential is 100 V. Six charges are placed at rest in the field, as shown. Rank from greatest to least the magnitude of the electric force on the six charges. Please explain your reasoning.

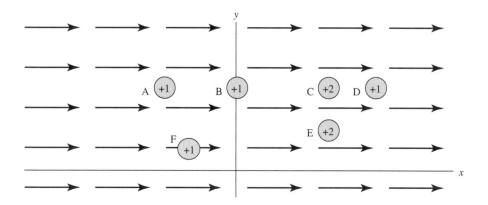

A correct response would be

$$C = E > A = B = D = F.$$

All ranking tasks follow a similar approach, so students quickly learn what is expected of them.

Although small computations are sometimes needed, ranking tasks focused on conceptual beliefs rather than equations. As such, they are a good tool for generating class discussion. The booklet by O'Kuma, Maloney, and Hieggelke contains well over 100 ranking task exercises from all areas of physics.

Tutorials (McDermott et al., 1994; Shaffer and McDermott, 1992; McDermott and Shaffer, 1998): For schools where little can be changed in lecture, due either to constraints or to instructor preferences, McDermott and her physics education research group at the University of Washington have developed a series of tutorials for use in recitation sections. The tutorials lead students through a series of conceptual and graphical questions known, from their research, to elicit errors based on students' alternative conceptions.

The tutorials are an integrated set of pretests, worksheets, and homework assignments. The sequence begins with a pretest that identifies areas of difficulty. Students interact with peers on worksheet exercises and receive instructor feedback at each stage. Tutorials for most introductory physics topics are now available in a booklet published by Prentice Hall.

This activity can be implemented independent of lectures by using graduate student TAs and one instructor who is in charge of recitations. It is very effective at addressing students' conceptual difficulties. TA training is an essential part of effective implementation, both because graduate students are not familiar with this style of instruction and because many TAs continue to hold some of the same alternative conceptions as the students.

Workshop Physics (Laws, 1991, 1997a): Priscilla Laws and her colleagues at Dickinson College have developed a successful new approach to teaching physics without lectures. The distinction between lecture, recitation, and lab is abolished, and students meet for three two-hour periods each week. Each pair of students sits at a computer station and has available a variety of laboratory equipment, most interfaced to the computer for rapid data acquisition. The goal, as befits the *workshop* title, is to learn physics by doing physics. Students are actively engaged in every aspect of the course.

Students work through a series of activities in the *Workshop Physics Activity Guide* (Laws, 1997b). Activities are varied, including collecting and analyzing data, working through conceptual issues, solving problems, using spreadsheets to do calculations, and—more recently—using video capture and analysis tools. Students interact with their peers and instructor, and occasionally the instructor stops the activity for a short mini-lecture or whole-class discussion. Nearly all activities involve the use of the computer. These include MBL tools for data acquisition, video capture and analysis tools, spreadsheets and graphics programs for modeling and displaying results, simulations, and special-purpose programs such as *Interactive Physics*.

Workshop Physics is an approach to teaching physics, not a curriculum unto itself. It has been very successful at a number of smaller colleges that have the flexibility to implement it. This approach requires a high instructor-to-student ratio, and it may not be feasible for large universities that teach hundreds or thousands of students each term.

Studio Physics (Wilson, 1994): Jack Wilson at Rensselaer Polytechnic has developed an instructional approach that is similar in many respects to *Workshop Physics.* It also abolishes the distinction between lecture, recitation, and laboratory, with students attending a two-hour session two or three times a week. *Studio Physics* is somewhat more formally structured than *Workshop Physics* and is intended to function with the large number of students in large universities.

A dedicated room (the *studio*) is outfitted with computers, and students (typically 60 at a time) sit two per computer. The room is arranged so that students normally have their backs to the instructor, but they sit on swivel chairs that allow them to pivot around to the instructor. The instructor can see the computer screens while students work in this arrangement, and students are forced to interrupt their activities if called upon to turn around to the instructor.

As with *Workshop Physics,* students have a variety of computer-interfaced experiments and computer tools. *Studio Physics* gives somewhat more emphasis to problem solving and numerical computation than does *Workshop Physics,* and there's somewhat more emphasis on mini-lectures by the instructor. These are difference of degree, however, and the individual instructor could move along a continuum between *Studio Physics* and *Workshop Physics.* Because *Studio Physics* was designed at a large university, it is more tightly time constrained than *Workshop Physics,* with new groups of students moving into the studio every two hours. Successful implementation requires training of both faculty and graduate TAs.

Research-Based Textbooks: Several new textbooks have been influenced by physics education research. Although stylistically quite different, each of these books attempts to

- Deal directly with students' conceptual beliefs.
- Provide a balance between qualitative reasoning and quantitative calculation.
- Reduce the pace of the course and the amount of material covered.
- Support an active learning classroom.

Although these textbooks could be used in a traditional lecture format, that was not the intent of the authors. Any of these books would nicely complement the active learning techniques outlined in this chapter.

Four research-based textbooks for calculus-based physics are

Ruth Chabay and Bruce Sherwood, *Matter and Interactions* and *Electric and Magnetic Interactions* (John Wiley & Sons, 1999). This is a two-volume set covering the major topics of introductory physics. Chabay and Sherwood's approach is based heavily on observations and simple experiments carried out by students, and in this regard it follows Arons' advice to work from the concrete to the abstract.

Randall Knight, *Physics: A Contemporary Perspective,* Preliminary Edition, and *Physics: A Contemporary Perspective Student Workbook* (Addison-Wesley, 1997; forthcoming in 2004 as *Physics for Scientists and Engineers: A Strategic Approach,* First Edition). My textbook is the most traditional of these four in terms of content. It has a strong emphasis on dealing with students' conceptual beliefs, multiple representations of knowledge, and on developing coherent problem-solving strategies. The *Workbook* is an extensive collection of qualitative and graphical exercises.

Thomas Moore, *Six Ideas That Shaped Physics* (McGraw-Hill, 1998). Moore's book (six slim volumes, one for each of the six ideas) is an outgrowth of the most successful new curriculum that was designed and tested under the auspices of the Introductory University Physics Project in the early 1990s. It has a strong emphasis on modern physics.

Frederick Reif, *Understanding Basic Mechanics* (John Wiley & Sons, 1995). This text and accompanying workbook cover only the subject of mechanics. It is particularly strong on knowledge structures, which is one of Reif's research specialties.

A research-based book for teaching physics to prospective pre-college teachers is

Lillian McDermott and the Physics Education Group at the University of Washington, *Physics by Inquiry* (John Wiley & Sons, 1996). This is a set of laboratory modules that develop conceptual understanding and scientific reasoning. Students work in collaborative groups.

Publishers are always willing to send complimentary copies to instructors. Publisher Web sites are listed in the Resources section of this guidebook.

Just the FAQs

This final section addresses some of the concerns and Frequently Asked Questions that instructors have about active learning.

Can I cover the same amount of material? Mazur, at Harvard, reports no change in coverage with his peer instruction methods, but he places a larger burden on the students to learn problem solving elsewhere. McDermott's tutorials are intended for use in recitation, independently of what's happening in lecture—although they may not be optimal if lectures are unchanged. Most developers of the interactive-engagement methods described above cite a 15–20% reduction in material covered to allow time for active learning. All think that this reduction is more than adequately compensated for by increased student learning.

How can I get students to cooperate? One important way to encourage cooperation is to establish a class atmosphere in which students feel free to express their ideas without fear of being criticized. Discuss ideas, but never be critical of them in a negative way. Let students know that others hold the same view. Whenever possible, let the correct answer be determined by an experiment or demonstration rather than simply presented by you in your role as an authority.

Another useful approach is to use an absolute grading scale, rather than to grade on a curve. This is intended to remove the concern of many students that they're competing with each other for grades. Most instructors, if they examine their grading over several terms, find there's really very little difference in their grading scale from one term to another. Thus it's not difficult to establish a fixed grading scale and to announce it at the beginning of the term. You can then say that, "Your grade is your responsibility, and someone else receiving a high score in no way affects your grade."

Won't many of these activities help the best students while leaving the poorer students behind? You might think that poorer students would try to coast along in cooperative groups, or that other activities would reward the better students who have more initiative while penalizing the poorer students. All evidence is to the contrary. The benefit of active learning appears to go disproportionately to the lower third of the class. The best students can likely learn under any circumstances—active learning, lecture, or just on their own. It is the weaker students who are most severely impacted by conventional lecture instruction, continually getting further and further behind with no understanding, and who are most assisted by active learning approaches that engage them and help them over the difficult hurdles.

Will students really read the text? Nearly all active learning methods place a greater responsibility on the students to read and learn from the text. Practitioners of active learning find that students can and will do this *if* you continually make your expectations clear and *if* you continue running class on the expectation that they have done the reading. If you relent and begin to lecture just because the students seem not to have read, then they'll know you weren't serious and will have little incentive. Reading quizzes, described earlier, are a good motivational device that take minimal time and effort on your part.

I don't have the right personality. Can I do this? Physics instructors have few, if any, role models for teaching an interactive-engagement course. It seems rather scary, but it really takes no special personality—just a willingness to give it a try. We've all interacted extensively with our peers and colleagues in graduate school and as faculty members, and we know that those interactions are an important part of our professional lives. Many of us have graduate students that we interact with one-on-one. So it's not that we lack the "right personality" to interact with students, it's simply that we don't know how to go about it in the classroom. The purpose of the preceding section was to outline a variety of possibilities. Pick one or more that look promising to you, check the references for more details, then just jump in and try it! Few instructors who switch to an active-learning classroom ever go back to a lecture mode, and most begin thinking about how to make their upper-division and graduate classes more interactive.

Will it work for me? Yes.

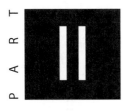

P A R T

II

Topics in Introductory Physics

Introduction

The Preface posed the question, "If I don't lecture, what else could I do?" Many instructors agree that the educational attainment of their students is less than expected or desired. Physics education research has characterized the difficulties that students face and provided new tools for teaching. But when it comes time to prepare class, there's still that fundamental issue: What should I do? It's easy to give a lecture, much harder to think of alternatives.

To address this issue, the chapters of Part Two are a practical, hands-on guide to teaching the major topics of introductory physics. Each chapter is structured the same:

- Background Information
- Student Learning Objectives
- Pedagogical Approach
- Using Class Time
- Sample Reading Quiz or Discussion Questions
- Sample Exam Questions

The Background Information section summarizes what is known from physics education research and highlights the major difficulties students are known to have. The Student Learning Objectives and Pedagogical Approach sections then outline a general approach that is built on the research base and is consistent with the constructivist philosophy of this guidebook.

Finally, the Using Class Time section is a collection of specific suggestions—demonstrations, exercises, discussion questions, and sample problems. This is the section that tries to give a coherent answer to the question, "If I don't lecture, what else could I do?" For most topics, I've indicated the number of days I devote to it. You could certainly "cover" the material faster by lecturing, but the pace I've suggested seems to be about the fastest at which typical university students can assimilate the material.

Nearly every suggestion is classroom tested. This is, in essence, how I teach my introductory physics classes, based on 20 years of "tinkering" to learn what works best. I've given more suggestions than you could possibly use in one course, but this provides flexibility and allows for variation from year to year. In any one year, I probably use half to two-thirds of the demonstrations and exercises that I've described. The important points are that

- There are more than enough student-centered activities to fill the time available, and
- Students learn physics from these activities as well as, and usually better than, they do from lectures.

The Sample Reading Quiz or Discussion Questions can be used either as reading quizzes, to provide an incentive for students to read the chapter before class, or as simple opening questions to launch a class discussion. I've used them both ways. They're intended as starting points, questions that students should be able to answer after a first look at a chapter.

Eventually, though, students have to be tested for mastery of the material. As students know all too well, you're not serious about a topic or an issue unless it's on the exam. Exhortations to think conceptually or to follow a coherent problem-solving strategy will be to no avail if your exams don't reflect this expectation. Test questions that can be worked by manipulation of equations will reinforce an equation-hunting approach to physics. The Sample Exam Questions mix qualitative analysis and graphical interpretation with quantitative problem solving. These questions quickly reveal when a student is adept at algebra but has little idea of the underlying physical concepts.

The suggestions in these chapters are, of course, simply suggestions, not a prescription for teaching physics. Instructors can adopt as much or as little as is appropriate for their students and the situation in which they teach. I expect that most readers of this guidebook will quickly think of many other effective uses of class time. But when it comes time to decide what to do in class, I hope the suggestions herein provide a good starting point.

5

Vectors and Mathematics

Background Information

Surveys (Knight, 1995) have found that only about one-third of students in a typical introductory physics class are knowledgeable enough about vectors to begin the study of Newtonian mechanics. Another one-third have partial knowledge of vectors (e.g., a student who can add vectors graphically but not use vector components), while the final one-third have essentially no useful knowledge of vectors. Surveyed students who were repeating the course generally displayed major gaps in their knowledge of vectors, and this was likely a contributing factor to their previous failure.

Students who can successfully add and subtract vectors are still often confused as to just what a vector *is*. When posed the open-ended question "What is a vector?" they may respond with "A vector is a force" or some similar answer. These students may have difficulty recognizing velocity or acceleration as vector quantities.

Although students have used Cartesian coordinate systems throughout high school, many have a hard time interpreting a statement such as "A vector 'points' in the $-x$-direction." These students are especially prone to making sign errors when decomposing vectors into components.

Many students, especially if they are starting calculus concurrently, are not sure what a *function* is. They don't really understand the notation $x(t)$ or our discussion of "position as a function of time." A not insignificant fraction of students interpret $x(t)$ as meaning x *times* t, as it would in an expression such as $a(b + c)$. Instructors need to give explicit attention to this issue.

Students are easily confused with changes in notation. Math courses tend to use functions $y(x)$, with x the independent variable. This includes graphing y-versus-x and taking derivatives dy/dx. In physics, we use functions $x(t)$, with x the dependent variable. We make x-versus-t graphs and take derivatives dx/dt. Despite how trivial this seems, instructors should be aware that many students are confused by the different notation and need assistance with this.

Finally, students at this stage often lack an operational understanding of differentials and integrals. They're not perturbed by writing expressions such as $dx = x^2$, in which they equate an infinitesimal to a finite expression. Faced with an integral such as $\int v\, dt$, students are likely to pull the v out of the integral, as if it were a constant, rather than recognize that v is an implicit function of t. Physics can help them solidify their understanding of calculus, but you should be cautious about assuming that students have a working knowledge of calculus.

Student Learning Objectives

- To understand the basic properties of vectors.
- To add and subtract vectors both graphically and using components.
- To decompose a vector into its components and to reassemble vector components into a magnitude and a direction.
- To recognize and use the basic unit vectors.
- To work with tilted coordinate systems.
- To understand and use the *basic* calculus of derivatives and integrals and, especially, to use a geometric interpretation in terms of slopes and areas.
- To understand and use proper significant figures.

Pedagogical Approach

Most textbooks represent vectors with boldface notion (e.g., **F**), although more are beginning to use an explicit vector notation (e.g., $\vec{F}$). Students pay little attention to the boldface type. I've found that students handle vectors better when the text uses an explicit notation, when you are extremely consistent about proper vector notation in work you do on the board, and when you insist that students label vectors correctly on all written work. An automatic 1-point deduction for failure to use proper notation is effective enforcement. This may seem to be merely a notation problem, but students who don't label vectors are much more prone to treat them as scalars.

Another notation difficulty is that most texts don't distinguish between F as the magnitude of a force vector (positive values only) and F as the *component* of a force vector (signed quantity) in one-dimension. Experienced instructors interpret the symbol properly by recognizing the context, but this is a major source of confusion to beginners. I highly recommend that you represent the *magnitude* of vector $\vec{F}$ by the explicit notation $|\vec{F}|$ and that you insist students do likewise. Although the $|\vec{F}|$ notation is admittedly cumbersome, students are much less likely to make errors when this notation is used consistently. This notation can be relaxed later, after students are more experienced, when you will want to use simply E and B to represent field strengths.

A difficulty with explicit vector notation is that some students will conclude that *everything* associated with a vector needs an arrow. You'll find that arrows over vector components (scalars) are not uncommon. This reflects an uncertain knowledge as to what constitutes a vector.

Textbooks differ as to whether the initial chapter on vectors includes the dot product and cross product. Because students have sufficient difficulties with basic vector arithmetic, and vector multiplication is not needed for many chapters, I recommend that you postpone this material and return to it when needed.

Using Class Time

As simple as the rules are, students need extensive practice to become familiar and comfortable with vectors. Two class days are desirable, although one will suffice if students have an opportunity for supervised vector practice in recitation or laboratory. Regardless of time, it is far preferable to spend class time with students *practicing* vector problems rather than listening to a lecture about vectors. Vector questions and problems are especially good for starting to teach students how to work with each other in think/pair/share activities or using cooperative groups.

The first examples should focus on basic vector problems and on the graphical addition and subtraction of vectors, without the use of a coordinate system. Good class examples are:

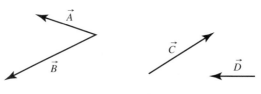

Draw the vectors

a. $\vec{A} + \vec{B}$
b. $\vec{A} - \vec{B}$
c. $2\vec{A} - \vec{B}$

Draw the vector $\vec{C} + \vec{D}$.

The question on the right stumps many students because they don't yet recognize that you can "slide" a vector around to another location as long as you don't change its length or direction. Textbook figures tend to draw the vectors in the "right places," as in the question on the left, so students need to face some less conventional situations.

Coordinate systems and vector components are then introduced. It is better to introduce vector components *without* reference to unit vectors. Once students are comfortable with the decomposition of a vector into components parallel to the axes, then the unit vector becomes a convenient way to express this.

Physicists are rather cavalier in the choice of the angle to call θ, leading to $F_x = |\vec{F}|\cos\theta$ in some cases but maybe $F_x = -|\vec{F}|\sin\theta$ in others. Unlike math books, which insist on defining θ as an angle measured from the positive x-axis, we tend to label and use angles based on their convenience in the problem. You'll want to *insist* that students use a figure to identify the angle they are using in their calculations.

Find the *x*- and *y*-components of vector $\vec{A}$.

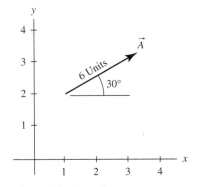

Start with a few examples of finding the components of a vector located at the origin but pointing into different quadrants. Then pose a question such as the figure shown above. This question returns to the issue of whether the location of the vector influences the properties of the vector, and many students will have initial difficulties with a vector not drawn at the origin. Next, draw the same vector in the second quadrant. Because the vector is drawn at a point where *x* is negative, many students will want to give A_x a negative value. Confusion of where a vector *is* with the direction it *points* is a big source of the difficulties students have throughout the study of motion and force. It's important to attack this problem early with numerous examples.

Once students can find components reliably, they can try vector arithmetic problems. For example, give them vectors $\vec{E}$ and $\vec{F}$ and ask them to find quantities such as $3\vec{E} - 2\vec{F}$. Next, practice going the reverse direction with questions such as "Vector $\vec{B} = 3\hat{i} - 4\hat{j}$. Describe this vector as a magnitude and a direction." Textbooks often define the direction of a vector as $\theta = \tan^{-1}(B_x/B_y)$, but this gives a negative angle if one component is negative and an angle in the wrong quadrant if both components are negative. Students are confused by this. I recommend first selecting an angle between 0° and 90° to specify the direction—which is what physicists usually do in practice—then using $\tan^{-1}(|B_y|/|B_x|)$ or $\tan^{-1}(|B_x|/|B_y|)$.

Conclude by asking students to find the components of vectors parallel and perpendicular to a tilted line. Even students already familiar with vectors find this difficult, but it's clearly a prerequisite to working successfully with forces on inclined planes. The figure below is a good example.

What are the components of vector $\vec{B}$ parallel to and perpendicular to the surface?

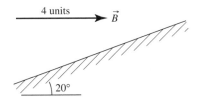

Significant figures: At some point fairly early you need a discussion about significant figures. Most students are aware of the rules, but likely they've never been required to follow them. We all know the students who write down ten digits from their calculator display. An equally serious problem is the student who keeps his or her calculator set to display two decimal points, leading them to give the one-significant-figure answer 0.02 when computing 2.87/123.

The usual rule in physics is three significant figures except when a number begins with 1, in which case four is acceptable. I find it worthwhile to enforce proper usage with an automatic one-point deduction on homework and a two-point deduction on exam problems for improper significant figures, although I usually accept two or four significant figures without penalty.

An interesting situation that occurs all the time in textbooks, and one that bothers some students, is the typical problem stating, "A block with a mass of 2 kg. . . ." Isn't this a mass known only to one significant figure, thus dictating a one significantfigure answer? You'll want to point out explicitly that textbook problems (and exam problems!) are written in a somewhat casual language but that there is an *implicit assumption* that all information is accurate to three significant figures unless otherwise stated.

Sample Reading Quiz or Discussion Questions

1. What is a vector?

2. What is the name of the quantity represented as $\hat{\imath}$?

3. This chapter shows how vectors can be added using
   a. graphical addition.   d. Both a and b.
   b. algebraic addition.   e. Both a and c.
   c. numerical addition.   f. Each of a, b, and c.

4. To *decompose* a vector means
   a. To break it into several smaller vectors.
   b. To break it apart into scalars.
   c. To break it into vectors parallel to the axes.
   d. To place it at the origin.
   e. This topic was not discussed in this chapter.

Motion and Kinematics

Of all the intellectual hurdles which the human mind has confronted and has overcome in the last fifteen hundred years, the one which seems to me to have been the most amazing in character and the most stupendous in the scope of its consequences is the one relating to the problem of motion.
Herbert Butterfield—The Origins of Modern Science

Background Information

The goal of the first part of the course is to find the connection between force and motion. An essential task is to establish and clarify just what we *mean* by the terms "force" and "motion." Students' ideas about force and motion are often non-Newtonian. They cannot begin to understand Newton's laws until they have a better grasp of what force and motion *are*.

As Butterfield notes in the above quote, the "problem of motion" was an immense intellectual hurdle. Galileo was perhaps the first to understand what it means to *quantify* observations about nature and to apply mathematical analysis to those observations. He was also the first to recognize the need to separate the *how* of motion—kinematics—from the *why* of motion—dynamics. These are very difficult ideas, and we should not be surprised that motion and kinematics are also an immense intellectual hurdle for students.

Student difficulties with the concepts of motion have been well studied (Trowbridge and McDermott, 1980, 1981; Rosenquist and McDermott, 1987; McDermott et al., 1987; Thornton and Sokoloff, 1990). Although to physicists kinematics says nothing more than $v = dx/dt$ and $a = dv/dt$, these are symbolic expressions for difficult, abstract concepts. These expressions are largely meaningless for students until they work through a large number of conceptual and mathematical barriers.

Difficulties with Concepts: Students have a rather undifferentiated view of motion, without clear distinctions between position, velocity, and acceleration.

In one study, students were presented with two balls on tracks. Ball A is released from rest and rolls down an incline while ball B rolls horizontally at constant speed. Ball B overtakes ball A near the beginning, as the motion diagram shows, but later ball A overtakes ball B. Students were asked to identify the time or times (if any) at which the two balls have the same speed. Prior to instruction, roughly 50% of students in a calculus-based physics class identified instants 2 and 4, when the balls have equal *positions,* as being times when they have equal speeds.

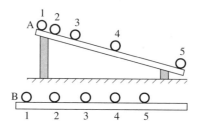

Confusion of velocity and acceleration is particularly pronounced. McDermott and her coworkers found that roughly 80% of students beginning calculus-based physics make errors when asked to identify or compare accelerations, and that the error rate was still roughly 60% after conventional instruction. Thornton and Sokoloff (1990) report very similar pre-instruction and post-instruction error rates for students' abilities to interpret acceleration graphically.

Issues to be aware of include:

- Students don't readily differentiate between position, velocity, and acceleration. They have a single, undifferentiated idea of "motion."
- Position and velocity are sometimes confused. If car B overtakes and passes car A on the freeway, both traveling in the same direction, many students will say the cars have the same *speed* at the instant when B is alongside A.
- Velocity and acceleration are frequently confused. When asked to draw velocity and acceleration vectors, students often draw acceleration vectors that mimic the velocity vectors. At a turning point (end of a pendulum's swing, top of the motion of a ball tossed straight up, etc.), nearly all students will insist that the acceleration is zero. This is an especially difficult belief to change.
- Acceleration is associated only with speeding up and slowing down. Very few students associate acceleration with curvilinear motion. This is not surprising, because a vector acceleration as we use it in physics is a *definition,* not a common-sense observation. Many students, from high school physics, may know that circular motion has a centripetal acceleration. But this is a memorized fact; almost none can tell you *why* the acceleration points to the center.
- Students interpret a positive acceleration as always meaning "speeding up" and a negative accelerating as "slowing down," rather than associating the sign with the direction of the acceleration vector. This is a difficult idea to change, and for many students it becomes a serious difficulty when they get to Newton's second law.

These difficulties are compounded by most students' lack of knowledge of vectors. Formal definitions, such as $\vec{v}_{avg} = \Delta\vec{r}/\Delta t$ are meaningless to students who can't interpret $\Delta\vec{r}$. Velocity and acceleration need to be introduced as *operational definitions,* and students need ample opportunities to apply the operations needed to find $\Delta\vec{r}$ and $\Delta\vec{v}$. This is especially true if the definition $\vec{a}_{avg} = \Delta\vec{v}/\Delta t$ is to make any sense.

Difficulties with Graphs: Nearly all students can graph a set of data or can read a value from a graph. Their difficulties are with *interpreting* information presented graphically. In particular:

- Many students don't know the meaning of the phrase "Graph *a*-versus-*b*." They graph the first quantity on the horizontal axis, ending up with the two axes reversed.
- Many students think that the slope of a straight-line graph is found from y/x (using any point on the graph) rather than $\Delta y/\Delta x$.
- Students don't recognize that a slope has *units* or how to determine those units.
- Many students don't understand the idea of the "slope at a point" on a curvilinear graph. They cannot readily compare the slopes at different points.
- Very few students are familiar with the idea of "area under a curve." Even students who have already started calculus, and who "know" that an integral can be understood as an area, have little or no idea how to use this information if presented with an actual curve.
- Many students interpret "slope of a curve" or "area under a curve" literally, as the graph is drawn, rather than with reference to the scales and units along the axes. To them, a line drawn at 45° *always* has a slope of 1 (no units), and they may answer an area-under-the-curve question with "about three square inches."
- Students don't recognize that an "area under the curve" has *units* or how the units can be something other than area units. We tell them, "Distance traveled is the area under the *v*-versus-*t* curve." But distance is a length. How can a length equal an area?

A recitation hour spent interpreting and using graphs is an hour well spent for all students.

Difficulties Relating Graphs to Motion: Nearly all students have a very difficult time relating the *physical* ideas of motion to a *graphical representation* of motion. If students observe a motion—a ball rolling down an incline, for example—and are then asked to draw an *x*-versus-*t* graph, many will draw a *picture* of the motion as they saw it. Confusion between graphs and pictures underlies many of the difficulties of relating graphs to motion.

Part of the difficulty is that we measure position along a *horizontal* axis (for horizontal motion), but then we graph the position on a *vertical* axis. This choice is never explained, as it seems obvious to physicists, but it's a confusing issue for students who aren't sure what a function is or how graphs are interpreted.

Confusion between position and velocity, and difficulty interpreting slopes, is seen with the simple example shown on the next page. The graph shows the motion of two objects A and B. Students are asked: Do A and B ever have the same speed? If so, at what time? A significant fraction will answer that A and B have the same speed at $t = 2$ s, confusing a common height with a common slope.

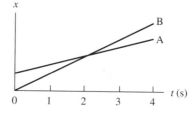

In another exercise (below), students are shown a position-versus-time graph and asked at which lettered point or points is the object moving fastest, at rest, slowing down, changing direction, etc. Students initially have difficulty with such exercises because they can't interpret the meaning of the graph. Fortunately, most students can master questions similar to these with a small amount of instruction and practice.

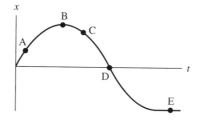

A much more difficult problem for most students, and one that takes more practice, is changing from one type of graph to another. For example, students might be given the x-versus-t graph shown below on the left and asked to draw the corresponding v-versus-t graph. When first presented with such a problem, almost no students can generate the correct velocity graph shown on the right. Many feel that a "conservation of shape" law applies and redraw the position graph—perhaps translated up or down—as a velocity graph. They need a careful demonstration, with several examples, of how the *slope* of the position graph becomes the *value* of the velocity graph at the same t. Changing from a velocity graph back to a position graph is even more difficult.

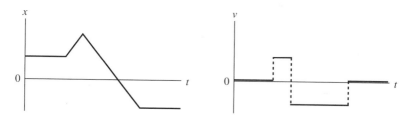

These examples give physical meaning to the slope of and area under curves, but they are still somewhat removed from the actual *situation* in which the motion occurs. To tie all aspects of a student's understanding of kinematics together, McDermott and her group presented students with situations of a ball rolling along a series of level and inclined tracks. The students are then asked to draw x-versus-t,

v-versus-t, and a-versus-t graphs of the motion, with the graphs stacked vertically so that a vertical line connects equal values of t on each of the three graphs. Students in a conventional physics class were presented—after kinematics instruction—with the simple track shown below. Only 1 student of 118 gave a completely correct response. Many students draw wildly incorrect graphs for questions like these, indicating an inability to translate from a visualization of the motion to a graphical description of the motion.

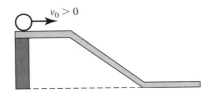

Difficulties with Terminology: Arons (1997) has written about student difficulties with the term *per*. Many students have difficulty giving a verbal explanation of what "20 meters per second" *means*—especially for an instantaneous velocity that is only "20 meters per second" for "an instant." Students will often say things such as "acceleration is delta v over delta t," but they frequently *don't* use the word "over" in the sense of a ratio but rather to mean "during the interval."

Another difficult terminology issue for students is our use of the words *initial* and *final*. Sometimes we use *initial* to mean the initial conditions with which the problem starts, and *final* refers to the end of the problem. But then we also use $\Delta x = x_{final} - x_{initial}$ and $\Delta v = v_{final} - v_{initial}$ when we're looking at how position and velocity change over *small* intervals Δt. Students often don't recognize the distinction between these uses.

Finally, students often don't make the same *assumptions* we do about the beginning and ending points of a problem. We interpret "Bob throws a ball at 20 m/s . . ." as a problem that starts with Bob releasing the ball. Students often want to include his throw as part of the problem. Similarly, a question to "find the final speed of a rock dropped from a height of 10 m" will get some answers of "zero," because that really is the *final* speed. These are not insurmountable issues, but you need to be aware that students don't always interpret a problem statement as a physicist would.

Student Learning Objectives

- To differentiate between the concepts of position, velocity, and acceleration.
- To identify velocity and acceleration vectors (both direction and relative magnitude) at different points in an object's motion.
- To recognize the relationship between $\vec{v}$ and $\vec{a}$ when an object is speeding up, slowing down, curving, or at a turning point.
- To interpret kinematic graphs and to relate position, velocity, and acceleration graphs to each other.

- To translate kinematic information between verbal, pictorial, graphical, and algebraic representations.
- To begin the development of a robust problem-solving strategy.
- To learn the basic ideas of calculus (differentiation and integration) and to utilize these ideas both symbolically and graphically.
- To understand free-fall motion, both linear and in two dimensions.
- To understand the basic ideas of circular motion.
- To solve quantitative kinematics problems and to interpret the results.

This list represents many new concepts and new mathematical techniques for students to learn. It can't be rushed! If students leave kinematics without getting past some of the conceptual hurdles about motion, there's little chance of acquiring a Newtonian understanding of forces and dynamics. I usually spend seven days on these topics, with many opportunities for in-class practice and feedback. This is probably twice as long as a typical course would spend on kinematics, but I feel it is time well spent.

Pedagogical Approach

A major goal is to provide students with both the conceptual foundations of kinematics and a systematic approach to analyzing problems. To this end, I emphasize *multiple representations of knowledge*. In particular, motion has the following descriptions:

- Verbal—as presented in typical end-of-chapter problems.
- Diagrammatic—with position, velocity, and acceleration shown in a motion diagram.
- Pictorial—showing beginning and ending points as well as coordinates and symbols.
- Graphical—in position-, velocity-, and acceleration-versus-time graphs.
- Mathematical—through the relevant equations of kinematics.

To acquire an accurate, intuitive sense of motion, students must learn to move back and forth between these different representations.

My approach to the subject of motion, influenced heavily by the work of Van Heuvelen (Van Heuvelen, 1991a, 1991b) is through the use of motion diagrams. Students pick up the motion diagram idea quickly, and this provides a powerful tool for *thinking* about motion.

A motion diagram is an ordered series of dots (representing the object as a particle) showing the position of a moving object at several points in time. You can have students think of these as different frames in a movie of the object (with the camera held steady, not "tracking" the object) or as an object lit by a flashing strobe light. If a position vector $\vec{r}$ is drawn to each dot, then the displacement vectors $\Delta\vec{r}$ are found to be vectors connecting each dot to the next. After introducing $\vec{v}_{avg} = \Delta\vec{r}/\Delta t$, students quickly recognize that (other than a scale factor) they can interpret arrows connecting the dots as velocity vectors.

The power of motion diagrams appears when you introduce acceleration $\vec{a}_{avg} = \Delta\vec{v}/\Delta t$. The acceleration at a point is found by taking the velocity vector $\vec{v}_i$ leading up to that point and the velocity vector $\vec{v}_f$ leading away, drawing these two vectors with their tails together, then finding $\Delta\vec{v}$ as the vector drawn from the tip of $\vec{v}_i$ to the tip of $\vec{v}_f$.

Because vector subtraction is not as visual as vector addition, I rewrite $\Delta\vec{v} = \vec{v}_f - \vec{v}_i$ as $\vec{v}_f = \vec{v}_i + \Delta\vec{v}$. The velocity change $\Delta\vec{v}$ is then the answer to the question "How do you *change* the initial velocity into the final velocity?" This sets the stage for interpreting acceleration as a rate of *change* of velocity. The graphical addition technique of drawing a vector from the tip of $\vec{v}_i$ to the tip of $\vec{v}_f$ is a straightforward operational procedure to obtain a specific answer to the question "What is $\Delta\vec{v}$?"

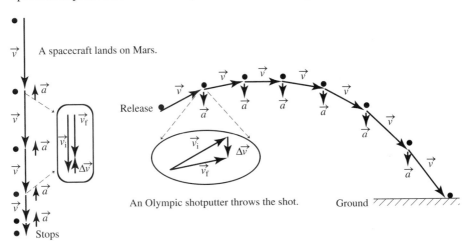

Because $\vec{a}_{avg}$ points the same direction as $\Delta\vec{v}$, motion diagrams are a practical tool for determining acceleration vectors. Indeed, this is the main point of using motion diagrams. However, students need extensive practice before becoming adept at this.

The figure shows two motion diagrams, one of linear and one of curvilinear motion.

As you proceed, the connection between motion diagrams and graphs can be emphasized. Students should be able to draw a motion diagram that goes with a kinematic graph or, conversely, draw a set of kinematic graphs that match a motion diagram. This is a good practice at interpreting a verbal description of motion, and it is particularly helpful for establishing the correct signs of v and a in one-dimensional motion.

Successful solution of trajectory and projectile problems requires understanding that the horizontal and vertical motion are independent of each other *but* that the time parameter t is common to both. Students either don't see the motions as separate, or they see them as *so* separate that they don't recognize the role of t. Projectile problems also introduce the need to solve simultaneous equations. Although they have "seen" simultaneous equations in high school algebra, I find that a significant fraction of students need remedial help with these. Fortunately, a couple of examples worked in excruciating detail is usually sufficient.

Most instructors are aware of the student tendency to latch onto the range equation, unfortunately highlighted in most textbooks as a major equation, and to use it in all projectile problems. I tell students—repeatedly—that although the range equation is interesting, I will *never* give them a problem where it is relevant. I then make sure that all class examples, all homework problems, and all exam problems consider projectiles starting and stopping at different elevations. This does get most students to actually *analyze* each projectile problem rather than looking for a canned formula.

Circular motion presents a whole range of potential difficulties, but you'll have circumvented many of these if you've been using motion diagrams. If so, students should now have little difficulty understanding that there is an acceleration even without a change of speed. A potential stumbling block, however, is the terminology "centripetal acceleration." Some students interpret this to be some *new* kind of acceleration. You'll need to convince them that the regular definition of acceleration $\vec{a} = \Delta\vec{v}/\Delta t$ is still being used and that the word "centripetal" is merely identifying the acceleration of a special kind of motion.

In addition to the conceptual and graphical hurdles students need to overcome, we also want students to start learning problem-solving skills. Much research has shown that most students don't "pick up" better skills simply by being assigned lots of problems. They *can* learn better problem-solving skills, but only if given specific instruction and feedback.

Again, I've been heavily influenced by the work of Van Heuvelen (Van Heuvelen, 1991a, 1991b). As part of his Overview, Case Study method, Van Heuvelen teaches a structured approach to problem solving and reinforces this technique with problem-solving worksheets. Kinematics problems are approached through a three-step strategy:

- Draw a motion diagram and determine the acceleration vector at different points in the motion.
- Draw a pictorial representation to establish known information, a coordinate system, and symbols for later use.
- Use kinematic equations to build and solve a mathematical representation of the problem. The equations should use only symbols defined in the pictorial representation.

The key to this approach is the pictorial representation. As instructors, we always tell students to "draw a picture," but neither we nor textbooks give them much guidance about how to draw one that is useful for solving the problem. Our own drawings, that we use in examples, are often inconsistent with each other or incomplete. This stems from the fact that we, as expert problem solvers, keep much of the information in our heads that students, as beginners, need to spell out explicitly. The pictorial representation is a *systematic* approach to drawing pictures that both clarifies the problem statement and helps set up the problem. As such, the pictorial representation forms an important bridge in the translation of a problem from words to symbols.

The pictorial representation I teach at this point in the course will carry students through Newton's laws. (A different representation is needed for conservation laws.) The basic steps are

Drawing a Pictorial Representation

- Draw a *sketch* of the situation, showing the object or objects at the beginning of the motion, at the end, and at any points in between where the nature of the motion changes (i.e., where the acceleration changes).
- Establish a *coordinate system*. Label the axes and the origin.
- Add *symbols* to represent the time, position, and velocity of each object for each time at which you've shown the object. Use subscripts to distinguish different times and different objects. Also include symbols for the geometry of the situation. This step *defines* the symbols to be used later in calculations.
- Draw *arrows* to show the acceleration of the object between each position and the next in your sketch (or write $\vec{a} = 0$). Check that these are consistent with your motion diagram!
- Make a *table* listing all the symbol values that are known from the problem statement or that can be determined from simple geometry. Check that signs are consistent with the motion diagram and the coordinate system you've chosen. Also list known relationships, such as $v_1 = v_2$, even if you don't yet know values. Do any unit conversions at this point. But don't start *working* the problem yet!
- Make a *list* of the unknown quantity or quantities you'll need to answer the question.

Because a picture is worth a thousand words, the example on the next page illustrates the three-step strategy being carried out on a Van-Heuvelen-style worksheet.

Needless to say, many students resist being "forced" to follow a structured problem-solving approach. They want to leap straight from the problem statement to an equation, and the fact that this strategy usually worked in high school makes it difficult to convince them of the need to improve their skills. I'm simply hard-nosed at insisting that homework problems be done on the worksheets, with all parts included. I also make sure that I model this approach in its entirety when working example problems. Finally, about the time we finish Newton's laws and have a "tough" midterm exam, most of the students begin to realize that they're successfully solving problems that they never would have solved just by trying to find the right equations. I've seen a marked improvement in students' problem-solving skills since I started teaching—and insisting on—such a structured approach.

Dynamics Worksheet   Problem _____

1) Physical Representation
   a. motion diagram
   b. force identification
   c. free body diagram

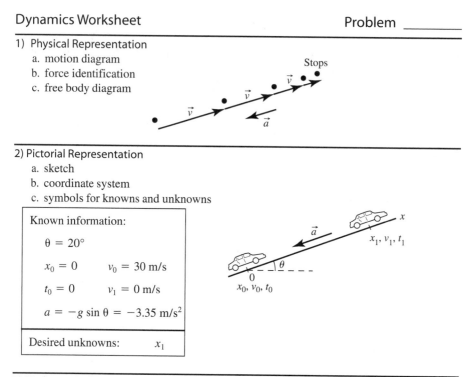

2) Pictorial Representation
   a. sketch
   b. coordinate system
   c. symbols for knowns and unknowns

Known information:

$\theta = 20°$

$x_0 = 0 \qquad v_0 = 30 \text{ m/s}$

$t_0 = 0 \qquad v_1 = 0 \text{ m/s}$

$a = -g \sin \theta = -3.35 \text{ m/s}^2$

Desired unknowns: x_1

3) Mathematical Representation

Constant acceleration motion. Assume no friction, so $a = -g \sin \theta = -3.35 \text{ m/s}^2$. Sign is negative since acceleration vector points in negative x direction.

From constant acceleration kinematics:
$$v_1^2 = v_0^2 + 2a(x_1 - x_0)$$

Thus $0 = v_0^2 + 2ax_1$

$$x_1 = -\frac{v_0^2}{2a} = -\frac{(30 \text{ m/s})^2}{2(-3.35 \text{ m/s}^2)} = 134 \text{ m}$$

Example of a Worksheet Approach to Problem Solving

The problem is:

A car traveling at 30 m/s runs out of gas while traveling up a 20° slope. How far up the hill will the car coast before starting to roll back down?

This is a simple problem to illustrate the worksheet idea. Worksheets are especially valuable for more complex problems, where there are more pieces of information to keep track of.

Using Class Time

I usually spend 7 class days on motion and kinematics, as follows:

Day 1:   Introduce motion diagrams, vector subtraction, and velocity.
Day 2:   Acceleration, complete motion diagrams, turning points.
Day 3:   Linear velocity. Position and velocity graphs.
Day 4:   Linear acceleration. Relating position, velocity, and acceleration graphs.
Day 5:   Kinematics problem solving, using the structured approach.
Day 6:   Projectile motion, both graphical and problem solving.
Day 7:   Circular motion, centripetal acceleration, and problem solving.

There's little quantitative work until day 5, after which quantitative problem solving assumes an ever increasing importance. This seems like an excessive amount of time on kinematics, but it's time well spent because motion is such an important aspect of all areas of physics. Well prepared students may find this a bit slow, but much of the class will be struggling to assimilate the ideas even at this pace.

Unless you're using my textbook, you'll have to introduce motion diagrams as a class activity. Have students imagine cutting apart the frames of a movie, stacking them, and projecting them onto a screen. Ask a student to walk steadily across the front of the room, then show how this gets converted to a motion diagram. Draw stick figures at first, and only consider the displacement vectors $\Delta \vec{r}$.

Once the basic idea is clear, have various students actually demonstrate speeding up, slowing down, or maybe both. To keep these initial motions as simple as possible, identify some points in the room (such as the edges of the lecture table) as being the edges of the camera's field of view. That way a student can speed up across the field of view, but the camera won't see his sudden deceleration before hitting the wall! After drawing stick figures for the first few, you can introduce the idea of a "particle" and switch to simply drawing dots to represent the object.

This is a good opportunity to use the think/pair/share approach, which you can use even in a large classroom. Have students watch a motion and draw their motion diagrams, then have them compare with their neighbor *and discuss any differences.* It's not the act of comparison that is as important as the discussion that follows! After any discussion, put your version on the board. The instructor's version is always important, even when most students seem to have it right, both to clarify subtle points and to give positive reinforcement. Many students who have it right will be very unsure about their answer until they hear it from you.

An initial goal is simply to have students *observe* motion carefully—something many have never done. Roll a ball down an incline so that it accelerates slowly, toss a ball in a high parabolic trajectory so that it visibly slows but doesn't stop, roll a cart across a table with enough friction that it slows and stops, and so on. (This is a good place for a digression on the "start" and the "end" of the motion.) Ask students to focus on the *shape* of the trajectory and on *how* the speed changes.

As a lead-in to velocity, draw the above figure on the board and ask students if the object is moving to the right, to the left, or if there's enough information to tell. If you're like most instructors (and books!), all your initial examples have been left-to-right motion. But after they all have agreed that they really can't tell, you can introduce the velocity vector to indicate both the speed and *direction* of the motion. Because $\vec{v} = \Delta\vec{r}/\Delta t$, the vectors they have been labeling $\Delta\vec{r}$ could just as well have been labeled $\vec{v}$. It's useful to write this definition in the form $\vec{r}_f = \vec{r}_i + \vec{v}\Delta t$, which you can interpret as saying that "Velocity is what changes an initial position to a final position. Without a velocity, the position wouldn't change."

Students can observe that the velocity also varies during some types of motion. If velocity changes the initial position to a final position, what changes an initial velocity to a final velocity? It's plausible that a parallel definition holds true: $\vec{a} = \Delta\vec{v}/\Delta t$, or $\vec{v}_f = \vec{v}_i + \vec{a}\Delta t$. Thus "Acceleration changes the initial velocity to the final velocity." This is an important prelude to understanding what happens at a turning point.

Initial examples of acceleration should first draw the velocity vectors, then isolate two of those vectors to find the acceleration by explicit vector subtraction. A couple of examples were shown above, in the Pedagogical Approach section. Note that three points (three frames) yield two velocity vectors and just a single acceleration vector, so $\vec{a}$ goes at the midpoint of the three. Have the students work on examples that show speeding up, slowing down, curving right, curving left. Even in a large class, you can draw the three points (number them so the time-sequence is clear), then have students determine $\vec{a}$ and compare their answer with a neighbor. An important conclusion you want them to reach is that an object is "speeding up" when $\vec{v}$ and $\vec{a}$ point in the same direction, "slowing down" when their directions are opposite.

Finally, you can begin asking students to draw complete motion diagrams: several points to represent the object, velocity vectors drawn between the points, and acceleration vectors drawn between the velocity vectors. Some simple starting examples might be:

- A car rolls down a hill after the parking brake fails.
- A space ship makes a soft landing on Mars.
- Bob runs once around a track with straight sides and semicircular ends.
- Ferris wheels, roller coasters, and other amusement park rides.
- An Olympic shot putter throws the shot.

After seeing centripetal acceleration for circular motion, many students draw the acceleration vectors of a projectile all pointing to a common "center." Well chosen examples can help clarify this difficulty. However, it's important to note the motion diagram of a projectile is *not* sufficiently accurate to conclude that the acceleration vector is perfectly vertical. A roughly drawn trajectory can be used to show that $\vec{a}$ is more or less vertical, but you need either accurate measurements, or the introduction of forces, to conclude that $\vec{a}$ really is vertical.

Motion diagrams are a good point to introduce Physics Jeopardy questions. Draw a motion diagram on the board, then ask students to write the description of a moving object that would have this motion diagram. Insist that they write about real objects, such as "A car slows down, makes a left turn, then speeds up," rather than simply "An object speeds up." Now if you ask them to compare with their neighbor, they'll have to think not only about the diagram you drew but also what their neighbor wrote. Students tend to like these exercises.

Time permitting, you can also ask students to use a motion diagram to answer a question. As an example, "A car drives over a hill at a steady 60 mph. Is it accelerating as it crosses the crest of the hill? Justify your answer."

Now is a good time to consider the acceleration at a turning point. If possible, demonstrate a turning point using *horizontal* motion. Push a fan cart away from you that rolls some distance, then reverses and rolls back. Or use a cart connected via string and pulley to a hanging mass; push the cart so that it rolls away (lifting the mass), then reverses and rolls back. Ask for a motion diagram from the time you release the cart until it returns, but don't initially call attention to the turning point. (For an object that returns along the same path, have students draw the second half of the motion diagram displaced to the side of the first half.) After seeing their responses, you can home in on the question of the velocity and acceleration at the turning point. Even with the motion diagram construction, students are reluctant to believe there is an acceleration at the turning point. But if you got them to agree earlier that "Acceleration changes the initial velocity to the final velocity," and if their diagram shows a velocity vector pointing toward and away from the turning point, then most will begin to agree that $\vec{a} = 0$ is an untenable position. Verbal arguments also help, such as "How would it get away from the turning point if $\vec{v}$ and $\vec{a}$ were both zero?" Follow up with the motion of a ball that rolls up and down an incline, then finally a ball tossed vertically upward.

Kinematics is the first serious test of an instructor's intent to use an active-learning teaching style. The temptation to start lecturing about slopes and derivatives is strong, but I urge you to jump right in with questions and exercises for the students to work on. You can make the necessary points about slopes, derivatives, and other matters as you go over the answers and underlying reasoning of the questions.

Students have a hard time getting the signs right in one-dimensional kinematics. Acceleration is particularly troublesome, since students want a positive acceleration to mean "speeding up" and a negative acceleration to mean "slowing down." A good exercise is to draw an x-axis across the board, mark the origin at the center, and draw a motion diagram showing an object speeding up toward the left. Draw the entire diagram on the *left* side of the origin. Ask students to determine the signs of x, v, and a. Most will get x and v, but they'll want a to be positive because the object is "speeding up." After a couple of examples like this, ask them to draw the motion diagram of an object such that x is positive, v negative, and a positive. These are hard for students until they can change their reasoning from "Is it speeding up?" to "Which way does the acceleration vector point?"

A particularly important point to make early is the role of Δ. Students tend to make no distinction between position and displacement (x and Δx) or between velocity and change of velocity (*v* and Δ*v*). Half-remembered formulas from high school, such as *v* = *d*/*t*, are more hindrance than help for coming to a solid understanding of kinematics. Even most college texts don't always distinguish between *t*, an instant of time, and Δ*t*, an interval of time. Equations such as *x* = *x*₀ + *vt* are actually using *t* to represent an interval, not an instant. I recommend writing this explicitly as $x = x_0 + v\Delta t$, with similar notation for other equations.

Kinematics gets off to a faster start if students have *already* had the opportunity to measure the motion of their own bodies in a microcomputer-based laboratory. Otherwise, you'll want to start with a number of examples asking students to draw a position-versus-time graph for an object they see moving, then draw the corresponding velocity-versus-time graph.

A good start is to establish a coordinate system across the front of the class, with a well-defined origin and with the "x-axis" pointing to the students' right. Ask a student to start at the origin, then walk across the room (left to right) at *constant* speed. Have the students first draw a motion diagram, then an *x*-vs-*t* graph, and finally a *v*-vs-*t* graph. This will give you an opportunity to talk about slopes and to note that the velocity vectors in the motion diagram are all equal length, pointing to the right. Then repeat the process with the student

- Walking right-to-left at constant speed, ending at the origin.
- Starting at a negative value of *x*, then walking to the right (or left) at constant speed.

These will provide an opportunity to discuss the role of signs for both *x* and *v*.

Next have a student, starting at the origin, *slowly* speed up until moving very fast at the far side of the room. (At this time, it's best to talk about *speeding up* and *slowing down* rather than to use the term *acceleration*.) Again, use motion diagrams, position graphs, and velocity graphs to illustrate the idea of instantaneous velocity. (For simplicity, consider the position graph to be parabolic and the velocity graph to be linear.) A good analogy is to ask what a speedometer would read at different points in the motion, if the student were carrying one.

Finally, have a student start very slowly on the far side of the room, gradually speed up while moving to the left, and reach the origin at top speed. Students find this one *much* more difficult. Focusing on the motion diagram helps.

Do A and B ever have the same speed? If so, at what time or times?

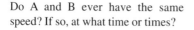
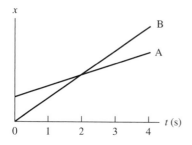

Do A and B ever have the same speed? If so, at what time or times?

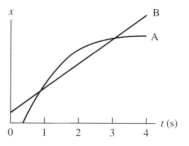

Once students seem to have the basic idea, the two questions shown at the bottom of page 84 are effective. For each, the issue is whether A and B ever have the same speed, and if so, when? Students who haven't practiced graph interpretation tend to confuse the crossing points (equal position) with points of equal speeds. The graphing exercise they've just completed should have most of them thinking about slopes, so error rates shouldn't be too high, but this exercise reinforces the message and catches a few more who are still confusing height with slope.

Another good question to pose is shown in the figure below. First, ask them to give a *verbal* description of the motion, Then, ask them to rank order the speeds at points A, B, and C, from fastest to slowest. Finally, ask them to draw a velocity-versus-time graph—*with a proper numerical scale*. Computing the slope at B will prove to be difficult for many students.

Rank order the speeds at points A, B, and C, from the fastest to the slowest.

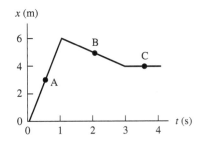

The following exercise illustrates the meaning of *per,* and it is a prelude to a similar acceleration exercise later.

A train is moving at a steady 30 m/s. At $t = 0$, the engine passes a signal light at $x = 0$. *Without using any kinematic equations,* determine the engine's position at $t = 1$ s. Also at $t = 2$ s and $t = 3$ s.

The objective is for students to realize the *meaning* of "30 meters per second" to be that "x increases by 30 meters during each second." The position increases by 30 m during the first second, to $x_1 = 30$ m, by 30 m more during the next second, to $x_2 = 60$ m, and so on. Some students will find this so obvious as to be trivial, but others—those who are very equation oriented—find this a difficult way to reason.

Have students graph both position and velocity for the train, then call their attention to the fact that the displacement Δx is exactly the same as the area under the velocity curve. The equation $x = x_0 + v \Delta t$ is merely giving algebraic expression to the observation that

$$\Delta x = \text{"area" under the velocity-versus-time curve.}$$

You can then direct them to the text for a proof that this result is true for *any* velocity, not just constant velocity.

For a nonconstant velocity, try a graph like the one shown on the next page and ask students to find—using the graph, not an equation—the position at $t = 1, 2, 3, 4,$ and

5 s. Then have them draw a position-versus-time graph. Now they've practiced going forward, from position to velocity, and backward, from velocity to position.

Find the position at times $t = 1, 2, 3, 4,$ and 5 seconds. Assume $x_0 = 0$.

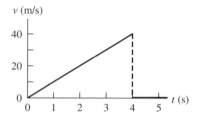

It's worthwhile to close this initial discussion with a simple kinematics example problem that uses only velocity, not acceleration, but that illustrates all the steps in the problem-solving strategy described in the Pedagogical Approach section. If you're using worksheets, encourage students to fill out a worksheet as they work along with you. A good first problem is:

EXAMPLE 1: Sally opens her parachute at a height of 2000 feet and descends at a steady 25 feet/s. How long does it take her to touch down?

This is not intended to be hard but to illustrate the problem-solving strategy. Start with a motion diagram. Then draw a pictorial representation that establishes a coordinate system and defines symbols. Finally—and for the first time—use the mathematical representation section of the worksheet to solve the problem. Call attention to the fact that all the symbols used in the mathematical solution, such as y_0 or t_1, were identified and defined in the pictorial representation. End by having them assess whether or not the result is "reasonable."

Note: Physicists often use a coordinate system for vertical-motion problems with the y-axis pointing down. However, students often find this more confusing than helpful. To minimize confusion, it's best to be consistent with your textbook and not adopt this coordinate system unless the textbook does so.

After first looking at velocity, you can now explore acceleration. It's good to use exercises and examples that parallel or build on the examples from velocity. A good place to start is with a demonstration of a ball rolling down a small incline, where the speed stays slow enough for students to observe the motion carefully. They can observe that the velocity increases continuously. A sonic rangefinder interfaced to a computer, or even a stop watch, can be used to establish that the velocity is increasing linearly with time.

Students have previously associated velocity with a changing position. Now, by analogy, you can associate acceleration with a changing velocity and can show that a is the slope of the velocity-versus-time graph. Remind students that they can judge velocity fairly easily when observing an object, but that it's much more difficult to judge acceleration. That's why the motion diagrams and graphical tools are so important.

Then roll the ball *up* the incline, catching it right at the top so that it doesn't roll back down. You want students to recognize that the velocity has changed sign but the acceleration is the same as for the ball rolling down. This is hard for most students. Graphs and motion diagrams can be used together to make this important point.

Sonic ranger probes are also good for showing that objects fall with a *constant* acceleration and that the acceleration is *independent of the mass*. Without a sonic ranger, there's not enough time in lecture to make the measurements that would be required to demonstrate that free-fall motion is one of constant acceleration, so you're forced to assert this without proof.

Students, for some reason, have a strong tendency to call g by the name "gravity." It's worthwhile to emphasize that g is "the acceleration due to gravity" and require them to use it correctly. You will also want to emphasize that g is always a *positive* value. The acceleration is negative (for an upward-pointing y-axis), given by $a_y = -g$, but g itself is positive.

A good exercise is:

A jet plane accelerates at 3 m/s^2 during take-off. *Without using any kinematics equations,* determine the plane's velocity at $t = 1, 2, 3,$ and 4 s.

This is analogous to the 30 m/s train exercise. You want students to reason that 3 m/s^2 = (3 m/s) per second, so the velocity increases by 3 m/s every second. A velocity graph is linear, and you can then easily use the area under the curve to find (and graph) the position at $t = 1, 2, 3,$ and 4 s. One goal of this exercise is for students to recognize that $\Delta x = \frac{1}{2}a(\Delta t)^2$ for constant acceleration can be understood from the geometry of the graphs, that it's not "just" a result derived from calculus.

A capstone to this mostly conceptual and graphical introduction is to return to the idea of a turning point. Roll a ball up and back down an incline, with a turning point at the top. (It's best to do this with the ball starting toward the right, so that v_0 is positive.) They've looked at motion up the ramp and motion down the ramp separately, so they should now quickly agree that the velocity graph looks as shown below. (If you have time, you might also have them draw a position graph and discuss its shape.) But then ask them to draw an acceleration graph. This shouldn't present too much difficulty now. Finally, ask them at what time the ball reaches the top and what its acceleration is at that instant. Many students will be surprised how "easy" it was to show that their earlier belief that $a_{\text{top}} = 0$ was wrong. Easy though it may seem now, this exercise doesn't have a full impact on their learning if they haven't first struggled with the idea using motion diagrams.

Draw the position-versus-time graph. Assume $x_0 = 10$ m.

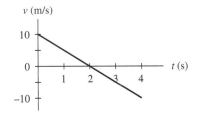

Now that the conceptual foundations have been laid, you can start to move into more typical kinematics problems. For the first few examples, students should be led step-by-step through the structured problem-solving approach. Here are three good examples:

EXAMPLE 2: Bob throws a ball straight up at 20 m/s, releasing the ball 1.5 m above the ground. What is the maximum height of the ball? What is the ball's impact speed as it hits the ground?

EXAMPLE 3: A ball is released at a height of 1 m on a frictionless 30° slope. At the bottom, it turns smoothly onto a 60° slope going back up. What maximum height does it reach on the right side? (Most students will be surprised that the answer is 1 m, and this gives you an opportunity to say a few initial words about energy.)

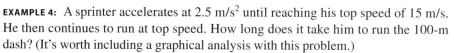

EXAMPLE 4: A sprinter accelerates at 2.5 m/s² until reaching his top speed of 15 m/s. He then continues to run at top speed. How long does it take him to run the 100-m dash? (It's worth including a graphical analysis with this problem.)

There are several possibilities for wrapping up linear kinematics. One is to continue example problem solving, having students first use pictorial representations, motion diagrams, and graphs to analyze the problem before getting into the mathematical analysis. Problems involving two moving objects are particularly challenging. Such problems point out the usefulness of the qualitative analysis tools and the problem-solving strategy because they can't be solved by equation-hunting.

EXAMPLE 5: Ball A rolls along a frictionless, horizontal surface at a speed of 1 m/s. Ball B is released from rest at the top of a 2-m-long, 10° ramp at the exact instant when A passes by. Will B overtake A before reaching the bottom of the ramp? If so, at what position? (The answer is *yes, at x =* 1.175 m.)

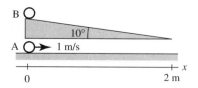

Another good possibility is to have students work through several examples of a ball rolling along a series of level and inclined tracks, as discussed earlier in the section on student difficulties relating graphs to motion. These exercises require students to *visualize* the motion and to relate it to graphs. Most students find these difficult, even after the previous graphing exercises. But quite a few "get it" after a few such examples, and their ability to connect visualized motion with graphs suddenly takes a quantum leap.

The track shown in the Background Information section is a good starting example. It's easy to invent more examples as needed, including ones in which the ball has an initial velocity. It's also good to do at least one example in which you give the graphs and ask students to draw the track. Many exercises of this type are available in an excellent computerized version called *Graphs and Tracks,* available from Physics Academic Software (Web address in *References and Resources* section).

Turning to trajectories, many students have never observed projectile motion very carefully. It's worth tossing a ball back and forth across the room in a high arc so that it slows perceptibly near the top, and asking about the shape, about where the motion is slowing down or speeding up, if the velocity is zero at the top, and so on. A good demonstration, if you have the facilities, is to capture the motion with a video camera, then play it back frame by frame. Computer software that plots the trajectory from the frames is especially good. There are also many films and videos that show slow motion and frame-by-frame trajectories of projectiles.

Trajectories other than projectiles are not so easily demonstrated, but one option exists for instructors with a large air table and overhead TV camera. Start a puck moving to the right, then use a compressed air gun to supply a small but constant (as best you can) force at right angles. Then repeat, but this time give the puck a large, short duration force (a "kick"). You want them to recognize the circumstances under which the puck is deflected in a straight line and when it follows a curved trajectory.

It's worth exploring these ideas with motion diagrams before becoming quantitative. Remind them that a constant force gives a constant acceleration. Give them an initial velocity vector $\vec{v}_0$, like the one shown, and an acceleration vector $\vec{a}$ that will be the same at all points. Then ask them to construct a motion diagram step-by-step by generating a series of about 5 velocity vectors. You may need to demonstrate finding $\vec{v}_1$, then let them do the rest. They should generate the curved trajectory shown, then compare it to the trajectory they observed for the ball.

Construct a motion diagram of the trajectory of a particle with the initial velocity and constant acceleration shown below.

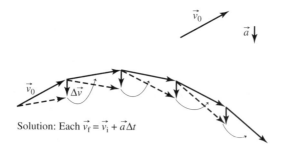

Solution: Each $\vec{v}_f = \vec{v}_i + \vec{a}\Delta t$

Before launching into problem solving, an exercise that I've found to generate intense discussions is the following.

A physics student on Planet Exidor throws a ball that follows the parabolic trajectory shown. The ball's position is shown at one-second intervals until $t = 3$ s. At $t = 1$ s, the ball's velocity is $\vec{v} = (2\hat{i} + 2\hat{j})$ m/s.

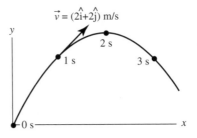

a. Determine the ball's velocity at $t = 0$ s, 2 s, and 3 s.
b. What is the value of g on Planet Exidor?

This appears to be straightforward, but I've found few students who can immediately recognize how to answer. Many begin by thinking that there's not enough information. Give them hints, but let them struggle for a while. One hint is to look at the horizontal and vertical components of the velocity—what can they learn from this? Another is to remind them of the earlier example—if you did it—of a jet accelerating at 3 m/s^2 and how they found this to cause the velocity to change in a regular series.

Projectile problems have many details to keep track of—vertical and horizontal components of positions, velocities, accelerations, and initial and final locations. Students who simply equation hunt often fail because they lose track of information or get pieces of information switched around. These are good problems for emphasizing and demonstrating the use of a pictorial representation for writing down information and associating it with symbols.

The properties of projectile motion—constant horizontal speed, equal fall time of dropped and horizontally fired balls, etc.—are surprising to most students, even those who have taken high school physics. It's worth doing as many demonstrations as you have available. Carts that fire a ball out of a vertical tube while they roll along, and then catch the ball as it comes back down, are effective. They can be used at constant speed, where they catch the ball, and they can also be accelerated by connecting them with a string and pulley to a falling mass. In both cases, you can conduct these as interactive lecture demonstrations, asking students to predict in advance whether the ball will land in front of the tube, behind it, or in it.

Many schools have the set-up to demonstrate the traditional "hunter and the monkey" problem—although a hunter trying to shoot a coconut that falls just as he pulls the trigger is more palatable to most students. This can make an extended interactive demonstration, where reasoning about what should happen is followed by the demonstration, and then followed by a numerical example of how far the coconut falls before the bullet hits it.

Finally, there's circular motion. Before getting into the acceleration, most students need practice with the basic description and mathematics of circular motion—namely period and frequency. The simple fact that $f = 1/T$ is not obvious to all students, so you may need to demonstrate this with simple numerical examples. ("If an object makes 10 revolutions in 1 s, how long does each revolution take?") Students are encouraged to remember the relationship between speed, radius, and frequency by recalling that speed = circumference/period. Thus $|\vec{v}| = 2\pi r/T = 2\pi rf$.

It's worth returning to motion diagrams to establish that there *is* an acceleration, even though the speed is constant, and that the acceleration points to the center. All textbooks derive the standard result that $|\vec{a}_c| = v^2/r$, so there's no reason to spend class time on this. Circular motion will be studied much more thoroughly after centripetal forces are introduced, so there's little need now for anything other than a couple of simple examples relating centripetal acceleration to speed, period, and frequency. Note that the pictorial representation for circular motion, if you're using worksheets, is simplified because there are no initial and final positions.

Sample Reading Quiz or Discussion Questions

1. What is a "particle"?

2. What quantities are shown on a complete motion diagram?

3. Draw the position graph and the acceleration graph that go with this velocity graph. The initial position is $x_0 = 0$ m.

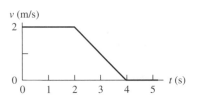

4. An acceleration vector
   a. tells you how fast an object is going.
   b. is constructed from two velocity vectors.
   c. points in the direction of motion.
   d. is parallel or opposite to the direction of motion.

5. The pictorial representation of a physics problem consists of
   a. a sketch. d. a table of values.
   b. a coordinate system. e. both a and b.
   c. symbols. f. all of a, b, c, and d.

6. The slope at a point on a position-versus-time graph of an object is
   a. the object's speed at that point.
   b. the object's average velocity at that point.
   c. the object's instantaneous velocity at that point.
   d. the object's acceleration at that point.
   e. the distance traveled by the object to that point.

7. The area under a velocity-versus-time graph of an object is
   a. the object's speed at that point.
   b. the object's acceleration at that point.
   c. the distance traveled by the object.
   d. the displacement of the object.
   e. This topic was not covered in this chapter.

8. At the turning point of an object,
   a. the average velocity is zero.
   b. the instantaneous velocity is zero.
   c. the acceleration is zero.
   d. all the above.
   e. This topic was not covered in this chapter.

9. A 10-pound sled and a 100-pound sled are placed side-by-side at the top of a frictionless hill. Each is given a very light push to begin their race to the bottom of the hill. In the absence of air resistance
   a. the 10-pound sled wins the race.
   b. the 100-pound sled wins the race.
   c. the two sleds end in a tie.
   d. there's not enough information to determine which sled wins the race.

10. A ball is thrown upward at a 45° angle. In the absence of air resistance, the ball follows a
    a. tangential curve.   c. parabolic curve.
    b. sine curve.   d. linear curve.

Sample Exam Questions

These questions cover kinematics in one and two dimensions. If you've been having students do homework on worksheets, you'll probably want to require their use on the quantitative problems on an exam.

1. A car drives through an S-shaped curve (seen from above) at constant speed. Draw a complete motion diagram of the car.

2. Mike falls out of a tree and lands on a trampoline. The trampoline sags 2 feet before launching Mike back into the air. At the very bottom, where the sag is the greatest, is Mike's acceleration upward, downward, or zero? Use the tools that you've learned in these first chapters to give a convincing *explanation* of your answer.

3. Is it possible for an object with a negative acceleration to be speeding up? If so, give an explicit example. If not, explain why not.

4. The figure shows a ball rolling along a smooth track. Each segment of the track is straight, and the ball can move from segment to segment with no loss of speed. The ball starts from the left edge with an initial velocity v_0 that is large enough for it to make it over the top. Draw position-, velocity-, and acceleration-versus time graphs for the ball until it rolls off the right edge of the track. (Position is measured along the track.) Your three graphs should have the same time scale.

(It's good on a question like this to supply them with three empty sets of axes stacked one above the other.)

5. An object moving horizontally has the acceleration-versus-time graph shown. At $t = 0$, the object has $x_0 = 0$ and velocity $v_0 = 10$ m/s.

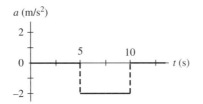

a. Draw a velocity-versus-time graph for the object. Include a numerical scale on the vertical axis.

b. Draw a motion diagram of the object's motion.

c. Write a description of a real object for which this is a realistic motion.

6. The figure below shows a ramp and a ball that rolls along the ramp. Draw vector arrows on the figure to show the ball's acceleration at each of the lettered points A to E, or write $\vec{a} = 0$ if there is no acceleration.

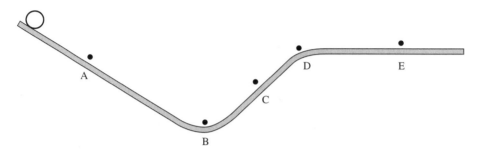

7. You are standing on top of the Empire State Building (height 330 m) when Superman flies past. He is headed *straight down* with a *steady speed* of 35 m/s. (Superman can do things like this.) At the instant he goes past, you drop a 10 kg lead ball over the edge. At what height above the sidewalk does the ball pass Superman?

8. A small block is placed at height h on a frictionless 30° ramp. When released, the block slides down the ramp and then falls 1 m to the floor. A small hole is located 1 m from the end of the ramp. From what height h should the block be released in order to land in the hole?

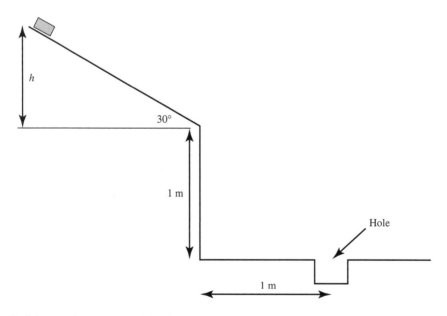

9. It is sometimes proposed that future space stations create an artificial gravity by rotating. Suppose a space station is constructed as a large cylinder 400 m in diameter. At what rotation frequency would an object on the rim of the space station have an acceleration equal to g? Give your answer in rpm.

7

Force and Newton's Laws

When asked to draw a force diagram for some simple situation, most students emerging from any level of introductory physics course are likely to draw objects which look like a porcupine shot by an Indian hunting party—the number and direction of pointed entities being essentially stochastic.

Arnold Arons (1979)

Background Information

There are three basic issues, each of which causes serious difficulties for students:

- What is a force?
- What is the connection between force and motion?
- How are forces related when two objects interact?

What Is a Force? Students don't have a clear idea of just what a force is. They tend not to distinguish between force, inertia, energy, power, or even velocity, often using these terms interchangeably. In addition:

- Many students believe that only animate objects can exert forces. They don't believe that a table exerts an upward force on an object; the table simply "gets in the way" of the object wanting to fall.
- Forces recognized by physicists are often seen by students as simply *influences* on an object's motion, not as forces. Thus friction is not a force but merely "what makes it stop." Gravity is not a force but simply "what makes it fall."
- A majority of students believe in an *impetus theory of motion.* In throwing a ball, the hand imparts a "force of the throw" to the ball. This is a property of the ball (an *inherent force*) and travels with the ball "to keep it moving." Typically 75% or more of students beginning calculus-based physics think that a ball tossed upward

has an upward "force of the throw" or "force of your hand" on it *after* it leaves your hand.

- Students tend to view forces from the perspective of the *applier* of force rather than from the perspective of the object experiencing forces. That is, they recognize the pushes and pulls *they* must apply to move an object, but they don't recognize that the object may experience additional inanimate forces of friction, gravity, and so on. This is one of the major reasons they think that motion requires a force.

How are Force and Motion Connected? The prevailing student belief is that *motion requires a force.* This belief is based on much common-sense evidence, and it is a belief that is *highly* resistant to change. More specifically, the "student version" of the laws of motion is

- If there's no force on an object, the object is at rest or will *immediately* come to rest.
- The converse is *not* true. An object at rest does *not* automatically imply no net force.
- Motion requires a force or, alternatively, force causes motion.
- In general, force is proportional to velocity.

Many student ideas about a "force of motion" take on new life in two dimensional situations. Clement (1982) posed the following problem to engineering students beginning calculus-based physics:

> A rocket is moving sideways in deep space, with its engine off, from A to B. It is not near any stars or planets or other outside forces. Its engine is fired at point B and left on for 2 s while the rocket travels from point B to some point C. Draw in the shape of the path from B to C. Then show the path from point C, after the engine is turned off.

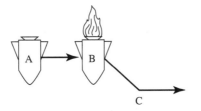

Most common response.

The figure shows the most common response. An incorrect linear trajectory between B and C was drawn by 89% of students. This is perhaps not too surprising because few students would have any reason to recognize that this situation gives a parabolic (or even curved) trajectory. Far more interesting is that 62% of students predicted that the rocket would return to its horizontal direction after the engine is shut off at C. Interviews with students found many giving explanations such as "Whatever was making it go to the right before will take over again after point C." This reflects a belief in a continuing "force of motion" to the right.

A force of motion often seems to have a *memory* of previous motions. Arons (1997) cites an experiment in which a marble is rolled around the inside of an incomplete circular hoop lying flat on a table. Students are asked to predict the marble's trajectory after passing point B. Some predict it will continue with perfect circular motion, reentering the hoop at A. Others predict a circular tendency that "runs down," so the marble curves to the left but gradually straightens out. Few predict the correct linear trajectory from B.

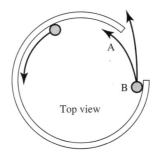

Common incorrect responses.

Although students may have learned the term *centripetal force* in high school, few have any sense of what it means. This is demonstrated by rolling a heavy ball (such as a bowling ball) across the floor and asking students to create a circular motion about a specified point by tapping on the ball with a rubber mallet. Most attempt to apply tangential forces, rather than centripetal forces—another example of thinking that force needs to be in the direction of motion.

Just the *term* "centripetal force" causes difficulty for many students. They consider it to be a new *kind* of force, like the gravitational force or the normal force, and they will dutifully add a force vector labeled "centripetal force" to their free-body diagrams.

These are common student beliefs, but not every student necessarily holds every belief. It is important to note that students are not at all consistent in their application of these beliefs; for example, they may apply some beliefs to vertical motion but not to horizontal motion. Even so, recognition of these prevailing alternative conceptions will give instructors better insight into student responses and the difficulties students face.

How Do Forces Behave During Interactions? Newton's third law is a subtle and difficult topic for students. However, if you spend sufficient time on the third law and provide students with ample opportunities for practice and feedback, most will end up finding this a rewarding topic where the concepts of force and motion suddenly begin to "make sense."

Suppose a large truck and a compact car have a head-on collision. During the collision, is the force of the truck on the car larger, smaller, or equal to the force of the car on the truck?

This question, and several similar questions on the Force Concept Inventory, is initially missed by 70–80% of students in a typical calculus-based physics course. Conventional instruction makes little improvement, with ≈60% still missing these questions on the posttest.

Halloun and Hestenes (1985b) have characterized student beliefs about interactions in terms of a *dominance principle:* the larger (or faster or more active) object exerts a larger force than the smaller (or slower or less active) object. Students tend to view an "interaction" as a "conflict" in which the stronger wins. It's not hard to understand how this common-sense view comes about. After all, the *effect* of the collision on the compact car is much larger than its effect on the truck. Different effects would seem to require different *causes,* hence different amounts of force. The difference in the masses does not appear to students as a significant factor in drawing conclusions about forces. This belief about interaction forces is likely the most persistent and hard to change of all the student misconceptions in mechanics.

Some of the more specific difficulties students have with Newton's third law and with interacting systems are

- Students don't believe Newton's third law. It's too contrary to common sense.
- Students have difficulty isolating systems from each other and from the environment.
- Students have difficulty identifying action/reaction force pairs:
  They match two forces on the same object.
  They attach forces to the wrong objects.
  They don't believe that long-range forces (e.g., gravity) have reaction forces.
- Students confuse equal force with equal acceleration.
- Students don't understand tension:
  They think that tension is the *sum* of the forces exerted at the ends of a string.
  They think that tension exerts a force only in the direction of motion.
  They think that tension passes through an object to another string on the other side.
- Students often don't recognize that objects connected by an inextensible string must have accelerations of equal magnitude.

Although the practice is thankfully declining, there are widely used textbooks with figures such as the one shown below. Students cannot tell from the figure that forces $\vec{N}$ and $\vec{N}'$ act on *two different objects*. Perhaps it's not surprising that students find it so hard to identify action/reaction pairs!

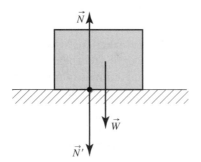

The major research paper on interacting systems is McDermott et al. (1994). As a first test (see figure below), they asked students in the calculus-based course at the University of Washington to compare the tension at the point at which the string in (a) is attached to the wall to the tension in the middle of the string in (b). *After* students had finished studying Newton's laws, only 50% answered correctly that the tensions are equal. Most students expected the tension in string (b) to be twice as large.

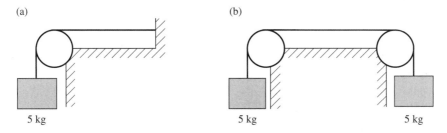

Two other questions involving tension are of interest. In the first (see figure below), a person pulling on a string accelerates blocks A and B, connected by a massless string, across a frictionless surface. Students were asked to compare the force exerted by string 1 on block A with the force exerted by string 2 on block B. Only 40% answered correctly that $F_{1 \text{ on } A} > F_{2 \text{ on } B}$. The other answers were divided between

- $F_{1 \text{ on } A} = F_{2 \text{ on } B}$ because the tension in string 1 is somehow "transmitted" through block A to string 2.
- $F_{2 \text{ on } B} > F_{1 \text{ on } A}$ because $m_B > m_A$, so it takes a larger force to accelerate block B.

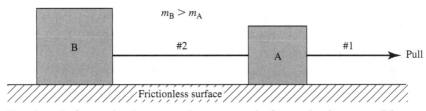

How does the force string 1 exerts on A compare to the force string 2 exerts on B?

Neither response is aware that the tension in string 2 exerts a "backward" force on block A. Indeed, students giving these responses failed to show a force $\vec{F}_{\text{string 2 on block A}}$ when asked to draw free-body diagrams. It's also of interest to note that 15% of students, in a preliminary question, failed to recognize that blocks A and B must have the same acceleration.

In the second tension question, shown on the next page, block A can slide across a frictionless table when pulled by falling block B and a massless string. Initially, block A was held in place by someone's hand. Students were asked how, or if, the tension in the string would change if the hand were removed. Only 25% of students, after Newton's laws, answered correctly that the tension would decrease. Nearly all predicted no change. In a related question, 55% of students predicted that block B would fall with an acceleration of 9.8 m/s².

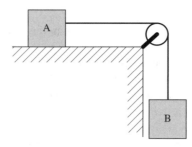

Interacting-system problems are difficult and frustrating for students. There are no magic formulas to apply, so students must correctly apply Newton's second law to each object. But even if they realize this to be the proper approach, they cannot succeed without first identifying all the interaction forces and then making proper use of the third law.

Free-Body Diagrams: Free body diagrams are the primary tool for working with forces. Student difficulties with free-body diagrams include:

- Identifying a force correctly, but not knowing which way it points.
- Including forces exerted *by* the object, not just forces exerted *on* the object.
- Not using a coordinate system, or using an inappropriate one.
- Placing the tips, rather than the tails, of the vectors at the origin, which makes it hard to determine $\vec{F}_{net}$ and the *x*- and *y*-components of the forces.
- Making errors in determining the components of the force vectors.
- Placing action-reaction force vectors on the wrong diagrams.

Fortunately, these are mostly technical errors rather than conceptual errors. Most can be overcome fairly quickly with focused practice and feedback on drawing free-body diagrams.

Student Learning Objectives

- To recognize what does and does not constitute a force.
- To identify the specific forces acting on an object.
- To draw and use accurate free-body diagrams.
- To understand the connection between force and motion.
- To learn a strategy for solving force and motion problems.
- To identify action/reaction pairs of forces.
- To understand and use Newton's third law.

Pedagogical Approach

Most students require a major reorientation of thinking to learn Newton's laws successfully. It is essential to lay the conceptual foundations before rushing in to solve problems. Students must recognize what a force is, correctly identify the forces on an object, and be able to express their knowledge about forces on a free-body diagram *before* they can successfully solve dynamics problems that go beyond simple plug-and-chug.

In approaching forces and Newton's laws, I strongly recommend that you delay the introduction of Newton's third law as long as possible. The conceptual hurdles that students face are high enough already, and the topic of interactions adds yet more. I've found that students do much better if instruction concentrates first on how a *single* object responds to forces. Only after these ideas begin to seem comfortable can students deal with the issue of how two objects interact with each other.

I begin by dividing a situation into the *system,* which is the object or objects of interest, and the *environment,* which is everything else. At the beginning, the system is a single object and all forces originate in the environment. Later, when it's time for the third law, you can consider systems interacting with each other and with the environment.

Identifying Forces: Forces are initially identified as pushes and pulls on the system. Every force must have an *agent*—an identifiable cause in the environment—and, with few exceptions, every force is a *contact force* that is applied at a point where the environment touches the system. The exceptions are *long-range forces.* Magnets are a good demonstration of a long-range force, although the only such force students will face for now is gravity.

With this "definition" of force, you can give students a step-by-step procedure for identifying forces:

Identifying Forces
- Divide the problem into "the system" and "the environment." The system is just the object whose motion you wish to study, and the environment is everything else.
- Draw a picture of the problem, showing the object and everything in the environment that touches the object. Ropes, springs, surfaces, and so on are all parts of the environment.
- Draw a *closed curve* around the system, with the object inside the curve and everything else outside the curve.
- Locate *every* point on the boundary of this curve where the environment touches or contacts the system. These are the points where the environment exerts *contact forces* on the object. Do not leave any out!
- Identify *by name* the contact force or forces (there may be more than one!) at each point of contact, then give each an appropriate symbol.
- Identify any *long-range forces* acting on the object. Name the force and write its symbol beside your picture. For now, the only long-range force is gravity.

The figure on the next page shows this procedure being used to find the forces on a car as it is towed up a hill.

Although this approach is simple, it immediately begins to address student difficulties with forces. In particular, it allows students to recognize which forces are actually being applied, and it helps them avoid nonexistent "force of the throw" forces.

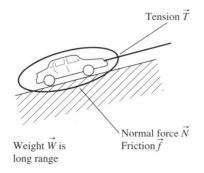

Tension $\vec{T}$

Weight $\vec{W}$ is
long range

Normal force $\vec{N}$
Friction $\vec{f}$

Identifying forces leads naturally into drawing free-body diagrams. I place far more emphasis on free-body diagrams than do most textbooks. The goal is for students to understand that they can "read" all the necessary information about forces off their free-body diagrams. You will want to emphasize that this part of the analysis is the *physics,* and that the subsequent solution for the acceleration and the motion is "merely" mathematics. Memorizing every equation in the book will be of little use if students can't identify forces and draw a correct free-body diagram.

Newton's Second Law: Once students can identify forces with fair accuracy, it's time to make the link between force and motion. It's vital to use whatever means you have at your disposal to demonstrate (or have students demonstrate) that force determines an object's acceleration, not its velocity. This is a new idea—*experimental evidence*—and it can be a powerful tool against the entrenched misconception that force is proportional to velocity. Specific ideas are given below. I prefer to "discover" Newton's second law in the form $\vec{a} = \vec{F}_{net}/m$, rather than the traditional $\vec{F}_{net} = m\,\vec{a}$, since this makes a better cause-and-effect statement.

You'll notice that this approach bypasses Newton's first law (and inertial reference frames) and moves directly to the second. Debating the relationship between Newton's first and second laws is an old and honorable tradition. On philosophical and logical grounds it may well be that the first law is needed as a definition of force ("... that which causes an object to deviate from motion at constant velocity ...") before the second law can be given meaning, but this sophisticated reasoning is much too subtle for beginning students and is not an appropriate teaching tool. At this stage, the best motto is "Keep it concrete and keep it simple." To define forces as tangible pushes and pulls and then to identify acceleration as the consequence of a net force is an operational procedure that will get students up and running. From this perspective, the first law is then a "special case" of the second law when $\vec{a} = 0$.

After this preparation, you're finally ready to focus on solving problems in dynamics. Even so, much of the emphasis needs to remain on *analyzing* problem statements. All of the analysis tools—pictorial representations, motion diagrams, and free-body diagrams—can now be gathered into a full strategy for analyzing force and motion problems:

Strategy for Force and Motion Problems

■ Analyze the problem statement by preparing a *pictorial representation*. The pictorial representation translates words to symbols, clarifies information that is known, and identifies what the problem is trying to find.

■ Analyze the motion by preparing a *motion diagram*. The motion diagram will determine the acceleration vector $\vec{a}$. It will also provide useful information to be used later in the kinematical analysis.

■ Identify all forces acting on the object and show them on a *free-body diagram*.

■ Invoke Newton's second law, usually in the operational form: $\vec{a} = \dfrac{1}{m} \sum_i \vec{F_i}$.

■ Determine the vector sum of the forces *directly* from the free-body diagram.

■ Solve for the subsequent motion, using appropriate results from kinematics.

■ Assess the results.

Notice that the Dynamics Worksheet, shown as an example in the previous chapter, is designed to accommodate force identification and free-body diagrams.

At this point you can start to work a wide variety of single-particle dynamics problems in one and two dimensions. Many students will still have difficulty using vector components or recognizing that $\vec{F} = m\,\vec{a}$ is a shorthand way to write three simultaneous equations involving the components. They will get the components mixed up, perhaps use magnitudes instead of components, and have difficulty reassembling components into a magnitude and direction. You'll need to demonstrate these steps in considerable detail as you work example problems.

The "big picture" you want students to be developing is the idea

$$\text{individual forces} \Rightarrow \text{net force} \Rightarrow \text{acceleration} \Rightarrow \text{kinematics.}$$

You can encourage this awareness by

■ Insisting that students use *all* the steps in the problem-solving strategy, preferably on Dynamics Worksheets.

■ Insisting that they begin the mathematical representation of each problem with a full statement of Newton's second law.

■ Insisting that they *use* the free-body diagram to obtain the force components for Newton's laws.

Equilibrium need not be singled out for special treatment. Statics problems fit perfectly well into the above problem-solving strategy just by recognizing that $\vec{a} = 0$. Many typical single-particle problems are, effectively, "in equilibrium" along one axis but not the other. For example, an accelerating car is in equilibrium along the vertical axis. This is an example of an issue that is so obvious to instructors that they may fail to call attention to it. Nonetheless, students don't always recognize this. You'll also want to point out explicitly the wisdom of choosing a coordinate system with $a = 0$ along one axis, such as in inclined plane problems.

Circular motion is an excellent application of single-particle dynamics, with most such problems needing only the second law and not the third. In my classes I consider only uniform circular motion. The goal is to understand how circular motion occurs as a consequence of center-directed forces. This is a very *non*-intuitive idea and one that students have a lot of trouble with. The additional complications of tangential acceleration are not needed at this time.

There have been many suggestions in recent years, especially in *The American Journal of Physics,* that textbooks and instructors *not* use the term "centripetal force"—at least not at the beginning of circular motion. Many students believe that this refers to a new and additional *kind* of force acting on the object. We don't label a tension force that causes linear acceleration "the linear force," so why should the same force get a special name if it happens to cause circular motion? The term seems to hinder rather than help many students in their efforts to understand a difficult topic. I do use the term *centripetal acceleration* as the acceleration of an object moving in a circle, but I avoid the term *centripetal force* at this time. The term can be introduced later, such as in the chapter on gravitational forces and orbits, after students are more familiar with the ideas and less likely to be confused by the terminology.

Because circular motion analysis requires looking at forces both in the plane of motion (the xy-plane) and perpendicular to the plane of motion, I introduce a special "rz-coordinate system." The origin is placed on the particle (for free-body diagram purposes), the r-axis is chosen to point toward the center of the circle, and the z-axis is perpendicular to the plane of the circle. This choice makes the radial component of acceleration always positive $(a_r = v^2/r > 0)$ and the perpendicular component zero $(a_z = 0)$. Likewise, the r-component of a force is positive if it points toward the center of the circle. The figure shows an example. In this case, $N_r = +|\vec{N}|\sin\theta$.

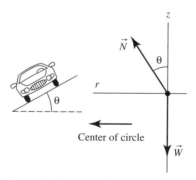

Example: Car on a frictionless
banked curve.

I've found that students do better when this coordinate system is made explicit, rather than merely inferred as it is in some texts, and when it is clearly different than the xy-coordinates used in the plane of motion. It's worth showing, in an example, that it makes no difference whether the diagram shows the particle to the right or the left of the center. Center-pointing forces have positive components in either case.

Friction is the first force to be introduced as a *model,* but many students don't distinguish between a model and a physical law, such as Newton's second law. This is a good point for a short digression about the role of simplifying assumptions in problem solving.

Kinetic friction is straightforward, but most students have a hard time with static friction. They don't understand how there can be a force without a well-specified value, and they often have difficulty knowing which way the static friction force points. It's good to ask them what would happen to the object if there were *no* friction, then note that the static friction force responds *as needed*—both size and direction—to prevent slipping. Note that because $|\vec{N}| = mg$ in so many problems, some students will try to apply $|\vec{f}_s| = \mu_s mg$ indiscriminately to any situation. Instructors should note when doing examples where $|\vec{N}| \neq mg$.

Interactions: Forces were introduced by dividing a problem into the system and the environment. That process becomes even more important as you turn to interacting systems. Student difficulties with interacting systems often stem from very basic failures to identify the relevant systems or the interaction forces between them. Consequently, the starting point for Newton's third law is a very basic, but systematic, approach to identifying action/reaction force pairs.

The initial procedure is:

- Draw a picture with *every* object shown as a *separate* system.
- Identify and draw every force acting on every object.
- Express each force in the form $\vec{F}_{A \text{ on } B}$.
- Identify action/reaction pairs by matching $\vec{F}_{A \text{ on } B}$ with $\vec{F}_{B \text{ on } A}$. Connect the two forces of a pair with a dotted line.
- Verify that *every* force is paired with *one* other force, that the forces of a pair act on two *different* objects, and that the two vectors point in opposite directions.

The figure below shows an example of this strategy applied to the forces when a truck tows a car with a rope. Few students can produce such a diagram without considerable assistance.

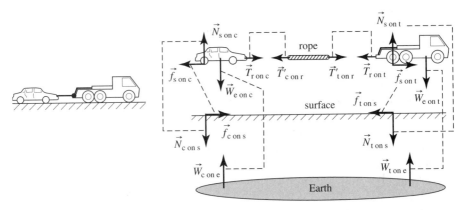

This initial procedure, although complete, is obviously cumbersome. After a few practice examples, students readily recognize that only some of the objects in the problem are of interest. These can be identified as "the systems." Others—the earth, the table, and so on—can be banished to "the environment" from where they exert "external forces" on the systems. Students can now recognize that there are no true external forces, since all forces are interactions, but that it is safe to treat some forces *as if* they are external forces. A revised procedure is then:

Analyzing Interacting Systems

- Separate the objects into "systems of interest" and "the environment."
- Draw each system *separated* from all other systems and from the environment. Separating the systems is a critical step.
- Identify and draw the forces on each system.
- Identify the action/reaction pairs between the systems. Verify that the two forces in a pair act on *different* objects.
- Draw a *separate* free-body diagram for each system. Use dotted lines to connect the two members of an action/reaction pair. Verify that the two forces connected by dotted lines are on two *different* free-body diagrams.

The emphasis on *separate* and *different* is important. Students have a strong tendency to identify the weight $\vec{W}$ and the normal force $\vec{N}$ on an object as an action/reaction pair. They will avoid this error if they first verbalize these as $\vec{F}_{\text{earth on A}}$ and $\vec{F}_{\text{table on A}}$. You should then reinforce the idea that the forces of an action/reaction pair *always* appear on two different free-body diagrams.

Newton's Third Law: Newton's third law is best given as a very explicit statement about forces. Almost all students know the phrase, "For every action, there is an equal and opposite reaction"—although their answers to questions asking them to "compare the forces" clearly show that they don't believe it! It's a catchy phrase, but misleading. Most significantly, the phrase doesn't indicate that the action and reaction must be on *different* objects. A better statement of the third law is:

1. *Every* force occurs as one member of a pair of interaction forces.
2. The two members of a pair of interaction forces act on two different objects.
3. The two members of an interaction pair are equal in magnitude but opposite in direction. That is, $\vec{F}_{\text{A on B}} = -\vec{F}_{\text{B on A}}$.

Early examples of third law reasoning can explore the idea of *propulsion forces*—how people walk, cars accelerate, rockets take off, and so on. Many students find this to be a difficult idea. They are not sure why the forces involved in walking and driving are static rather than kinetic friction, and they have difficulty understanding the directions. Ask them to imagine what would happen on a frictionless surface. Also ask them to imagine what direction loose gravel would be "kicked" as they sprint forward or a car accelerates forward.

The third law allows a closer look at tension and at the motion of objects connected by strings or ropes. As the research of McDermott, Shaffer, et al. (1994) has shown, tension is a difficult concept for students. A useful approach is to make an imaginary cut through the string, then to ask how the left and right sides of the string "hold on" to each other. The idea of an *internal* force is new to most students, but this approach is generally effective.

A standard application of tension is the Atwood machine or the "modified Atwood machine" of a hanging mass connected to a mass on a horizontal surface. Both involve pulleys, and pulleys introduce another complication. Most textbooks assume, with no discussion, that the tension forces at the two ends of the rope are the same—i.e., that the tension forces are an action/reaction pair of forces. But this directly contradicts the third law, since these forces aren't in opposite directions! A useful exercise is to consider both masses, the rope, and the pulley as separate systems and show *all* the forces between them. Then students can be led through the reasoning behind the massless string and massless, frictionless pulley assumptions, finally concluding that the tension forces at the ends of a massless rope act *as if* they are an action/reaction pair, even when the forces are not in opposite directions.

Students also have a difficult time with the *acceleration constraint* of connected objects. Texts rarely note that the acceleration constraint is an independent equation needed to solve the simultaneous equations typical of third-law problems. Although most students quickly recognize that $|\vec{a}_A| = |\vec{a}_B|$ if A and B are connected by a string (at least for simple connections; pulleys can make for more complex relationships), most need several practice examples to learn how to express this in terms of the vector components used in the problem. For example, a modified Atwood machine with B falling (negative acceleration) and A moving to right has $a_B = -a_A$.

Although the physics of interacting systems is difficult, it is not above the capability of introductory students. However, true understanding rarely occurs spontaneously on the basis of expository readings and lectures. These are very counterintuitive ideas, and they hit all of the misconception hot buttons of most students. Coming to an understanding takes explicit instruction and practice with the techniques of identifying and using interaction forces. Students need opportunities to reason their way to an explanation, based on Newton's laws, of observed phenomena. Providing them with such opportunities and practice is a powerful teaching tool for coming to grips with the essence of Newtonian mechanics.

But the time is well spent. By the time you finish Newton's laws, most students will be coming to realize that this systematic approach to reasoning and problem solving is paying off. They are beginning to solve difficult problems that they could never have solved with an equation-hunting strategy. It is an empowering experience for many students to realize they are solving "hard" problems that seemed far out of their reach just a few short weeks ago.

Using Class Time

I usually spend 8, or even 9, days on force and Newton's laws, as follows:

Days 1–2: Identifying forces, free-body diagrams, "discovering" Newton's second law.

Days 3–4: One-dimensional dynamics problems involving tension, friction, and "external" forces, such as the "propulsion force" on a rocket.

Days 5–6: Two-dimensional trajectories (brief) and circular motion.

Days 7–8: Interaction forces and Newton's third law.

Forces are a topic for which instructors need to have a wide variety of simple props available—springs, ropes, sticks, masses, blocks, rubber balls, etc. An effective starting tactic is to apply a force to a student, asking him or her to verify from sensations that there is a push or a pull. Then apply the same force to an inanimate object, such as a block, and ask if the block also experiences the same force.

If you pull on a student's arm (gently!), he'll agree that he feels a pulling force. If you now hand him one end of a rope, pull on the other end, and ask what he experiences, more than likely he'll reply, "You're pulling on me." This is an opportunity to distinguish the *immediate cause* at a point of contact, namely the tension in the rope, from the *ultimate cause* of whatever pulls the other end of the rope.

Hand a student one end of a spring and pull the other end. Have her agree that she feels a pulling force as the spring stretches. Then hang a mass from a spring. Because the spring stretches, most will now agree that the spring exerts a force on the mass. Then hang the same mass from a string. They don't see the string stretch, so does it exert a force? You can ask students to imagine if *they* were hanging from a rope tied around their waist—would they feel a force from the rope? Once they agree that the string does exert a force on the block, you have a good opportunity to talk briefly about molecular bonds as atomic-level springs; the string does stretch ever so slightly as the molecular bonds stretch, and that's what tension is.

The Normal Force: These activities are good preparation for what I think is a critical demonstration.

Place a heavy block in the center of the lecture table and ask students what force or forces are acting on the block. You'll get lots of responses of "gravity." If you inquire about other forces, a few will say "the normal force." Reply, "OK, so you learned in high school about this thing called the normal force, but how many of you *really* believe* that the table is exerting an upward force on the block." My experience is that less than one-third of the students will raise their hands. While their high school physics books and teacher may have told them about normal forces, their doubts arise because they don't see any *mechanism* by which the table exerts a force on the block. They need to be convinced, with evidence, that such a force is really there.

To begin a series of demonstrations, hand out several compression springs (fairly stiff ones) and ask students to squeeze them. They'll agree that the spring pushes back when it's squeezed. Stand a fairly soft spring on the table and set the block on it, giving a very visible compression. (You'll need some way to stabilize the block on top of the spring.) They'll agree that the spring exerts an upward force on the block. Then switch to a stiffer spring that barely compresses. This leads to the conclusion that the *amount* of compression is not the issue. Then place the block on a thin board that is supported at the ends, causing the board to sag. They'll agree that the board is also springy and exerts an upward force. Finally, return to the block sitting on the table. No discernible sag or compression, so is there a force?

The *coup de grace* is to place a mirror flat on the table, then reflect a laser beam from it at grazing incidence (laser beam almost parallel to table surface) with the reflected laser spot striking the side wall of the classroom. Ask a student to place a pencil tip next to the laser spot, then climb up and *stand* on the lecture table! Virtually all lecture tables will flex enough under the weight of the instructor to deflect the laser spot up a couple of millimeters. (It's good to try this in advance to make sure it's going to work. Also note that lecture tables have cross bracing underneath, and you don't want to stand right over one of the stiffer braces.) *Voila!* The table really does compress, even if only a microscopic amount, and it exerts the same upward force as a spring. (Those molecular bonds again, as it's good to note.)

This sequence of demonstrations takes some time, but it's well worth it. They make a memorable impression on students. I've received numerous unsolicited comments on evaluation forms that this demonstration was what really convinced students about the reality of forces exerted by inanimate objects. Note, by the way, that the usual approach of appealing to $\vec{F}_{net} = 0$ in static equilibrium to *infer* the existence of a normal force is not in the least convincing to most students. Many don't yet accept that an object at rest must have $\vec{F}_{net} = 0$.

You can consolidate what they've seen about forces to this point: a force is a push or pull, it occurs at a *point of contact* between the object and some *identifiable agent* that exerts the force, and a force can be exerted by either an animate or inanimate agent. This is a good opening to consider friction. Ask them to imagine dragging their hand across a very rough surface. Is there a force? What direction? Then what about a box sliding across the surface?

Then turn to gravity. Drop a ball—why does it fall? This is a different type of force, a long-range force. You might want to have some magnets to demonstrate other long-range forces, but emphasize that gravity is the only long-range force that will be considered for quite some time. Every other force *must* be a contact force.

Finally, toss a ball straight up into the air and inquire about the force or forces on the ball after it leaves your hand but before it reaches the top. If you had opened class with this question, a majority of the students would likely assign an upward "force of the throw" to the ball. Some may be doubting this answer after the sequence of demonstrations you've been through, but many others will still want an upward "force of the throw." You can play devil's advocate, first getting them to agree that there's no contact with anything in the environment, then asking "But how can it go up unless there's an upward force?" Don't answer! This is where you really want them to be confronted by difficulties with their alternative conceptions of force and motion. Ask them to think about it, and promise to resolve the issue during the next class.

Once students understand how to *identify* forces, they've already overcome the largest hurdle to drawing free-body diagrams. Even so, they'll need some explicit instruction and practice. Incorporating free-body diagrams into your explanations reinforces their importance as an *analysis tool.* Call attention to the fact that free-body diagrams show only the forces acting *on* the object. Many students will try to include forces exerted *by* the object on other systems. An object sliding down an inclined plane, with friction, is an excellent example for first identifying all forces, then showing them in the correct orientation on a free-body diagram with a tilted coordinate system.

The objective of day 2 is to "discover" Newton's first and second laws. What are the *consequences* of a force on an object? You'll want to use whatever demonstrations you have available to show that an object continues to move at constant velocity in the absence of a net force and that force causes an object to accelerate. Convincing demonstrations here go a long way toward changing the common student beliefs that motion requires a force and that force is proportional to velocity.

Gliders on air tracks (the longer the better) are good for demonstrating the first law. Also effective are dry ice pucks that slide down a long sheet of glass on a cushion of CO_2 vapor.

Microcomputer-based force probes and motion sensors are especially effective for demonstrating Newton's second law. Attach a force probe to a low-friction cart on a track, then use a string and pulley to connect the force probe to a hanging weight. When the weight is released, the force probe measures the string tension while the motion detector measures the motion. This gives excellent results, showing that a constant force produces a linearly increasing velocity and a constant acceleration. Doubling the tension force doubles the acceleration while doubling the cart mass halves the acceleration. These results can all be shown in just a few minutes with MBL tools.

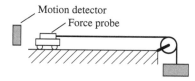

Priscilla Laws (1997b) has developed an alternative approach that can be shown in a large lecture hall, using volunteers, though it is better suited to lab. One student sits on a cart with low-friction wheels (a Kinesthetic Cart is available from PASCO) and holds one end of a spring scale. Another student pulls the cart by pulling the other end of the spring scale, with instructions to "keep the reading as constant as possible." Nearly all students expect, before trying this, that they're going to walk or run at a steady speed; they are very surprised to find that they must *accelerate* to keep the force constant. Once the cart is moving, they can let the spring scale reading drop to zero but the cart keeps moving!

These demonstrations lead to the introduction of Newton's second law in the *operational form* $\vec{a} = \vec{F}_{net}/m$. Although $\vec{F} = m\vec{a}$ looks more elegant, the operational form conveys a better sense of cause and effect and helps students to develop better ability to *reason* with Newtonian concepts. These demonstrations should lead to the explicit conclusion that motion does not need a *cause*. The question is not "Why does an object move?" but "Why does it *change* its motion?" Remind students that this is easy to say, but it takes effort and practice to begin to *think* this way. They shouldn't be surprised if it all seems confusing at first, but promise them that they will "get it" if they persevere.

Several examples are worthwhile at this point.

EXAMPLE 1: An elevator is going up at a *steady* speed. First have students identify tension and weight as the only two forces. Then ask, "Is $|\vec{T}|$ greater than, equal to, or less than $|\vec{W}|$? Or is there not enough information to tell?" Many will answer "greater" because "motion requires a force."

EXAMPLE 2: Push a block across the table at *steady* speed. Since you're exerting a force on it, why isn't it accelerating? Ask them to identify all the forces and to draw a free-body diagram. Finally, ask them to compare the size of the pushing force and the size of the friction force.

EXAMPLE 3: Push the same block fairly quickly, then release it so that it slides some distance before stopping. Have them analyze the forces and reach the conclusion that the acceleration vector points backwards. Remind them of what they learned in kinematics about situations in which the acceleration vector is opposite the velocity vector. When done, congratulate them on having performed a Newtonian analysis to *explain* why a block coasts to a stop after you release it.

EXAMPLE 4: Return to the issue of the ball tossed straight up—where you left them hanging on day 1 as to how it moves up without an upward force. Now you can complete the analysis, using Example 3 as a horizontal analogy. Remind them that no *cause* is needed for the ball to move upwards—inertia takes care of that. The proper question is not "Why does the ball move upward?" but "Why does the ball slow down and eventually fall?"

As you work through these examples, point out that you are giving an *explanation* of an event in terms of physical principles (Newton's laws) and logical inference. Students find this kind of reasoning *very* hard to do.

The next day can be spent developing the problem-solving strategy, on mass and weight, and on simple equilibrium and non-equilibrium problems. The following day can then introduce friction and solve more complex examples. It's best to work through as many examples as possible, making points within the context of a problem rather than lecturing about them. You'll want to strike a balance between work that you do, as an example, and work that you ask students to do. When students work problems, it's usually best to ask them to do one part at a time—such as "draw a pictorial representation" or "draw the free-body diagram"—then stop and wait for feedback and discussion before going on to the next part.

The following suggested example problems are chosen to visit all the major issues.

EXAMPLE 5: A 1000 kg block hangs on a rope. Find the tension in the rope if the block is stationary, then if it's moving upward at a steady speed of 5 m/s, finally if it's accelerating upward at 5 m/s^2. Although this is a simple problem, students should draw a motion diagram, identify the forces, draw a free-body diagram, then explicitly use Newton's second law.

EXAMPLE 6: Show a mass m as a point on the board, and draw a downward weight vector $\vec{W}$. Then ask students to find the components W_x and W_y in each of the three coordinate systems shown. You may first want them to prove that the angle θ between the weight vector and the negative y-axis is the same as the "tilt angle" of the coordinate system.

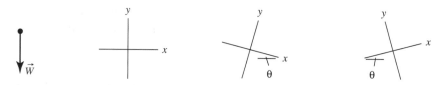

Find the components of $\vec{W}$ in each of the three coordinate systems shown.

EXAMPLE 7: "Sammy Skier (75 kg mass) starts down a 50-m-high, 10° slope on friction-less skis. What is his speed at the bottom?" This is a good first opportunity for a full-blown worksheet solution, from the pictorial model all the way to the kinematics. Along the way you can demonstrate that the acceleration on a frictionless incline is $a = g \sin \theta$. This problem also provides an opportunity to talk about modeling and making approximations. Is it reasonable to treat the skis as frictionless? Is it reasonable to ignore air resistance?

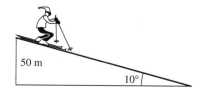

50 m

10°

EXAMPLE 8: Burglars are trying to haul a 1000 kg safe up a *frictionless* ramp to their get-away truck. The ramp is tilted at angle θ. What is the tension in the rope if the safe is at rest? If the safe is moving up the ramp at a steady 1 m/s? If the safe is accelerating up the ramp at 1 m/s²? Do these answers have the expected behavior in the limit $\theta \rightarrow 0°$ and $\theta \rightarrow 90°$? Most students have no experience with checking the limiting behavior of solutions, so this is a worthwhile exercise for comparison with your answers to Example 5.

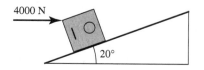

θ

EXAMPLE 9: The same burglars push the 1000 kg safe up a 20° frictionless slope with a *horizontal* force of 4000 N. What is the safe's acceleration?

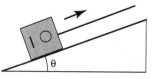

4000 N

20°

EXAMPLE 10: A 50 kg student gets in a 1000 kg elevator at rest. As the elevator begins to move, she has an apparent weight of 600 N for the first 3 s. How far has the eleva-tor moved, and in which direction, at the end of 3 s?

It is worthwhile to revisit some of these examples with friction added. For exam-ple, you can repeat Example 7 after giving them $\mu_s = 0.12$ and $\mu_k = 0.06$. You can tell them that the safe in Example 8 has $\mu_s = 0.8$ and $\mu_k = 0.5$, then ask them how much tension is needed in the rope to start the safe moving. If that tension is maintained after the motion starts, does the safe move with constant velocity or constant acceleration? What is the value of v or a?

EXAMPLE 11: A car traveling at 20 m/s stops in a distance of 50 m. Assume that the deceleration is constant. The coefficients of friction between a passenger and the seat are $\mu_s = 0.5$ and $\mu_k = 0.3$. Will a 70 kg passenger slide off the seat if not wearing a seat belt?

This is an excellent example, but one that students find very difficult. First of all, the question doesn't tell them what to compute. Second, the passenger's mass is extraneous information. Third, it confronts their difficulties with static friction. All the analysis tools that have been developed so far are needed to understand and solve this problem. Walking students through it is a valuable exercise. You need to establish that *static* friction is the only horizontal force on the passenger, that the acceleration (as seen from a motion diagram) points backward, and so static friction is the force that decelerates the passenger. But since static friction has a maximum possible value, there's a maximum possible deceleration. Finally, they have to realize that the question will be answered by comparing the car's deceleration to the maximum possible deceleration of the passenger. In this case, the passenger stays on the seat.

Projectile motion is important, but it shouldn't be overemphasized. It's mostly kinematics, which you've already covered, and all you can add at this point is a "proof" that the acceleration really is $\vec{a} = -g\,\hat{j}$. However, an example with acceleration along both axes is worthwhile. A nice example to consider is a rocket of mass m that takes off with, and maintains, a 45° angle above the ground. The thrust force is $\vec{F}_{\text{thrust}}$, and it is assumed that the mass and thrust remain unchanged. First ask students what the motion would be in the absence of gravity. Most will agree that it accelerates along a straight line. Then ask them to predict the shape of the trajectory with gravity. Most will expect some kind of curve, but they won't be very sure of themselves. Rather surprisingly, the analysis shows that the trajectory is still a straight line, but with slope $= 1 - (\sqrt{2}mg)/|\vec{F}_{\text{thrust}}|$. This has the expected behavior for $g \to 0$. The minimum thrust needed to get off the ground is found in the limit of slope $\to 0$.

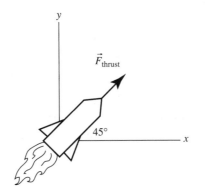

Before getting into the dynamics of circular motion, it's worth returning to motion diagrams to establish that there *is* an acceleration, even though the speed is constant, and that the acceleration points to the center. This is a good point to introduce the rz-coordinates that were described in the Pedagogical Approach section.

Then, turn to the issue of force. It's tempting to assert that the net force points toward the center because $\vec{a}$ does, but this is actually backward reasoning. You need to consider situations of circular motion, analyze the forces present, and conclude *from the situation* that the net force points to the center. In other words, you want students to understand the chain of reasoning:

- The circumstances make $\vec{F}_{net}$ point toward a fixed point.
- Therefore, $\vec{a} = \vec{F}_{net}/m$ also points toward a fixed point (a centripetal acceleration).
- Therefore the motion is uniform circular motion.

You can show this with two related demonstrations. First, swing a ball or block on a string around on the table top. (Don't swing it in a horizontal plane over your head, because that adds extra complications.) Have students identify the forces: $\vec{T}$, $\vec{W}$, and $\vec{N}$. Then have them recognize that $\vec{W}$ and $\vec{N}$ cancel, leaving $\vec{F}_{net} = \vec{T}$. Now they see, from the circumstances, that the net force points to a fixed point. It's worth pointing out that tension can also cause linear accelerated motion. It's the circumstances of *how* the force is applied, not the force itself, that determines the motion.

Next, roll a ball or a marble around the inside of a horizontal circular hoop and repeat the same procedure. Now there are *two* normal forces, a novel situation to consider, but again $\vec{N}_{hoop}$ points toward a fixed point. An important thing to note is that $\vec{N}_{hoop}$ is a *pushing* force toward the center, whereas $\vec{T}$ was a *pulling* force. Many students think that circular motion occurs *only* with a "pull" toward the center of the circle.

A good problem to work in detail—and to demonstrate if you can—is a block placed on a turntable that rotates at fixed frequency f. If given values for f, μ_s, μ_k, and m, how far can the block be placed away from the center before it slips? (μ_k and m aren't needed, but it's good to give values so that students have to separate relevant from irrelevant information.) Alternatively, if you're doing a demonstration, you can measure the radius at which the block slips and ask students to find μ_s.

Another demonstration that can be analyzed is to spin a marble around the inside of a cone. (A large glass funnel works well.) How does the height h above the bottom depend on the marble's speed and frequency? The speed for circular motion at height h is $v = \sqrt{gh}$, independent of the cone angle. But $f = v/2\pi r$, which leads to an interesting situation: To increase the height of the circle requires a larger speed but a *smaller* frequency.

In analyzing a marble in a cone, you'll want to note that the plane of the motion does *not* correspond with any physical surface and that there's no physical object at the center of the circle. You'll also want to note that the rz-coordinate system is *not* tilted, even though the surface is. This is different than the tilted axes used for linear motion, but in both cases the axes are chosen so that the acceleration along one axis is zero.

A demonstration well worth an extended analysis is swinging a ball on a rope in a horizontal circle over your head. Many students think that the rope will be parallel to the ground. You can easily show, with a slow swing, that the rope is tilted down at an angle and that the angle depends on the rotation frequency. No matter how fast you swing the ball, you can't get the rope to move in a horizontal plane. A good problem for the class is to find the angle of the rope as a function of the frequency. Students want the net force to point to the hand, so you need to have them recognize explicitly where the center of the circle is.

A final task is to reach an understanding of how water can stay in a bucket swung over your head. For demonstrations, you'll want a ball on a rope and a bucket with both blocks and water. A marble or cart that can do a loop-the-loop on a track is also worthwhile.

Swing the ball on a rope in a vertical circle and focus on the bottom point. Ask students if the tension at that point is less than, greater than, or equal to the ball's weight. Most students are still likely to think it will be equal. You can then work through the analysis. You can also remind them that they "feel heavy" at the bottom of a roller coaster dip.

Then you can turn their attention to the top point of the circle and the issue of when the ball's rope goes slack. It is hard for many students to understand this, so they may need multiple explanations of the fact that the vertical tension component T_y can't be positive (a rope can't *push* the ball upward at the top of the swing). You can then have students find the formula for the minimum frequency for which the motion remains circular.

Put some loose objects (blocks, erasers, etc.) in a bucket and swing it over your head. Is this really any different than the ball at the top of the circle? What is the condition that the objects stay with the bucket rather than beginning a parabolic free fall from the bucket? You want students to realize that bucket has to be *pushing* on the objects to create a centripetal acceleration, so you must have $|\vec{N}| > 0$. $|\vec{N}| = 0$ is the limiting case at which the objects are on the verge of leaving the bucket. Finally, put water in the bucket and repeat the discussion.

This is very tricky reasoning, and students often have a lot of questions. It's good (with the loose objects, not the water!) to go slow enough that the objects *do* fall out, asking students to think about how the trajectory of the objects differs from the trajectory of the bucket. It's important to focus student comments on Newtonian ideas— forces, acceleration, velocity, cause and effect, and so on. They tend not to use this terminology spontaneously, so you need to prompt them with "Could you state that in terms of specific forces?" or "Can you use one of Newton's laws to tell me why B should follow A?" Nonetheless, this is time well spent because it gives students an excellent opportunity to focus on the principles and logic of Newtonian mechanics.

Finally, you can turn your attention to Newton's third law. Most students readily accept that if A pushes/pulls B, then B pushes/pulls back on A. If you ask a student to stretch a spring, she can "feel" that the spring pulls on her hand at the same time she pulls on the spring. Long-range forces are more troublesome because students don't yet understand the role of mass in the "outcome" of an interaction. The earth clearly pulls down on a ball that is dropped, but there's little evidence of the ball exerting a force on the earth. You can make this idea plausible by demonstrations with magnets. If a student holds two fairly strong magnets, he can feel that each is pulling (or pushing) on the other. If you attach magnets in the repulsive orientation to two gliders on an air track, both gliders move apart when released. More important, if you now weight the gliders differently students can see that there *is* a mass effect and that the lighter glider does most of the motion. Now it's a much smaller step to accepting that the ball really does exert a reaction force on the earth.

With this as a starting point, it's good to spend an entire class asking students to follow the strategies summarized in the Pedagogical Approach section for identifying *and labeling* forces, for identifying action/reaction pairs, and for drawing free-body diagrams. As trivial as it seems, it's worth starting with a block sitting at rest on the floor (ask them to consider the floor to be part of the earth) and have students draw a picture showing the block, the surface, and the earth as a whole. Although somewhat artificial, it's important to distinguish "the surface," where contact forces occur, from "the earth" that exerts the gravitational force. Have them draw all force vectors, label them as $\vec{F}_{\text{A on B}}$, and connect all action/reaction pairs with dotted lines. This will rapidly lead to conflict for students who want $\vec{W}$ and $\vec{N}$ to be an action/reaction pair.

Then place a second block on top of the first. The lower block now experiences *two* normal forces, one from above and one from below. Again, this apparently trivial situation is initially difficult for many students. Fortunately, most will "get it" after just a few such examples.

Have a cart or some other large object that you can push as you walk. Have students draw pictures of you, the cart, the surface, and the earth and identify all the forces. The different roles of friction for the person and for the cart is a particularly troublesome point: kinetic friction in the reverse direction for the cart, but *static* friction in the *forward* direction for you! You'll need to talk about how you push against the ground—what would happen to loose gravel on the ground?—and, in reaction, the ground pushes forward on you. A surprisingly difficult idea for many to grasp, but essential for understanding rocket propulsion.

After showing *all* the forces, it's worth moving quickly to the simplified view of showing free-body diagrams of just the "systems of interest"—you and the cart in this case. You should still have students connect any action/reaction pairs between the systems with dotted lines.

These exercises show that the members of an action/reaction pair are always opposite in direction, but they say nothing about the size of the forces. You're now ready to give a full statement of Newton's third law. Unfortunately, it isn't easy to provide a demonstration of the third law. By far the most convincing demonstration I know is the colliding force probe experiment from Sokoloff, Thornton, and Laws' *RealTime Physics* (1999). This gives instant and dramatic confirmation that the forces between two colliding carts are *always* equal in magnitude, regardless of the masses or the initial velocities of the carts. A simpler demonstration is to have two students of different size push against each other with bathroom scales, each calling out the reading on "their" scale as they move forward or backward. Lacking these, you're forced to assert the third law as a hypothesis that will be tested on the basis of its predictions about the motions of interacting systems.

An important application of the third law is an analysis of tension. The idea of tension as "pulling equally in both directions" can be introduced by asking students to compare the tension in two ropes: one is tied to the wall and pulled by a student with a 100 N force, the other is pulled by two students in a tug-of-war as each pulls with a 100 N force. Research has shown that most students expect the tension in the tug-of-war rope to be twice as large because "it's being pulled twice as hard." It's good to demonstrate the equal tensions in a set-up such as shown below, using spring scales. Then have students go through the force analysis, making an imaginary division of the rope into two parts so they can see the tension force as "holding the rope together."

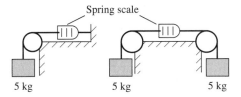

| 5 kg | 5 kg | 5 kg |

It's worth ending the day with a discussion of the massless string approximation. Arons (1997) suggests a good example in which an initial mathematical analysis using the third law helps to reveal the role of the string's mass. A string of mass m_S is used to accelerate a block of mass m_B across a frictionless surface.

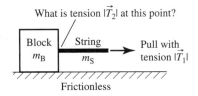

Here, in a nonequilibrium situation, you first need students to recognize that the tension at the two ends of the string is *not* the same. It is straightforward to show that

$$\frac{|\vec{T_1}|}{|\vec{T_2}|} = 1 + \frac{m_S}{m_B},$$

although students would never think of casting the result in this form. Now it's easy to see what happens as $m_S \to 0$. This can lead into a discussion of how two objects interacting through a massless string act *as if* they exert action/reaction forces on each other.

Finally, you can turn to working standard third-law example problems. These will be easier if you've spent time identifying action/reaction pairs and drawing free-body diagrams, but you should still discuss and show these steps as you work. Because third-law problems tend to be lengthy, you will probably want to do more of the work yourself rather than having students do it. Even so, it's good to have them do one piece of each example—draw the free-body diagrams for one, identify the acceleration constraint of another, and write down Newton's second law for each system in a third.

It's not unusual for a third-law problem to have four equations of motion (*x*- and *y*-components of the second law for each of two moving objects), one or two third-law relationships, one or two friction models, and an acceleration constraint. Eight or nine independent equations and relationships is not uncommon. Students are not used to problems with this level of complexity, and many become overwhelmed. Sympathize with them, but keep pointing out that the problem-solving strategies and the work-sheets are a systematic way of keeping track of all the information. Equation hunting won't help them now; they *have* to understand and follow the procedures.

Simple substitutions usually reduce everything to two simultaneous equations. Even though the substitutions are simple, you need to work through *all* the details in at least one example. Many students get lost with assertions that "If you substitute the first equation into the second you end up with . . . " Also be aware that some students have had little or no experience with simultaneous algebraic equations. Once again, they'll need to see one solution worked and explained in detail.

This is the high point of Newtonian mechanics. Although the concepts of upcoming chapters may be difficult, students won't face this level of detail and complexity again until they reach electricity and magnetism. Ample praise for their efforts and for the amount they've learned to this point is certainly called for.

Sample Reading Quiz or Discussion Questions

1. What is a "resultant force"?
2. List at least three of the steps used to identify the forces acting on an object.
3. What is the difference, or is there a difference, between *mass* and *weight?*
4. What is the purpose of the *massless string approximation?*
5. Is the tension in rope 2 greater than, less than, or equal to the tension in rope 1?

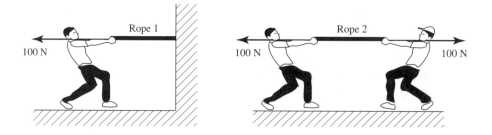

6. Which of these is *not* a force discussed in this chapter?
    a. The tension force.   c. The orthogonal force.
    b. The normal force.   d. The propulsion force.

7. The coefficient of static friction is
    a. smaller than the coefficient of kinetic friction.
    b. equal to the coefficient of kinetic friction.
    c. larger than the coefficient of kinetic friction.
    d. not discussed in this chapter.

8. The force of friction is described by
    a. the law of friction.   c. a model of friction.
    b. the theory of friction.   d. the friction hypothesis.

9. If a car turns a corner too quickly, you seem to be "thrown" against the door because of
    a. inertia.   c. the centrifugal force.
    b. the centripetal force.   d. the friction force of the seat.

10. An Atwood machine
    a. is a lever and a fulcrum used to lift a block.
    b. is two hanging masses connected by a rope over a pulley.
    c. is a rope-and-pulley device used to lift a block.
    d. is any labor-saving device that works because of Newton's third law.
    e. was not discussed in this chapter.

Sample Exam Questions

1. Give an explanation of *how* a horizontally thrown tennis ball bounces off a wall. (Assume that the ball is traveling horizontally when it hits the wall.) Your explanation should include
    - A motion diagram of the ball during the short time it is in contact with the wall (made with a very-high-speed movie camera).
    - A free-body diagram of the ball during its contact with the wall.
    - Several well-written sentences that use the laws and principles of physics to explain the observed behavior of the ball.

2. A 10 g bullet is fired with a speed of 400 m/s toward a solid wood door. The force on the bullet as it goes through the door is shown by the graph.

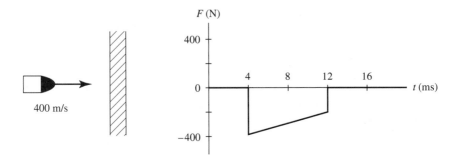

a. Why is the force negative?

b. Find the speed with which the bullet exits the door on the other side.

3. You have been hired to measure the coefficients of friction for the newly discovered substance jelloium. Today you will measure the coefficient of kinetic friction for jelloium sliding on steel. To do so, you pull a 200 g chunk of jelloium across a horizontal steel table with a constant string tension of 1.00 N. A motion detector records the motion and displays the graph shown. What is the value of μ_k for jelloium on steel?

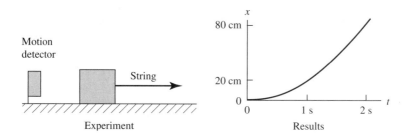

Experiment   Results

4. A car drives at steady speed $|\vec{v}|$ over the top of a circularly shaped hill of radius 50 m. What's the maximum speed the car can have without "flying off" the hill at the top?

5. Is the acceleration of Cart 1 greater than, equal to, or less than the acceleration of Cart 2? Explain your reasoning.

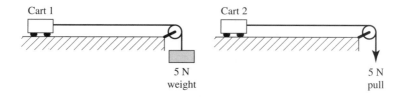

6. A 200 kg horse is hooked up to a 400 kg wagon, and you signal the horse to pull forward. The horse turns to you and says, "Why bother? However hard I pull, the wagon is just going to exert an equal but opposite force. The wagon won't move." Will the wagon move? If not, explain why not. If so, what kind of explanation would you give the horse to convince him to pull? You might want to use pictures as part of your explanation.

7. Block A has a mass of 1 kg and block B has a mass of 2 kg. They are being accelerated upward by massless ropes 1 and 2. Tension $\vec{T}_1$ is pulling on rope 1.

a. Draw separate free-body diagrams of block A and block B. Label all forces. Connect any and all action/reaction pairs with dotted lines.

b. Rank order *all* the forces on blocks A and B, from the largest magnitude to the smallest. *Explain* your rankings.

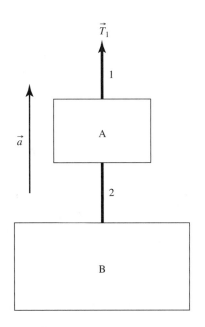

8. A 200 kg wood crate sits in the back of a truck. The coefficients of friction between the crate and the truck are $\mu_s = 0.9$ and $\mu_k = 0.5$. The truck starts off up a 20° slope. What is the maximum acceleration the truck can have without the crate slipping out the back?

8

Impulse and Momentum

Background Information

To the experienced physicist, conservation laws seem the "obvious" way to tackle many problems in physics. Yet students find conservation laws to be rather mysterious, and they are reluctant to use a conservation law unless explicitly asked to do so. Most instructors find that a majority of students will elect to use Newton's laws to solve (or try to solve) an exam problem that could have been solved more easily with a conservation law. My students, in informal questioning, report that they feel "comfortable" using the tangible ideas of forces and acceleration, but they find conserved quantities—especially energy—to be too abstract to hold any meaning for them. Conservation laws are at the center of our modern understanding of physics, so an important pedagogical task is to provide students with a learning environment in which they come to "own" the ideas of conservation of energy and momentum.

Give students a choice of a rubber ball or a sticky clay ball (of equal mass) that they are to throw against an upright wood block in order to knock it over. A significant majority will choose the clay ball, feeling that sticking to the block is somehow more effective at applying a force than is bouncing off the block. Students have little intuition for the idea of an *impulse,* and most are very surprised at the outcome of a demonstration showing that the elastic collision has more "effect" on the block than the inelastic collision.

There has been little systematic research on students' understanding of the ideas of impulse and momentum. The only paper I know of is that of Lawson and McDermott (1987). They used a stream of compressed air to accelerate two dry-ice pucks of significantly different mass through a fixed distance, then they asked students (who had completed instruction in energy and momentum) to make comparisons of the final momenta. The goal was for students to recognize that the heavier puck experienced the force for a longer time and thus received a larger impulse. Thus the heavier puck—although slower—had a larger momentum.

Only 25% of *honors* students responded correctly, and *no* students in an algebra-based course gave the correct answer. Students who answered incorrectly were given hints. Following the hints, the correct response rate of honors students went up to 58%, but the correct response rate of students in the algebra-based course remained at 0%. The investigators did not use students from a regular, non-honors section of calculus-based physics, but their responses could be expected to fall between the two groups tested.

The most common incorrect response was that the final momenta would be equal. Students explained this choice with a *compensation argument:* the heavier puck has more mass but it goes slower, so—since the forces were equal—both pucks should have the same value of *mv*. The investigators' conclusion was that even honors students, who could often apply conservation laws successfully in problem solving, had difficulty seeing how the ideas of momentum and impulse applied to real moving objects.

Many students are hindered in their learning of these concepts because of language. *Momentum,* in particular, is often thought of as being synonymous with *inertia*—things with a lot of momentum want to keep going. Students easily fall back upon their loose definitions of these terms rather than using the very specific meanings of physics.

Other students have difficulty distinguishing between p and Δp. The new perspective of conservation laws requires students to compare the situation "before" and "after" an interaction. Many want to interpret the momentum-impulse theorem as $p_{\text{final}} = J$ rather than $\Delta p = J$—that is, that "impulse causes momentum" rather than "impulse causes a *change* in momentum." Note the similarity to the common misconception that "force causes velocity" rather than "force causes a *change* of velocity." Instructors, as they introduce impulse and momentum, will want to remind students of previous discussions about Newton's second law.

A final difficulty is with the *signs* of p and J. Although momentum is a fairly intuitive idea, many students see it as an inherently positive quantity, similar to mass. This is particularly devastating when calculating Δp for an object that changes directions. Instructors need to give very explicit attention to signs during class discussions and examples.

Student Learning Objectives

- To understand interactions from the new perspective of *impulse* and *momentum*.
- To learn what is meant by an *isolated system*.
- To apply conservation of momentum in simple situations.
- To understand the basic ideas of inelastic collisions, explosions, and recoil.

Pedagogical Approach

I highly recommend introducing momentum before energy. A few texts are beginning to do this. There are several reasons for this choice:

- Momentum is a more intuitive concept than energy. In addition, momentum more readily lends itself to simple demonstrations. Teaching momentum conservation first allows students to gain experience with the conservation law perspective before getting into the more abstract ideas of energy.
- Momentum conservation follows directly from Newton's third law.
- Understanding impulse requires thinking about a force exerted over an interval of time. Students find this easier than work, which is a force exerted over a distance.

The chapters on momentum and energy introduce no new physics. The basic principles of mechanics remain Newton's laws of motion. Instead, these chapters introduce a *new perspective* for looking at problems of force and motion. This is not an easy transition for students who are still very insecure about their grasp of Newton's laws!

The limited research on this topic suggests that instructors should begin with an emphasis on the impulse-momentum theorem and not rush into using conservation of momentum. The application of impulse and momentum to single-particle dynamics helps students to see that this is just a new way of talking about what they already know. Students need to be asked to *explain* simple events—such as throwing a ball, bouncing a ball off a wall, or jumping straight up off the ground—first in terms of forces and Newton's laws, then in terms of impulse and momentum. The goal is make impulse and momentum as tangible and "real" as are force and acceleration. Along the way, you can discuss issues such as the signs of p and J and the fact that an impulse causes a *change* in the momentum.

Conservation of momentum is introduced as a way of comparing the situations "before" and "after" an interaction. It seems commonplace to instructors, due to long familiarity, but it's really quite surprising that there exists a quantity not changed by the complex interactions of collisions and explosions. Although the "proof" is easily made from Newton's third law, class demonstrations are equally important.

Adopting the before-and-after perspective is difficult for most students. They need to see a number of examples—preferably involving demonstrations—devoted simply to identifying the initial and final states. A common error in more complex problems—such as the ballistic pendulum or cars that collide and slide—is to try to include too much. That is, students will try to apply momentum conservation from "before" the cars collide until "after" they finish sliding.

A related issue is choosing the correct system for analysis. Consider dropping a ball and letting it fall. If the ball alone is chosen as the system, its motion can be analyzed with the impulse-momentum theorem because gravity is an external force. If the ball+earth system is chosen, then total momentum is conserved. Students need to practice choosing systems and seeing the implications of their choice.

Students also need to begin learning a new pictorial representation for conservation problems. The goal, with a two-part drawing, is to show the situation before and after the interaction. It's still necessary to establish a coordinate system and define symbols. Two-step problems, where a momentum conservation situation is coupled to a dynamics situation, require careful attention and a free-body diagram. The figure below shows an appropriate pictorial representation for the problem of two cars sliding after an inelastic collision.

Two cars collide and slide. The skid marks are 3 m long and $\mu_k = 0.8$. What was the initial speed of the Volkswagen?

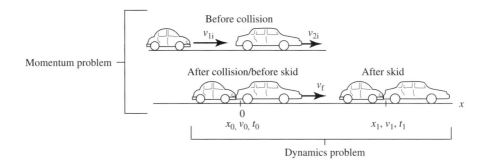

Known: $m_1 = 1000$ kg $m_2 = 2000$ kg $v_{2i} = 5$ m/s

Find: v_{1i}

Skid: $x_0 = 0$ $v_0 = v_f$ $x_1 = 3$ m $v_1 = 0$ $\mu_k = 0.8$

The one drawback to introducing momentum before energy is that you need to defer elastic collisions, which conserve energy as well as momentum, to the energy chapters.

Using Class Time

I focus on impulse and momentum as applied to single particles and to interactions between two particles, skipping the usual extension to center-of-mass and systems of many particles. This takes two days.

It's worth starting with conservation of mass. A nice demonstration is to measure out 100 g of water and 100 g of alcohol, then ask students to predict the mass after you mix them together. (You want two substances that are *different* but that do *mix,* such as alcohol and water.) After everyone correctly predicts 200 g (you should measure to verify!), you can ask them *why* they believed that mass would be conserved.

You can also have students articulate, in this familiar context, what it *means* to be conserved and what the significance of "before" and "after" is. This discussion can lead you to raise the possibility that there might be other less obvious quantities that are also conserved during an interaction.

The majority of day 1 is well spent on impulse and momentum as applied to single particles. As noted above, it's good to have students *explain* simple events—such as bouncing balls or jumping off the ground—first in terms of forces and Newton's laws, then in terms of impulse and momentum. A dropped ball has a "long" impulse due to gravity as it falls, then a "short" impulse due to both the normal force *and* gravity (most students leave out gravity) as it bounces. It's worth asking students if $\vec{F}_{\text{floor on ball}}$ is the same as, or different than, the normal force $\vec{N}$ they are familiar with. They may need help seeing that $\vec{N}$ is a force that grows from zero to a magnitude that far exceeds $\vec{W}$, then shrinks back to zero. Motion diagrams can still be useful!

You'll want students to think about impulse as an area under a curve. This reinforces the geometric interpretation of integrals they learned in kinematics and prepares them for a similar interpretation of work. The force probes from Vernier Software or PASCO can be used to capture a force-versus-time curve during a collision of carts, and the software then allows you to integrate the curve numerically to obtain the impulse. This is a worthwhile lecture demonstration if students are unable to do this in a laboratory.

Some students have difficulty believing that collisions between "hard" objects, such as steel spheres, take a finite amount of time. The spheres seem so incompressible that the collision "must" be instantaneous. It's worth a short digression to remind them that the molecular bonds inside the steel compress ever so slightly during a collision, just as the table compresses ever so slightly as it exerts the normal force on an object.

This initial discussion of impulse and momentum change can lead you to pose the "Should you throw the clay ball or the rubber ball?" question that was given earlier. An extended class discussion is worthwhile before revealing (or, better yet, demonstrating) the answer.

Possible numerical exercise: A 200 g rubber ball is released from a height of 2.0 m. It falls to the floor, bounces, and rebounds. The force of the floor on the ball is shown in the figure. How high does the ball rebound? (Answer: 1.62 m. Don't forget the 10 ms impulse due to the weight force $\vec{W}$, which is small but not entirely negligible.)

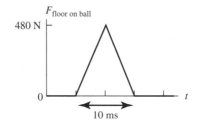

Collisions between a ball and the earth lead naturally to a consideration of collisions between two objects of similar mass. Gliders on an air track or low-friction carts on a track provide many opportunities to show that *each* object experiences an impulse and a change of momentum. You can then pose the question: What is the *total* change of momentum, $\Delta p_1 + \Delta p_2$? It is best to answer this question experimentally if you can, using motion detectors or photogates or whatever equipment you have for measuring speeds. You probably won't get exactly zero, but you can note that the sum is *much less* than Δp_1 and Δp_2 individually and thus—for all practical purposes—consistent with zero.

You can easily extend this idea to the more general law of conservation of momentum. The derivation from Newton's third law of momentum conservation for two interacting objects is so short and so important as to be worth making an exception to the general rule of "don't lecture." However, the derivation should *follow* demonstrations rather than precede them.

This can be followed by examples of inelastic collisions, "explosions" (such as two gliders of different mass pushed apart by a spring), and recoil. Lecture demonstrations are especially important, and most departments have a wide range to choose from. Note that only inelastic collisions can be analyzed by momentum conservation alone; elastic collisions have to await the introduction of energy conservation.

Conservation laws are a good place to use Jeopardy questions. For example, give students the equation

$$(1000 \text{ kg})(2 \text{ m/s}) = (1000 \text{ kg} + 2000 \text{ kg})v_f,$$

then ask them to "write a problem" for which this is the relevant equation.

It's worth ending day 2 with two or three full examples of problem solving with conservation of momentum, emphasizing the new pictorial representation that shows the "before" and "after" situations. Be sure to include at least one two-part problem that requires both conservation of momentum *and* a second-law analysis of the subsequent motion. One I like is

A 10 g bullet is fired into a 1 kg wood block, where it lodges. Subsequently, the block slides 4.00 m across a wood floor ($\mu_k = 0.2$ for wood on wood). What was the bullet's speed?

This requires a careful distinction between the collision and the subsequent slide, with many students wanting to apply momentum conservation for the entire process. Other students are not yet comfortable with the need to "work backwards." A careful analysis of the *reasoning* behind the solution leaves students better prepared for typical end-of-chapter homework. It is also an important prelude to the ballistic pendulum. The answer, by the way, is 400 m/s.

Sample Reading Quiz or Discussion Questions

1. What is an "isolated system"?
2. What is the "impulse approximation"?

3. Impulse is
    a. a force that is applied at a random time.
    b. a force that is applied very suddenly.
    c. the area under the force curve in a force-versus-time graph.
    d. the time interval that a force lasts.

4. The total momentum of a system is conserved
    a. always.
    b. if the system is isolated.
    c. if the forces are conservative.
    d. never; it's just an approximation.

5. In an inelastic collision,
    a. impulse is conserved.
    b. momentum is conserved.
    c. force is conserved.
    d. energy is conserved.
    e. elasticity is conserved.

6. Tommy and his skateboard have a combined mass of 50 kg. Joe and his skateboard total 100 kg. You push both of them with the same force—Tommy for 2 seconds and Joe for 1 second. After the pushes,
    a. Tommy is moving faster than Joe.
    b. Joe is moving faster than Tommy.
    c. both have the same speed.
    d. there's not enough information to tell about their speeds.

Sample Exam Questions

Sample exam questions on momentum are included in the next chapter on energy.

9

Energy

Background Information

Energy is a very abstract concept. The full importance of energy was not recognized until Joule's experiments of the mid-nineteenth century, nearly two hundred years after Newton's flash of genius. And unlike the well defined idea of momentum, $p = mv$, we keep "inventing" new forms of energy—kinetic energy, potential energy, thermal energy, chemical energy, nuclear energy, and so on. Energy, at least to beginning students, is an amorphous, ill-defined concept. It's not at all obvious how $\frac{1}{2}mv^2$ has any connection to chemical energy or nuclear energy, yet we lump them all under the same heading. And students are certainly not helped by the everyday use of the terms *work* and *energy*. Why, after all, should we be worried about "conserving energy" if energy is always conserved?

Yet out of this, we want students to understand that conservation of energy is one of the most hallowed and best established truths of physics!

Energy and its conservation *are* difficult ideas. There's no *thing* you can put your finger on and say, "Here, this is energy." It's just some number, calculated by adding a little of this and a little of that, that for some hard-to-fathom reason never changes. Perhaps, if we can put ourselves back in the mind of a freshman, we shouldn't be surprised at the difficulties most students have learning to use energy conservation.

There is surprisingly little research on students' concepts and understanding of energy. The paper of Lawson and McDermott (1987) cited in the last chapter asked students to compare the kinetic energies of two dry-ice pucks that had been accelerated equal distances by compressed air. Their performance on the energy comparison was slightly better than on the momentum comparison, but, even so, only 50% of *honors* students could answer correctly without being given hints. The investigators concluded that the majority of students—including those who can solve standard energy problems—cannot use the concepts of work and energy to reason about a real situation.

A different issue that has recently attracted the attention of several writers (see especially Arons (1997) and Mallinckrodt and Leff (1992)) is the applicability of work and the work-kinetic energy theorem to "real world" situations involving extended objects. The definition of work is the work done on a *particle* by a force that displaces the particle. This definition of work, and the work-kinetic energy theorem, remain valid when applied to a *rigid* body with purely translational motion— that is, to an object for which the particle model is strictly valid.

But difficulties quickly arise when these concepts are applied indiscriminately to non-rigid, deformable objects. Consider, for example, a person jumping straight up into the air. Although the floor exerts a normal force that accelerates the jumper's center of mass, this force does *no work* because the point of contact between the force and the jumper undergoes no displacement. The correct energy analysis is that the jumper's chemical energy is transformed, through internal forces, into the center-of-mass kinetic energy. The jumper is a deformable object, and this fact invalidates a simple-minded use of the work-kinetic energy theorem.

Nonetheless, no shortage of introductory texts have treated this and similar situations within the context of a particle model, calculating the "work" as the product of the force and the center-of-mass displacement (even though that's not the point where the force is applied). In some cases, this procedure does happen to give a correct numerical value for the final kinetic energy. But obtaining the kinetic energy is not the only goal; we also want to understand how energy is transferred from one system to another and is transformed from one kind to another.

Work needs to be introduced not simply as "force times distance" but as an *energy transfer.* For an extended object, work is the mechanical transfer of energy between a system and its environment by pushes and pulls (forces) at the boundary where forces are applied. Indeed, that is how a piston does work on a gas. But *no* energy is transferred from the floor to the jumper. Instead, the jumper has an internal *energy transformation* from chemical to kinetic energy.

Our ultimate goal, which will not be reached until thermodynamics, is for students to understand energy as a quantity that can be either transferred to or from the system or transformed within the system. Hence it becomes important to distinguish clearly between these two concepts early in the teaching of energy. In fact, looking ahead to thermodynamics, we always refer to the first law as the ultimate statement of conservation of energy. But with the usual presentation, students don't see any connection between the first law of thermodynamics and the "law of conservation of energy" they learned in mechanics. If we want students to understand and use the concepts of energy, we need to introduce the subject in such a way that the idea of mechanical energy is gradually and coherently generalized until we reach thermodynamics.

As an example, most textbooks state the "law of conservation of energy" in a form equivalent to

$$\Delta K + \Delta U + \Delta E_{\text{therm}} = 0,$$

where the change in thermal energy is related to the work done by nonconservative forces: $\Delta E_{\text{therm}} = -W_{\text{nc}}$. This statement, as usually presented, is not true. It hasn't properly distinguished transfer of energy from transformation of energy.

Simple examples are easily found. If you push a block across a table at constant speed, then $\Delta K = 0$, $\Delta U = 0$, and $\Delta E_{\text{therm}} > 0$. If you pick up a book from the floor and place it on a shelf, then $\Delta K = 0$, $\Delta U > 0$, and $\Delta E_{\text{therm}} = 0$. In neither case is the above sum equal to zero.

The difficulty is that "the push" and "the lift" are not nonconservative, dissipative forces associated with thermal energy. They are *external* forces that act on the system and *transfer energy* to the system. The above statement doesn't allow for energy transfers. This defect becomes critical when we get to thermodynamics. A piston compressing a cylinder of gas is an external force doing work on the system. The thermal energy of the gas increases $(\Delta E_{\text{therm}} > 0)$, but the center-of-mass kinetic and potential energy do not change $(\Delta K = \Delta U = 0)$.

Now the above statement of the "law of conservation of mechanical energy" would apply to an *isolated system,* one on which no work is done by external forces. Interestingly, only one of the four leading physics textbooks even mentions isolated systems, and even that text doesn't give a clear discussion as to the significance of isolation. But many systems of interest aren't isolated. What then?

My preferred approach is illustrated in the figure below. The basic idea, as it has been for many chapters, is to distinguish between a *system* and the *environment.* Action/reaction forces within the system may be either conservative (gravitational forces, spring forces) or nonconservative (friction forces). Equally important, the environment may exert external forces on the system.

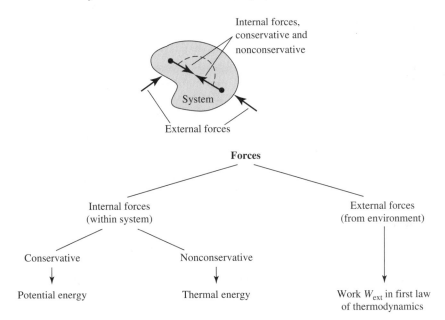

The work-kinetic energy theorem summed over all the particles in the system gives

$$\Delta K = W_c + W_{nc} + W_{ext} = -\Delta U - \Delta E_{therm} + W_{ext}$$

$$\Rightarrow \Delta K + \Delta U + \Delta E_{therm} = W_{ext}.$$

This is now a correct statement of the law of conservation of energy. Now students can see that mechanical energy $K + U$ is conserved only for systems that are both isolated ($W_{ext} = 0$) and nondissipative ($W_{nc} = 0$). The internal forces *transform* energy (kinetic to potential via W_c, kinetic to thermal via W_{nc}) whereas the external forces *transfer* energy to the system.

Furthermore, this statement is now easily generalized to the first law of thermodynamics. A thermodynamic system, such as a stationary cylinder of gas, usually has no change in center-of-mass kinetic or potential energy, so $\Delta K = \Delta U = 0$. The work of interest in thermodynamics, there called simply W, is the *external* work W_{ext} that transfers energy to or from the system. If we merely add heat as another form of energy transfer, we have the first law: $\Delta E_{therm} = W + Q$.

These ideas, which will unfold over many chapters, are essential for understanding the modern concept of energy. Although some of the discussion here is more than students need to know at the beginning, they can certainly start to learn the "big picture." And they *need* to learn this expanded view to place their concept of energy on a firm foundation. Thus much of my pedagogical approach to energy is based on energy transfer and energy transformation. It is an approach that requires thinking ahead to thermodynamics so that you can present students with a coherent view of energy rather than isolated facts about energy.

Student Learning Objectives

- To begin developing a concept of energy—what it is, how it is transformed and transferred.
- To learn about work, kinetic energy, and their relationship through the work-kinetic energy theorem.
- To learn Hooke's law for springs and the new idea of a *restoring force*.
- To learn and to use the gravitational potential energy and the elastic potential energy.
- To use and interpret energy diagrams.
- To understand the law of conservation of energy.
- To recognize transformations between kinetic, potential, and thermal energy.

Pedagogical Approach

Before leaping into details, it is important to give students a broad overview of where we're trying to go. This will help to motivate many of the details to come. The *basic energy model* shown here is a good starting point. This model will be expanded and generalized until becoming the *knowledge structure of energy* shown in the chapter on thermodynamics.

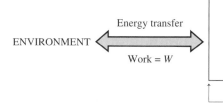

SYSTEM

Energy transfer

ENVIRONMENT

Work = W

Motional energy = "kinetic energy" = K
Stored energy = "potential energy" = U

Mechanical energy = E_{mech} = $K + U$

System boundary

The Basic Energy Model

I like to begin by using *money* as an analogy for energy. Money, like energy, has no single, unique definition. The simple, though inadequate, definition of energy as "the ability to do work" is analogous to a definition of money as "the ability to purchase goods." And like energy, money can be transformed and transferred in various ways. Kinetic energy is analogous to cash, potential energy to a savings account, and work to getting a paycheck or paying bills. Because students are familiar with money, and with at least some of the ways that money can be transferred and transformed, this analogy provides them with a mental image to begin thinking about energy.

With the monetary analogy, it makes sense to students that there must be ways of adding energy to a system or removing energy from a system. Thus the concept of *work* can be introduced as a *mechanical transfer of energy* to or from a system by the pushes and pulls of external forces. Only after establishing a need for the idea of work should you begin to inquire how one calculates the quantity of work. (This is in contrast to most textbooks, which start right in with "Let's define work as . . . " without having provided any motivation or rationale.)

Thinking of work as an energy transfer helps students understand the sign convention: W (and later Q) is positive when energy is added to the system, negative when energy is removed. The signs are a consequence of the fact that energy is a *property* of a system (the system *has* a certain amount of energy) whereas work is a *process,* an event that *changes* the energy of the system. Thus we use ΔE, ΔK, and ΔU to show how the system changes, but never ΔW. Work is a quantity of transferred energy, not the change of a state variable. This is an important point to emphasize.

It is best initially to talk only of work done *by the environment*—which may be negative—and avoid "work done by the system." Students have enough difficulty with this material; there's no need to add an additional complication by switching between two different perspectives. The work done *by the system* will enter naturally when you get to thermodynamics.

After establishing a conceptual framework for energy, you then have a rationale to look at the pieces in more detail. The first chapter is always on work, kinetic energy, and their relationship through the work-kinetic energy theorem. I prefer to start by defining kinetic energy. If you then ask, "What causes the kinetic energy to change?" you're led by simple reasoning with Newton's laws to the standard definition of work.

But starting with the definition of work, as most books do, is equally viable *if* the students have a clear understanding of the purpose.

As noted in the Background Information section, you'll want to limit examples to rigid objects that are adequately described by a particle model. Reasonable choices include throwing balls or pushing on objects with springs. In the latter case, note that the *object* is "the system" while the spring is simply a variable external force. The fact that the spring is not rigid is of no concern.

The vector dot product usually enters the textbook at this point. This mathematics is new to almost all students. They will need focused practice computing dot products before they're comfortable using $\vec{F} \cdot \Delta \vec{r}$ to calculate work. In particular, most students find it hard to see that $A_x B_x + A_y B_y$ and $|\vec{A}| |\vec{B}| \cos \theta$ are the same thing, so some class discussion is warranted.

Students quickly accept kinetic energy as an *energy of motion*. That is a fairly tangible idea. But how do you get students to understand the concept of potential energy?

To begin, potential energy needs to be seen as an energy of *interaction*. Gravitational potential energy arises due to the gravitational interaction between two masses. Elastic potential energy applies to two objects interacting through the force of a spring. Potential energy enters only when we expand the system to include more than one interacting particle. Potential energy is a convenient way to describe the work done by conservative *internal forces,* forces between particles inside the system, as opposed to the external work done by forces that originate in the environment.

As a consequence, potential energy is an energy (a state variable, if you wish) *of the system,* not an energy of a specific object or particle. We, and most textbooks, lose sight of this important idea when we talk about "the gravitational potential energy of the ball." The gravitational potential energy is really an energy of the ball+earth system, and the elastic potential energy is an energy of the two objects connected by the spring. This is an important idea to understand.

Because of its definition in terms of forces and distances, potential energy is an *energy of position*. Potential energy depends on height or distance or displacement. Consequently, the transformation of potential energy to kinetic energy (and vice versa) as the particles move inside a system can be seen as a connection between position and motion. This, of course, is the idea behind potential energy diagrams.

I start with the *idea* of stored energy and of energy transformations. The monetary analogy has already shown that "savings" can be transformed into "cash." If a ball is tossed into the air, we can treat it as a single particle whose kinetic energy changes because of work done by the gravitational force. But there are advantages to expanding the system to be ball+earth. Then it becomes convenient to think of the energy being "stored" as the ball rises and "released" as the ball falls—an energy transformation. The goal is to establish a rationale for potential energy before leaping in to give a definition. Once the rationale is in place, the questions become:

- How do you calculate the potential energy?
- Under what conditions is the mechanical energy $K + U$ conserved?

The derivation of a potential energy function in terms of the path independence of work is quite a sophisticated piece of reasoning. I rely more on plausibility arguments than on mathematical rigor since I want students to understand and be able to use the ideas of energy without getting sidetracked by the fairly sophisticated mathematics necessary to do energy "right." Only two specific potential energies are introduced at this time—"flat earth" gravitational energy and the elastic energy of an ideal spring. The results are simple, even if the derivations aren't. I certainly urge you to *use* the results without class derivation, referring students to the text for the details.

The "zero of potential energy" is a source of confusion for many students. They expect the potential energy to have some absolute significance. Arguments that "We've only defined ΔU, not U" are less effective than several explicit examples showing that the choice of a zero point doesn't have any effect on the outcome of a problem.

I prefer to give a fairly explicit discussion of thermal energy as I introduce non-conservative forces. Students know about atoms and molecules, but few have ever thought about the fact that atomic-level motion must have energy associated with it. They find it plausible, once you point it out, that friction can increase the molecular motion. They can also see that many common situations are an *energy transformation* from kinetic or potential energy to thermal energy. Once students recognize the connection between thermal energy and nonconservative forces, you can defer a more detailed treatment of thermal energy until later. If you have an especially inquisitive class, you can even provide them with a teaser at this point by calling their attention to the fact that macroscopic kinetic energy is easily transformed into thermal energy, but the reverse isn't true. Why the one way street?

Finally, you arrive at the point where you can introduce the law of conservation of energy as it was presented in the Background Information section: $\Delta K + \Delta U + \Delta E_{\text{th}} = W_{\text{ext}}$. Ask students if they can come up with examples where two of the terms are zero. For example, what is a situation where $\Delta K = 0$ and $\Delta U = 0$? Or where $\Delta K = 0$ and $W_{\text{ext}} = 0$? These exercises help clarify the distinction between energy transfer and energy transformation.

To put all these ideas to use, students need a well defined problem-solving strategy:

Strategy for Energy Problems
- Analyze the problem statement by preparing a pictorial representation that shows the situation "before" and "after."
- Analyze the energy transfers and transformations with an energy bar chart.
- Invoke energy conservation in the form $K_i + U_i + W_{\text{nc}} + W_{\text{ext}} = K_f + U_f$.
- Draw a free-body diagram and make any necessary calculations of work.
- Solve the energy equation.

Energy bar charts were introduced by Van Heuvelen (2001). The slightly modified version I use is shown below. Like motion diagrams, they are a tool to help students visualize the situation before rushing into the mathematics. Students can draw vertical bars to represent initial and final energies and the work done on the system.

Notice that the set-up exactly matches the form of the energy conservation equation used in the strategy for energy problems. Students can read information from the bar chart as they set up the equation, just as they've learned to read information from free-body diagrams as they use Newton's second law.

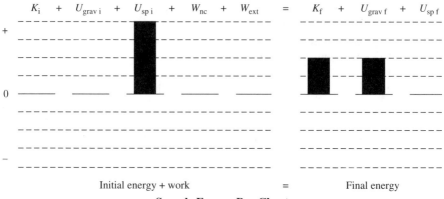

$$K_i \quad + \quad U_{grav\,i} \quad + \quad U_{sp\,i} \quad + \quad W_{nc} \quad + \quad W_{ext} \quad = \quad K_f \quad + \quad U_{grav\,f} \quad + \quad U_{sp\,f}$$

Initial energy + work   =   Final energy

Sample Energy Bar Chart

I have students solve energy problems on a Conservation Worksheet that is analogous to the Dynamics Worksheets they used with Newton's laws. The figure on the next page, worth a thousand words, shows all these pieces in action.

Potential energy diagrams are an afterthought in most texts. I prefer to give them considerably more emphasis for several reasons:

■ It is in keeping with an overall emphasis on graphical representations of knowledge.
■ Energy diagrams are a precursor to equipotential surfaces in electricity.
■ Energy diagrams will be used explicitly in quantum physics.

Although the idea seems straightforward to experienced physicists, students find energy diagrams to be quite difficult. They need several opportunities to go step-by-step through the interpretation of a potential energy graph before they can do so successfully on their own.

Finally, students should have at least some familiarity with the idea of power. The typical definition of power as "the rate of doing work" is clearly inadequate for later uses of power, especially in electric circuits. It's best to start with a more general definition of power as "the rate of energy transfer or transformation." Energy may be transferred into the system as work. But it may also happen that one kind of energy is transformed into another—the chemical energy in a runner's muscles is changed into runner's kinetic energy, or the kinetic energy of electrons in a current is changed into the thermal energy of a resistor. This more general definition of power allows students to understand the runner's power output or the power dissipated by the resistor.

Conservation Worksheet   Problem _____

1) Pictorial Representation
   a. sketch, showing "before" and "after"
   b. coordinate system
   c. symbols for knowns and unknowns

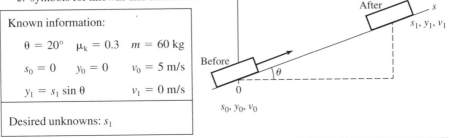

Known information:

$\theta = 20°$ $\mu_k = 0.3$ $m = 60$ kg

$s_0 = 0$ $y_0 = 0$ $v_0 = 5$ m/s

$y_1 = s_1 \sin \theta$ $v_1 = 0$ m/s

Desired unknowns: s_1

2) Energy Bar Chart
   a. What is the system? _skateboarder + earth_   c. Nonconservative forces? __friction__
   b. Potential energies? _____gravity_____   d. External forces? _____none_____

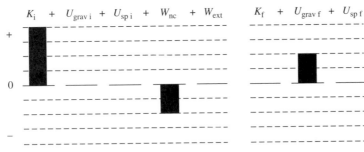

K_i + $U_{grav\,i}$ + $U_{sp\,i}$ + W_{nc} + W_{ext} K_f + $U_{grav\,f}$ + $U_{sp\,f}$

3) Mathematical Representation

Work calculation (FBD and details not shown here): $W_{nc} = -|\vec{f}_k|s_1 = -\mu_k(mg \cos \theta)s_1$

Energy equation: $\frac{1}{2}mv_0^2 + mgy_0 - \mu_k mgs_1 \cos \theta = \frac{1}{2}mv_1^2 + mgy_1$

$\frac{1}{2}mv_0^2 - \mu_k mgs_1 \cos \theta = mgs_1 \sin \theta$

Solution: $s_1 = \dfrac{\frac{1}{2}v_0^2}{g \sin \theta + \mu_k g \cos \theta} = 2.12$ m

Example of a Worksheet Approach to Problem Solving

The problem is:

A 60 kg skateboarder starts up a 20° slope at 5 m/s, then falls and slides up the hill on his kneepads. The coefficient of kinetic friction is 0.30. How far does he slide before stopping?

This problem could equally well be worked with Newton's laws, but you could specify an energy solution if doing such a problem while teaching energy.

Using Class Time

The basic introduction to work and energy needs 5 or 6 days. The longer period allows you to give a qualitative overview of the basic energy model, to deal with conceptual issues, and to have students practice *describing* simple situations in terms of work and energy before having to use the ideas quantitatively.

It's good to start with some simple demonstrations of the ideas of work, kinetic energy, and change of kinetic energy. A low-friction cart attached by a string to a hanging mass is good because students should now be quite comfortable identifying all the forces (assume it to be frictionless).

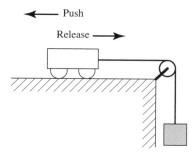

First release the cart from rest, then ask students:

- Does K increase or decrease?
- What is the *sign* of ΔK?
- What forces act on the cart?
- For each of the forces, is the work positive, negative, or zero?
- Is W_{net} positive, negative, or zero? Does this agree with ΔK?

Next, push the cart *away* from the pulley so that it slows as the hanging mass is lifted. Go through the same questions, asking them to focus on the motion after you release the cart and before it reverses.

Finally, have them state in their own words the fundamental idea that the work done by the tension force is a *transfer of energy* (motional energy in this case) to or from the cart. Let them know that this is *the only idea* introduced in the chapter on work and kinetic energy. There may be procedural difficulties in learning how to compute the work done by various forces, but they need to stay focused on the basic fact that they're simply learning how to compute the energy transfer to a system.

As a follow-up demonstration, roll a ball up and down a ramp and ask the same questions. Typical discussions of work and energy tend to not mention the presence of the normal force, so students need to see and be fully convinced that the normal force is neglected for rolling or sliding motion because it never does any work.

The ramp—especially if you use one where you can adjust the angle—also allows you to introduce the dot product definition of work. Students can see that ΔK increases as the ramp angle increases, so the amount of energy transferred must somehow depend on the angle between the weight vector and the displacement vector. The dot product is new mathematics to nearly all students, so practice calculations using both components and $|\vec{A}||\vec{B}|\cos\theta$ are worthwhile.

Next, swing a ball on a string in a horizontal circle around your head. What forces act on it? Do they do any work? Students tend to latch onto the first definition they see—that "work is force times distance"—and apply it to every possible situation. You want to dissuade them of this incorrect thinking before it becomes a habit.

Then use the string to lift the ball straight up from the floor, first at constant speed and then at increasing speed. This gives you a situation where *two* forces do work, but one work is negative. So $W_{net} = 0$ for lifting at constant speed but $W_{net} > 0$ for an increasing speed. You can ask them if there is a *lifting* situation in which $W_{net} < 0$. Throughout these demonstrations you want to keep emphasizing the basic idea of work as an energy transfer.

This last demonstration leads nicely into an initial numerical example using the work-kinetic energy theorem.

EXAMPLE 1: A 200 g ball is lifted upward on a string. It goes from rest to a speed of 2 m/s in a distance of 1 m. What is the tension (assumed to stay constant) in the string?

You can have students solve this first with Newton's second law, then with the work-kinetic energy theorem. You can admit that the work-kinetic energy theorem doesn't offer any great advantages in this problem—but that will soon change. Here are two other possible examples, one using the dot product and one requiring the work of a variable force to be calculated as an area under a curve.

EXAMPLE 2: A 1000 kg car is rolling slowly across a level surface at 1 m/s, heading toward a group of small innocent children. The doors are locked, so you can't get inside to use the brakes. Instead, you run in front of the car and push on the hood at an angle 30° below horizontal. How hard must you push to stop the car in a distance of 2 m?

EXAMPLE 3: A 1 kg block moves along the x-axis. It passes $x = 0$ with a velocity $v = 2$ m/s. It is then subjected to the force shown in the graph.

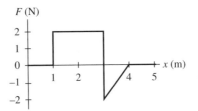

a. Which of the following is true:

   i) The block gets to $x = 5$ m with a speed greater than 2 m/s.

   ii) The block gets to $x = 5$ m with a speed of exactly 2 m/s.

   iii) The block gets to $x = 5$ m with a speed of less than 2 m/s.

   iv) The block never gets to $x = 5$ m.

b. Calculate the block's speed at $x = 5$ m. (Don't reveal this part of the question until they've answered part a.)

Most textbooks don't introduce springs and Hooke's law until the chapter on work and energy. Students have many difficulties understanding springs, and a full day devoted to springs is worthwhile.

First, a surprising number of students don't readily grasp the significance of the *spring constant.* It's good to have several springs of similar size but different stiffness to use in asking students how they would characterize the differences between the springs.

Second, many students don't distinguish clearly between the *length* of a spring and the *displacement* of the end of the spring.

Third, the usual way of writing Hooke's law as $F_{sp} = -kx$ is misleading and causes much unnecessary grief for students. The restoring force of a spring depends on the displacement. But x is a *position,* not a displacement. Writing Hooke's law as $F_{sp} = -kx$ carries a *hidden assumption* that the origin of the coordinate system is located at the end of the unstretched spring. This assumption is often not stated. Even if it is, students quickly forget or ignore it. And there are many problems in which this is *not* a good choice for the origin. Physicists, through long familiarity, find no difficulty interpreting x as a displacement and making adjustments as needed for different coordinate systems. But this is an unnecessary stumbling block for students. Consequently, I prefer to write Hooke's law as $F_{sp} = -k\Delta x = -k(x - x_0)$, where x_0 is the *equilibrium position* of the end of the spring. This will lead naturally to $U_{sp} = \frac{1}{2}k(\Delta x)^2$. Only after explicit examination of the coordinate system do some problems set $x_0 = 0$ and arrive at the more usual expressions for F_{sp} and U_{sp}.

Fourth, students have a hard time understanding the significance of the minus sign in Hooke's law. A good exercise is to draw a series of pictures of springs and coordinate axes, some with $x_0 = 0$ but others not. Then ask them:

- If the spring is stretched (or compressed), is the displacement Δx positive or negative?
- Is Δx the same as x?
- What direction does the force vector $\vec{F}_{sp}$ point?
- Is the scalar force quantity $\vec{F}_{sp}$ positive or negative?
- How does the sign of $\vec{F}_{sp}$ compare with the sign of Δx?
- What role does the minus sign play in Hooke's law?

In doing these, you'll want to emphasize that the force in Hooke's law is $\vec{F}_{spring\ on\ object}$. Some students, especially when an object is compressing a spring, tend to think that Hooke's law applies to the force of the object on the spring.

The following suggested examples are intended to work through many of these issues.

EXAMPLE 4: A 20-cm-long spring is attached to a wall. The spring stretches to a length of 22 cm when pulled horizontally with a force of 100 N. What is the value of the spring constant?

EXAMPLE 5: The same spring is now used in a tug-of-war. Two people pull on the ends, each with a force of 100 N. How long is the spring while it is being pulled?

After Example 5 you may want to note that students answered a similar question about the tension in a string. You can even note that tension is really a large number of "molecular springs," so there is a similarity between tension and stretched springs.

EXAMPLE 6: The same spring is now placed vertically on the ground and a 10.2 kg block is balanced on it. How high is the compressed spring?

EXAMPLE 7: Finally, the same spring is placed vertically on the ground and a 10.2 kg block is held 15 cm above the spring. The block is dropped, hits the spring, and compresses it. What is the height of the spring at the point of maximum compression?

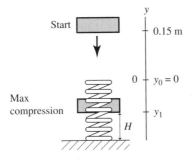

Example 7 is a hard problem that few, if any, students are likely to solve on their own. You can start by noting that they do *not* know how to solve this problem using Newton's laws. They could write down Newton's second law, but they don't know how to handle the kinematics for the nonconstant acceleration. They *can* solve it using work and energy, so they'll see their first example of the power of this new perspective.

You can approach the solution several ways. I prefer to use free-fall kinematics to find the velocity of the block as it hits the spring (1.715 m/s) and its kinetic energy $K_0 = 15$ J at that point. Use a coordinate system with the origin at the initial top of the spring. You'll want to use a clear pictorial model that identifies before and after. The ending point (with $K_1 = 0$) occurs at y_1, which will be a *negative number* in this coordinate system. The height of the spring at this point is

$$H = 20 \text{ cm} + y_1.$$

Note the plus sign. Students need to see these issues of setting up the problem explained in considerable detail. Then ask them what forces do work on the block as the spring compresses and the block slows. Many will forget about gravity and need an explicit reminder to include it. The work-kinetic energy theorem is then

$$[\Delta K = K_1 - K_0 = 0 - K_0 = -K_0] = W_{\text{net}} = W_{\text{grav}} + W_{\text{spring}}$$

$$\Rightarrow -K_0 = (F_{\text{grav}})_y \Delta y + \int_{y_0}^{y_1} (F_{\text{spring}})_y \, dy$$

$$\Rightarrow -K_0 = (-mg)(y_1 - y_0) + \int_{y_0}^{y_1} (-k(y - y_0)) \, dy$$

$$= -mgy_1 - k \int_0^{y_1} y \, dy.$$

Students will need help getting all the signs straight for both W_{grav} and W_{spring}. Call their attention to the fact that $-mgy_1$ is actually a positive number because $y_1 < 0$. Is that what they expected for W_{grav}? Also note that the full and correct statement of Hooke's law was used for the spring force, then simplified by using $y_0 = 0$. Because $y < 0$, the spring force $-ky$ is an *upward* force, as it should be. Students will not notice these issues unless you point them out.

Integrating then gives

$$-K_0 = -mgy_1 - \tfrac{1}{2}ky_1^2$$

$$\Rightarrow \tfrac{1}{2}ky_1^2 + mgy_1 - K_0 = 2500y_1^2 + 100y_1 - 15 = 0$$

$$\Rightarrow y_1 = +0.060 \text{ m} = 6 \text{ cm or } -0.100 \text{ m} = -10 \text{ cm}.$$

We know on physical grounds that y_1 is negative, so the solution is $y_1 = -10$ cm, leading to a spring height $H = 10$ cm.

This is likely the first problem students have faced in which they really must *do* an integral, so it's worth working in detail. The difficulty for most students is not with carrying out the integration itself but with knowing *what* to integrate. Hence the careful attention to setting up the integral. If time permits, consider redoing this problem with the origin of the coordinate system placed at ground level. This is somewhat more logical in that the positions are all positive and that y_1 is then equivalent to H, but the integral becomes more difficult because $y_0 \neq 0$. You also may want to revisit this example in the next chapter, doing it then with conservation of energy.

Potential energy can be introduced by focusing on the *transformation* of energy. Toss a ball into the air. The ball had kinetic energy when you released it, and it has the same kinetic energy at the end. Although you can calculate ΔK based on the work done by gravity, it certainly *seems* as if the energy was "stored" somewhere as the ball rose, then "released" as it falls. The same is true for a cart that compresses a spring and is then shot back with the same kinetic energy.

All textbooks describe how potential energies are calculated. I'm more interested in students being able to apply energy ideas than I am with their following (or reproducing) the derivation, so I leap right into reasoning about energy transformations with $U_{grav} = mgy$ and $U_{sp} = \tfrac{1}{2}k(\Delta x)^2$. Here are some sample questions you can pose. In addition to answering the question as stated, you also want to ask students to answer explicitly: "What is the system?" and "What energy transformations take place?" Students will find these more difficult than you might expect.

EXAMPLE 8: A block slides down a frictionless ramp of height h. It reaches speed v at the bottom. To reach a speed of $2v$, the block would need to slide down a ramp of height

  a. 1.41h b. 2h c. 3h d. 4h e. 6h

EXAMPLE 9: A block is shot up a frictionless 40° slope with initial speed v. It reaches height h before sliding back down. The same block is shot with the same speed up a frictionless 20° slope. On this slope, the block reaches height

a. $2h$ b. h c. $\frac{1}{2}h$
d. Greater than h, but I can't predict an exact value.
e. Less than h, but I can't predict an exact value.

EXAMPLE 10: Two balls, one twice as heavy as the other, are dropped from the roof of a building. Just before hitting the ground, the heavier ball has

a. one-half   b. the same   c. twice
d. four times the kinetic energy of the lighter ball.

EXAMPLE 11: A spring-loaded gun shoots a ball 12 m straight up into the air. The ball is shot again, but this time the spring is compressed only half as far. If air resistance and friction are negligible, the new height of the ball will be

a. 3 m   b. 6 m   c. 12 m   d. 24 m   e. 48 m

EXAMPLE 12: A block sliding along a frictionless horizontal surface with speed v collides with a spring. The far end of the spring is fixed in place, and the spring is compressed by the block. The maximum compression is 1.4 cm. If the block then collides with the spring while having a speed of $2v$, the spring's maximum compression will be

a. 0.35 cm   b. 0.7 cm   c. 1.0 cm   d. 1.4 cm   e. 2.0 cm   f. 2.8 cm   g. 5.6 cm

EXAMPLE 13: If the spring in Example 12 is replaced by a spring whose spring constant is twice as large, a block with speed v will compress the new spring a maximum distance

a. 0.35 cm   b. 0.7 cm   c. 1.0 cm   d. 1.4 cm   e. 2.0 cm   f. 2.8 cm   g. 5.6 cm

Student difficulties with the zero of potential energy are best addressed with specific examples. A good numerical example, during which you can illustrate all the different pieces of the Strategy for Energy Problems, is simply to find the impact speed of a 200 g ball dropped from a height of 2 m. Work this first with $y = 0$ at the floor, then again with $y = 0$ at the ceiling. In both cases, calculate U_i and U_f explicitly, not just ΔU. Even though the potential energies are negative in the second case (another trouble spot for students), both yield the *same* value of ΔU—which is all that ultimately matters.

Other straightforward examples of energy conservation include finding the speed of a block pushed away from a spring, finding the height of a spring-launched projectile, and finding the speed at $\theta = 0°$ of a pendulum released from, say, $\theta = 60°$. Students are quite surprised that the pendulum speed yields so easily to an energy analysis, but you will need to have them discuss and resolve the issue of why the string tension doesn't play a role. If you did Example 7, dropping a block on a spring, you might want to redo that one using energy conservation.

Following some simple examples where mechanical energy is conserved, you'll want to introduce the more complete statement of the law of conservation of energy, including both energy transfer due to W_{ext} and energy transformation into thermal energy due to W_{nc}. Students all know that friction makes surfaces hot, so there's no reason not to go ahead and note that the nonconservative forces transform kinetic energy into thermal energy. This is even a good first opportunity to note—as you'll have to many times—that the energy does *not* "go into heat," as students usually say.

Examples you can ask students to consider are

- Pushing a block across a table (with friction) at constant speed.
- The block sliding to a halt after you release it.
- Picking the block up and placing it on a high ledge.
- Tossing the block straight up, with the "toss" and the subsequent motion considered separately.

For each of these, and others you might want to add, ask students to do several things:

- Specify which objects are included in the system.
- List conservative interaction forces, nonconservative interaction forces, and external forces.
- For each of the terms ΔK, ΔU, W_{nc}, and W_{ext}, specify if the value is positive, negative, or zero.
- Describe any energy *transformations* that take place.
- Describe any energy *transfers* that take place.

These are excellent exercises to focus student attention on the essential ideas of energy.

The second day on potential energy is a good time to walk the students through interpreting an energy diagram. The figure below shows a good potential energy to use. Begin by having them draw the corresponding force diagram, including numerical values as well as shape. Have them identify regions where the force is to the right and to the left. Then have them discuss and describe the motion of a particle with a total energy of 1.5 J. Focus on where the particle speeds up and slows down, how energy transformations are taking place, where K is minimum and maximum, and where the turning points are. Try $E = 0.5$ J, then talk about the points of stable and unstable equilibrium. Next try $E = 2.5$ J.

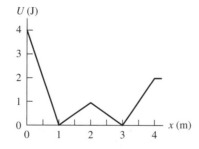

Finally, ask them to consider a particle with $E = 1.5$ J that is moving to the left. Just as it passes $x = 2$ m it is hit from behind by a marble that had been launched from a slingshot. The collision increases the particle's energy by 1 J. What happens to it? This example leads nicely into a discussion of molecular bonds, molecular vibrations, and photodissociation.

If you covered impulse and momentum prior to energy, you'll need to backtrack briefly to treat elastic collisions. The concepts shouldn't be a problem at this point, but some care is needed to avoid problems that are merely plug-and-chug exercises with equations derived in the textbook.

As time permits, it's good to close with more extended sample problems that combine energy conservation with either dynamics or momentum conservation. If you have a good ballistic pendulum demonstration, now is the time to use it and analyze it. A somewhat easier version is as follows.

EXAMPLE 14: A 1 kg wood block on a frictionless surface is attached to a 20-cm-long spring that has a spring constant 6000 N/m. A 10 g bullet is fired straight into the block where it sticks, causing the block to compress the spring a maximum of 5 cm. What is the speed of the bullet? (389 m/s)

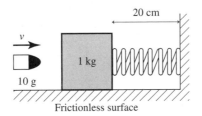

Sample Reading Quiz or Discussion Questions

1. What new mathematical idea about vectors was introduced in this chapter?

2. Write the work-kinetic energy theorem. (Accept any version, if you're just trying to see if they've read the chapter.)

3. Blocks A and B, of equal mass, slide down the two frictionless ramps shown in the figure. Their speeds at the bottom are v_A and v_B. Is $v_A > v_B$, $v_A = v_B$, or $v_A < v_B$? Write an explanation of your choice.

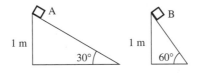

4. What information is shown on a potential energy diagram?

5. Energy is a physical quantity with properties somewhat similar to
   a. money.
   b. a liquid.
   c. momentum.
   d. heat.
   e. work.

6. Hooke's law describes the force of
   a. gravity.
   b. a spring.
   c. a collision.
   d. tension.
   e. none of the above.

7. This chapter introduced the potential energy for interactions involving
   a. gravity.
   b. tension.
   c. springs.
   d. both a and b.
   e. both a and c.
   f. all of a, b, and c.

8. The potential energy of a spring
   a. is proportional to the distance stretched.
   b. is proportional to the square root of the distance stretched.
   c. is proportional to the square of the distance stretched.
   d. was not discussed in this chapter.

Sample Exam Questions

These sample exam questions cover both impulse/momentum and energy.

1. A 0.2 kg plastic cart and a 2 kg lead cart both roll without friction on a horizontal surface. Equal-size forces are used to push the carts forward a distance of 1 meter, starting
   from rest.
   After the force is removed at $x = 1$ m, which of the following statements is true?
   a. The kinetic energy of the plastic cart is less than the kinetic energy of the lead cart.
   b. The kinetic energy of the plastic cart is equal to the kinetic energy of the lead cart.
   c. The kinetic energy of the plastic cart is greater than the kinetic energy of the lead cart.
   d. There is not enough information to compare the kinetic energies of the two carts.
   Write an *explanation* of your choice.

2. A 0.2 kg plastic cart and a 2 kg lead cart both roll without friction on a horizontal surface.
   Equal-size forces are used to push each cart forward for a time of 1 second, starting from rest.
   After the force is removed at $t = 1$ s, which of the following statements is true?
   a. The momentum of the plastic cart is less than the momentum of the lead cart.
   b. The momentum of the plastic cart is equal to the momentum of the lead cart.
   c. The momentum of the plastic cart is greater than the momentum of the lead cart.
   d. There is not enough information to compare the momenta of the two carts.
   Write an *explanation* of your choice.

3. A 0.5 kg cart and a 2.0 kg cart are attached and are rolling forward with a speed of 2 m/s.
   Suddenly a spring-loaded plunger pops out and blows the two carts apart from each
   other. The smaller cart shoots backward at 2 m/s.

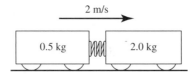

   a. What are the speed and direction of the larger cart?
   b. If the spring constant of the plunger is 25,000 N/m, by how much was the spring ini-
      tially compressed?

4. Workers at a packing factory shove a 10 kg crate against a horizontal spring. The crate has a speed of 1 m/s as it hits the spring. If the spring constant is 50 N/m and the coefficient of kinetic friction between the crate and the floor is 0.10, what is the maximum compression of the spring?

5. Sammy Skier, whose mass is 70 kg, stands at the top of a 10° slope on his new frictionless skis. A strong *horizontal* wind blows against him with a force of 50 N. *Without using Newton's laws of motion,* find Sammy's speed after traveling 100 meters down the slope.

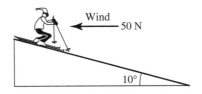

6. A 100 g rubber ball is thrown horizontally with a speed of 5 m/s toward a wall. It rebounds with the same speed. The collision force on the ball is shown on the top set of axes below.

  a. What is the value of the maximum force F_{max}?
  b. Draw an acceleration-versus-time graph for the collision on the middle set of axes. Make your graph align vertically with the force graph, and provide an appropriate numerical scale on the vertical axis.
  c. Draw a velocity-versus-time graph for the collision on the bottom set of axes. Make your graph align vertically with the acceleration graph, and provide an appropriate numerical scale on the vertical axis.

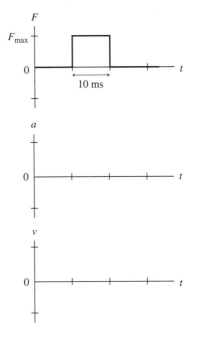

7. A 1 kg object has the potential energy shown in the graph.

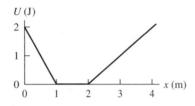

a. The object starts from rest at $x = 0$. What is its speed at $x = 3$ m?
b. Draw a force-versus-position graph showing the force on the particle as it goes from $x = 0$ to $x = 4$ m. Include an appropriate numerical scale on the force axis.

c. Use the work-kinetic energy theorem to find the object's speed at $x = 3$ m if it starts at rest at $x = 0$ m.

8. A 200 g rubber ball is tied to a 1-meter-long string and released from rest at angle θ. It swings down and at the very bottom has a perfectly elastic collision with a 1 kg block. The block is resting on a frictionless surface and is connected to a 20-cm-long spring of spring constant 2000 N/m. After the collision, the spring compresses a maximum distance of 2 cm. From what angle was the rubber ball released?

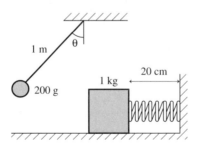

10

Oscillations

Background Information

Oscillations occupy a middle ground between mechanics and wave motion. Simple harmonic motion, the basic model for oscillations, is still single-particle dynamics. It plays a critical role in all of physics and engineering. But simple harmonic motion is also the starting point for the subsequent development of harmonic waves. Adequate time spent here, especially on the mathematics of sinusoidal functions, will help the chapters on wave motion go more smoothly.

Oscillations of springs, pendula, and other objects are familiar to students, but most have never observed the motion systematically. Thinking about oscillatory motion immediately reawakens the difficulties students have with concepts of motion, especially the concept of acceleration. Although kinematic graphs and motion diagrams will certainly have benefited students, don't expect to find them free of difficulties with acceleration. In addition, the concept of *phase* adds a whole new layer of difficulties. Oscillations will provide them with another opportunity, in a new context, to develop correct mental models of motion.

Students can easily become distracted by the mathematics of oscillatory motion. It is important to keep the mathematical representation connected to graphical representations, to pictures, and to demonstrations of real oscillating systems. All students studied trigonometry in high school, but many never got beyond thinking of trigonometry as simply dealing with triangles. A significant fraction of students don't really understand sines and cosines as oscillatory *functions,* and this becomes a severe hindrance to their understanding of oscillations and waves. Many students don't know the term *sinusoidal function* and think that it refers only to the specific function $\sin(x)$.

Common student difficulties with the mathematics of oscillations include

- Unfamiliarity with radians.
- Not knowing that the argument of an oscillatory function needs to be an angle, which has no *physical* units. Student expressions such as $\sin(t)$ are common.

- Unfamiliarity with trigonometric identities, such as $\sin(-x) = -\sin(x)$ or $\sin(x + \pi/2) = \cos(x)$.
- Thinking that sine and cosine are two distinct functions, whereas we want them to understand that sine and cosine are the *same* oscillatory function with different phase constants.
- Unfamiliarity with drawing or interpreting functions that are not pure sines or cosines, such as $\sin(\omega t + \pi/2)$.

The concept of *phase* is new to virtually all students and a major source of difficulty for most. Phase is an essential concept for talking about oscillations or waves in a meaningful fashion, but it will take a significant number of examples and exercises before students catch on to the idea.

Another idea new to most students is the *small angle approximation*. Few have gotten far enough in calculus to see Taylor series, and even those will not have realized that the idea is actually useful! The small angle approximation is so widely used in science and engineering, including several times later in this course, that it is worth a small class discussion.

Oscillations and waves begin to introduce quite a bit of new terminology. Physics instructors are often heard to say that there's very little to memorize in physics, but that is a rather extreme exaggeration. Students *do* need to learn the definitions, or else they will not follow discussions in class or in the text. This is a good place in the course to begin adding vocabulary terms to reading quizzes or exams.

Student Learning Objectives

- To understand the physics and mathematics of oscillations.
- To draw and interpret oscillatory graphs.
- To learn the concepts of *phase* and *phase constant.*
- To understand and use energy conservation in oscillatory systems.
- To understand the basic ideas of damping and resonance.

Pedagogical Approach

Simple harmonic motion serves as the model for many kinds of oscillations. You can help motivate the chapter with demonstrations of torsional oscillations, oscillations of a blade (or meter stick) clamped at one end, and any others that are available. Mention vibrating machinery and oscillating circuits, to get the interest of engineers. Although the specific details depend on the situation, you want students to understand that all these vibrations and oscillations are described by the same *mathematical language* that we'll develop for the case of a simple oscillating spring.

It's especially important to keep the mathematical description connected to real physical situations of springs and pendula. This is a chapter where it's easy to let mathematical analysis take over and obscure the physical ideas. A vertical spring with hanging masses is adequate for many demonstrations, although some students may be worried about the presence of the weight force. A better demonstration, for those who have it, is an air-track glider attached to a soft spring that gives a slow, easily observed oscillation.

You'll want students to practice *interpreting* oscillatory motion before they get into the detailed mathematics. Simple classroom demonstrations of oscillations provide a good opportunity for interactive activities with the class. You can ask students where in the motion does the position, velocity, or acceleration have specific signs or values—such as "Where is the velocity positive?" or "Where is acceleration zero?" You can ask students to predict what will happen to the period if the mass (or amplitude or length or ...) is changed. Most students will predict that the period of a spring depends on the amplitude and that the period of a pendulum depends on the mass. Interactive demonstrations, where they have a vested interest in the outcome, make a lasting impression on them.

You can then move from the actual demonstration to graphs of the motion, letting students practice converting one graph into another (e.g., "draw the velocity graph from the position graph"). You want them to see how the answers to the questions are related to the shapes of the graphs.

Although most students have seen radians defined elsewhere, few really understand the idea. Several simple class exercises involving arc lengths are worthwhile. These can lead into a discussion of *angular frequency,* another new idea for most students. Many have a difficult time distinguishing frequency from angular frequency. In introducing the small angle approximation, you'll want to emphasize that it's valid only if the angle is in radians.

The most difficult idea of the chapter for most students is the idea of *phase* and of the *phase constant.* This is such an important idea for later understanding of wave phenomena, especially interference, that I urge you to spend significant time on exercises about phase and the phase constant. Point to an oscillator at different points in its cycle and ask the students to determine the phase at that point. Help them make the mental links between the three related ideas of i) the physical position and velocity of the oscillator, ii) a point at angle ϕ as it moves around a reference circle, and iii) where the oscillator is on a position graph. Students need to recognize that position alone doesn't determine the phase, because an oscillator passing through $x = 0$ could have either $\phi = \pi/2$ or $3\pi/2$. Use different amplitudes and different frequencies so that students come to realize that phase measures "where" the oscillator is in a generic sense, as a fraction of a cycle, that is independent of the amplitude or frequency of a specific oscillator. Once they begin to understand phase, then it's a fairly easy step to recognize that the position and velocity at $t = 0$—the initial conditions—are characterized by a specific value of the phase called the *phase constant.*

Although the oscillator differential equation is easily found from Newton's second law, there is no expectation that students know anything at all about differential equations. It's sufficient merely to *confirm* that a sinusoidal function satisfies the equation of motion. You do want to call their attention to the fact that ω is determined by the physical characteristics of the oscillator—k and m for a spring, L and g for a pendulum—but that the amplitude A and phase constant ϕ_0 can have any values. It's good to have them verbalize the significance of this fact, a task which many will find hard to do but which is important for them to connect the mathematics to the physics.

Using Class Time

Although the mathematics of oscillation can be written down quickly, the conceptual density of the chapter on oscillations is high. Many students have not seen simple harmonic motion in high school, and those that did usually carried away few lasting memories. Consequently, three days are recommended for this topic. The first day can focus on basic observations of oscillations, day two on the mathematical description of oscillation and phase, and day three on energy, damping, resonance, and example problem solving.

I like to start day 1 with a variety of demonstrations. Either a mass hanging on a spring or a low-friction horizontal oscillator, such as an air-track glider connected to a spring, is the main demonstration. At the beginning, it's good to start all oscillations from maximum *positive* displacement so that the oscillation is described by a cosine with $\phi_0 = 0$. Different initial conditions can be considered later. If you're using a vertical spring as your main demonstration, this means that you'll want to start it each time by *lifting* and releasing rather than pulling and releasing.

Ask students, on the basis of their observations, questions such as "Where is the velocity positive?" or "Where is acceleration zero?" After a few such observations, draw both a cosine graph and a triangular-shaped oscillation graph on the board, then ask them which graph is consistent with their observations. After some discussion, most will finally agree that the motion they see is described by the cosine graph. If you have a motion detector for making real-time graphs, it's good to *confirm* this conclusion, but don't do it too soon.

Once you have the basic position graph, you can take derivatives (they know the derivatives of sine and cosine from calculus) to produce velocity and acceleration graphs. Then ask if the graphs are consistent with their observations that $v = 0$ when x is a maximum, v is a maximum when x is zero, a and x always have opposite signs, and so on. A good exercise is for students to identify specific points on the position graph (erase the velocity and acceleration graphs) where, for example, $v = 0$ while a is positive or $a = 0$ while v is positive. Also ask if there are any points where both $v = 0$ and $a = 0$ simultaneously.

After establishing how x, v, and a are related, you can switch to demonstrations about amplitude, period, and frequency. Use a timer to measure the period of an oscillating spring with a small amplitude. (Measure the time for ten cycles in order to have sufficient accuracy.) Then ask students to predict what will happen to the period if you increase the amplitude. You can give them the choices increase, decrease, or stay the same. Even if they've read the chapter, not many students correctly predict "same," so the result comes as a surprise to most.

Then ask for a prediction of the period if you double the mass. Most predict either "equal" (they're not going to be fooled twice!) or "double." After it turns out to be in between, ask a student with a calculator to take the ratio. If you're using a high-quality spring whose mass is significantly less than the mass of the object, the ratio should be quite close to 1.4. The question "Does this number have any significance?" gets mostly blank looks, but eventually someone usually, and timidly, ventures "Square root of two?" You can also do similar experiments for the mass and length of a pendulum.

It's good to start day 2 with a brief discussion, and perhaps a couple of simple exercises, of radians, the small angle approximation, and angular frequency.

A nice demonstration to motivate the idea of phase is to place an object on the edge of a rotating turntable (the larger the better), then tilt the turntable so that the class sees the motion from the edge. An alternative is to swing a ball on a string around your head, asking them to see it as a "one-dimensional motion" back and forth. You'll want them to verbalize the fact that the motion "hesitates" at the ends and "speeds up" through the center—that is, it's simple harmonic motion. Once they recognize this, you can associate each point in the oscillation with a point on a reference circle, thus defining the idea of *phase*.

Then return to your main oscillating spring. Point to the mass at different points in the cycle and ask students to identify the phase. Suggest they visualize a circular motion and think about where the mass would be on the circle. Start with the easy ones—multiples of $\pi/2$. Ask how they know whether the mass at $x = 0$ has $\phi = \pi/2$ or $3\pi/2$. Then point to $x = \frac{1}{2}A$, both as it's moving out and then as it's moving back in.

Once students get the idea, you can *start* the oscillation at some point other than $x = A$ and ask for the *initial phase*. Associating the phase constant with initial conditions is not too hard once they grasp the meaning of phase. You can then easily establish that the general equation for the position of an oscillator is $x = A\cos(\omega t + \phi_0)$.

As an exercise, ask students to draw a graph for an oscillator with an amplitude of 2 cm and a phase constant of 60° (or $\pi/3$). This will not be easy for students who are weak in trigonometry. Then ask them to *interpret* the graph, stating the initial position and whether the mass is initially moving in, moving out, or at rest. As a follow up, ask them to draw the graph of an oscillator with an amplitude of 2 cm that starts at $x = 1$ cm and is moving out. Then ask for the value of the phase constant (300° or $-60°$).

Having covered all the basic ideas of oscillations, you can now move into a more conventional description of the mathematics. One of the most common student errors in working with the mathematics of oscillation is having their calculators set to degree mode rather than radian mode. (Some students don't even know that there *is* a radian mode on their calculator because they've never used it!) You'll want to warn them of this, and suggest that they make a big note card to use as a bookmark that says "Set calculator to radian mode!"

A numerical example to end day 2 is as follows:

A 500 g block on a frictionless horizontal surface is attached to a spring with spring constant of 12.5 N/m. The block is pulled to $x = -20$ cm and released at $t = 0$.

a. What is the period of oscillation?
b. Draw a graph, showing three cycles of oscillations.
c. What is the phase constant?
d. *Where* is the block at $t = 2$ s?
e. What is the block's velocity at $t = 2$ s?
f. *When* does the block first pass through $x = +10$ cm?

In working part f, note that the block takes 4/8 of a period to move from -20 cm to $+20$ cm, but *not* 3/8 of a period to move to $+10$ cm. You'll also want students to verify that their numerical answers to parts d and f are in agreement with their graph.

Energy issues involving springs usually present no major problems because students have just recently completed the chapters on energy, including the potential energy a spring. Even so, many students are still reluctant to use energy conservation in a problem. Many will try to use the equations of motion, even though that entails messy inverse cosines. One typical energy problem, such as finding the speed at a specific displacement, is worth working through in detail.

It is also worth having students interpret a potential energy diagram, describing the motion and finding the turning points for an oscillator with a specified total energy. You want them to recognize the "signature" of simple harmonic motion as a parabolic potential energy graph.

An informative exercise is to ask students to give a verbal interpretation of the energy equation $\frac{1}{2}k(x_{max})^2 = \frac{1}{2}m(v_{max})^2$. A common response is, "It means that potential energy is equal to kinetic energy," revealing still-common misconceptions about energy. (This makes a good exam question.)

Demonstrations of damping and resonance are worthwhile. It's good to ask students to observe and graph an oscillation that damps out after a few cycles. The main point you want to convey, of course, is the idea of an exponential decay and of a *time constant*. It takes most students several exposures to the idea of a time constant before they begin to catch on.

Sample Reading Quiz or Discussion Questions

1. What is the name of the quantity represented by the symbol ω?
2. What term is used to describe an oscillator that "runs down" and eventually stops?

3. The starting conditions of an oscillator are characterized by
   a. the initial position.   d. the phase constant.
   b. the frequency.   e. the phase angle.
   c. the amplitude.

4. *Wavelength* is
   a. the time in which an oscillation repeats itself.
   b. the distance in which an oscillation repeats itself.
   c. the distance from one end of an oscillation to the other.
   d. the maximum displacement of an oscillator.
   e. not discussed in this chapter.

Note: Answer e is the correct response for a reading quiz on oscillations.

Sample Exam Questions

1. The graph shows the position-versus-time graph for a block oscillating on a spring.

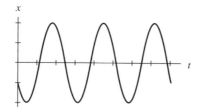

   a. On the first set of axes, draw the velocity-versus-time graph for the block. Make sure your graph aligns with the position-versus-time graph above it.
   b. What is the phase constant ϕ_0 for this oscillation?
   c. On the second set of axes, draw the position-versus-time graph for an oscillation in which the value of ϕ_0 is replaced by $-\phi_0$.
   d. On the third set of axes, draw the position-versus-time graph if the spring constant is increased by a factor of four.
   (Note: Provide three sets of empty axes below the figure.)

2. A 500 g block is attached to a spring on a frictionless horizontal surface. The block is pulled to stretch the spring by 10 cm, then gently released. A short time later, as the block passes through the equilibrium position, its velocity is 1 m/s.

   a. What is the block's period of oscillation?
   b. What is the block's speed at the point where the spring is compressed by 5 cm?

3. Astronauts visiting Mars measure the period of a pendulum to be 4.0 seconds.

   a. What will the period be if they make the string four times as long?
   b. What will the period be if they make the mass four times as large?

Waves

Background Information

Waves are one of the two great models of classical physics, and wave phenomena appear throughout science and engineering. With waves, we move from the physics of particles to the physics of continuous media. Many of the ideas introduced during the study of waves will later be important for understanding electric fields and quantum-mechanical wave functions.

Student difficulties with wave motion have recently been studied (Wittman et al., 1999). In one question, students were shown a pulse traveling on a stretched string after flicking one end. Students were asked what could be done to cause the pulse to reach the wall sooner. The most common response was "Flick the string harder." This and other questions have found that most students hold a *particle-pulse model* of wave propagation in which wave pulses travel through a medium as if they were particles. After all, the way to make a ball reach the wall sooner is to "Throw it harder." The medium is seen merely as something that "the wave goes through," like the ball going through the air, while the propagation speed depends on the initial conditions. This view of wave motion is incompatible with superposition, so it's perhaps not surprising that few students could correctly sketch the waveform as two wave pulses passed through each other.

These conceptual difficulties are compounded by the fact that waves are the first place students meet functions of two variables. Although this introduces mathematical complications, the more significant issue for most students is a difficulty *visualizing* what a wave looks like and how it moves through space. Textbook pictures are static images, and more often than not they show only one-dimensional, transverse waves. Students find it hard to imagine what the wave will look like at a later time or in a different position. Longitudinal waves and waves in two or three dimensions are even more difficult to visualize.

Unfortunately, it's difficult to demonstrate traveling waves in the classroom. Nearly all demonstrations reflect at the end to set up standing waves. Consequently, this is one area where commercially available videos and computer simulations can be especially helpful. Computer simulations are also useful for showing how the motion of the wave is different than the motion of particles in the medium.

Wave fronts and wave front diagrams are another source of difficulty, likely because pictures show them at only a single instant of time. As a result, some students end up thinking that the wave fronts are fixed in space. Other students don't understand how wave fronts are related to a graph of a wave. Because all students are familiar with ripples moving out from the splash where a rock hits the water, you can associate wave fronts with the moving crests of the ripples. You can then draw a displacement-versus-position graph along a line through the two-dimensional wave to show how the wave fronts align with the maxima on the graphs. You want students to understand that the midpoint between two wave fronts is a trough, or minimum. This will be an essential piece of information for understanding interference.

For standing waves, the major conceptual difficulty is to understand that a standing wave is the superposition of two traveling waves. After all, a simple observation of a standing wave gives no hint of the two counter-propagating waves. A *graphical* demonstration of adding two counter-propagating waves is more convincing than a mathematical analysis, although the latter is also important.

The standard textbook picture of a standing wave, such as the one shown below, is confusing to many students. They're not sure how to interpret this. As drawn, the wave appears to be *simultaneously* at its maximum and minimum displacements. Students need some help to see this picture as representing the extrema of the displacement. I like to use colored chalk to highlight the wave at a single instant of time.

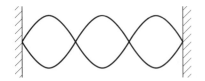

Another graphical source of confusion involves standing sound waves in air-filled tubes. A common practice is to draw sinusoidal waves that fit exactly inside a "picture" of the tube, with the peak-to-peak displacement matching the tube's diameter. Students already have a difficult time understanding the relationship between a longitudinal oscillation and a graph of the oscillation. Standing waves drawn to match the tube convey the impression that the amplitude of the wave is equal to the radius of the tube.

As an example, I gave an exam problem to an honors class based on the following figure of a standing wave in a tube. I asked several questions about the wave, one of which was "Can you determine the amplitude of the wave?" Approximately 75% of the class answered that the amplitude was 1 cm. I now either draw standing waves with an amplitude much larger than the picture of the tube or, even better, draw the picture of the tube above or below the graph of the wave.

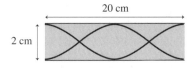

Student Learning Objectives

- To visualize wave motion and develop intuition about waves.
- To work with functions of two variables, using both graphical and mathematical representations.
- To become familiar with the properties of sinusoidal waves, such as wavelength, wave number, and frequency.
- To apply the ideas of phase and phase difference to waves.
- To understand and use the principle of superposition.
- To understand that standing waves are the superposition of two traveling waves.
- To understand the basic properties of standing waves.
- To study the properties of common waves—waves on strings, sound waves, and light.

Pedagogical Approach

Although many textbooks derive the wave equation for a string, it is a pointless exercise. Few students have reached partial derivatives in their math courses, much less partial differential equations! A hand waving derivation of the wave speed on a string is quite sufficient.

Sinusoidal waves are best delayed until late in the discussion. Any disturbance that propagates through a medium is a wave, but the early introduction of sinusoidal waves convinces many students that all waves *must* be sinusoidal. Pulses of various shapes that travel along strings are a better way to introduce many of the properties of waves. Sinusoidal waves are then seen as just one particular type of wave.

At the beginning, I make heavy use of what I call *snapshot graphs*—displacement versus position at a specific instant of time—and *history graphs*—displacement versus time at a specific position in space. These are two alternative ways to represent a function of two variables. Both graphs represent the same wave, so a snapshot graph can be converted into a history graph and vice versa. These turn out to be excellent exercises for thinking about how waves propagate.

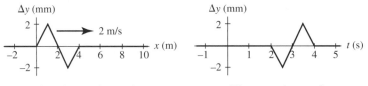

Snapshot graph at $t = 0$ s   History graph at $x = 8$ m

The figure shows an example. As an exercise, you would give students one and ask them to draw the other. Even experienced instructors find this tricky if they've not tried it before, but it's highly educational and gives you a better appreciation of wave motion. Nearly everyone finds it much easier to translate a snapshot graph to a history graph than vice versa, although I have no explanation for the asymmetry. Students find these exercises very challenging, but also rather fun.

Another interesting graph-picture combination turns out to be very useful for visualizing longitudinal waves and understanding graphs of longitudinal waves. The wave equation is usually written, and graphed, as $y = A \sin(kx - \omega t)$. Unfortunately, this x-y notation merely reinforces the common misconception that all waves are transverse. How should we write the wave equation when the displacement is parallel to the propagation? Clearly, an "x-versus-x" graph is meaningless. What we really want to show is the displacement Δx from an equilibrium position x—thus a Δx-versus-x graph. It helps students if you write and graph transverse waves in terms of the *displacement* Δy, with Δx then used for longitudinal waves.

To show a longitudinal wave, you can represent the molecules in a gas, or the coils of a slinky, by equally spaced positions along an axis. Displace each point by the amount shown on the graph, then draw a dot at that position on the axis. As the figure below shows, the result is a "picture" of the longitudinal wave pulse. Even experienced teachers are frequently surprised that the wave doesn't look like what they expected from the graph.

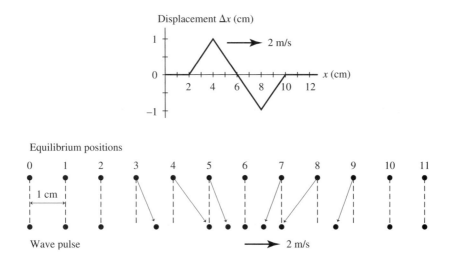

The motion of sinusoidal waves turns out to be more difficult for many students than you might expect. Computer simulations that show sinusoidal waves moving *slowly* down a string are very helpful. An essential idea you want students to see is that "a wave moves forward a distance of one wavelength during a time interval of one period." That is the *meaning* of the important relationship $v = \lambda f$, which is best derived from speed = distance/time = λ/T. If students can understand this relationship, rather than just memorizing it, then they will have made major progress toward understanding waves.

The full expression for sinusoidal waves—$\Delta y(x, t) = A \sin(kx - \omega t + \phi_0)$—is initially confusing for most students. To begin with, they get confused with the two variables and don't recognize that $kx - \omega t + \phi_0$ is just a *number*—some number of radians—at each point in space and time. Two types of exercises help students become comfortable with this expression. First, ask them to draw snapshot and history graphs of waves with specified values of wavelength, period, and phase constant. Second, draw a graph on the board (either a history or a snapshot graph), tell them the wave speed, and ask them to determine the values of A, k, ω, and ϕ_0. Although valuable, these exercises are time consuming and are perhaps best done in recitation.

The idea of the *phase* of a wave is difficult, even after emphasizing phase in the chapter on oscillations. It will be, of course, essential for understanding interference. Phase can be introduced in conjunction with wave fronts. Since a wave front corresponds to the crest of a wave, the value of $\sin(kx - \omega t + \phi_0) = \sin(\phi)$ must be equal to 1, so $\phi = \pi/2 \pm n \cdot 2\pi$. Thus any two neighboring wave fronts have a phase *difference* of 2π. There's a phase difference of π between a crest and a trough, and so on. It is useful if students can recognize that two points separated by Δx have a phase difference of $(\Delta x/\lambda) \cdot 2\pi$.

Using Class Time

I spend 4 or 5 days on traveling and standing waves. The first day should be focused on a basic understanding of wave propagation, leaving the mathematical aspects until the second day. Demonstrations of transverse and longitudinal pulses make a good opening. You want students to immediately begin to recognize that the motion of a wave pulse is not the same as the motion of the particles in the medium. Demonstrations can be used to show that the medium of a transverse wave has displacement Δy, which is a function of both x and t, while longitudinal waves have displacement Δx.

Stretch a long rubber hose or spring across the room and send a transverse pulse down it. Ask students what you should do to make the pulse travel faster. As the Background Information indicated, many students will want you to "flick the string harder" or faster. Demonstrate that the initial conditions don't affect the speed, but the tension does. This leads naturally into a discussion of wave speeds.

The linear density of a string, which appears in the expression for the wave speed, may be the first use in the course of the concept of *density*. This is a surprisingly difficult idea for many students, likely due to a difficulty with the whole concept of ratios. All students will have encountered the usual mass density in high school, even if they didn't really understand it, but this is likely the first time they've encountered linear density. It's important to understand that the *same* value of the linear density characterizes *any* piece of a particular string, regardless of its length. You may need a couple of examples to get this idea across.

EXAMPLE 1: A 5 kg block is hung from the ceiling on a 2-m-long metal wire with a mass of 4 g. The wire is "plucked" at the very bottom, where it connects to the block. How long does it take the pulse to reach the ceiling?

A good fraction of day 1 is well spent having students work with history and snapshot graphs. Here are some possibilities.

EXAMPLE 2: The figure shows a snapshot graph made at $t = 0$ of a wave moving to the right at the leisurely pace of 1 m/s (to keep the numbers simple).

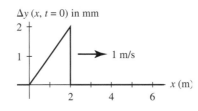

Draw:

a. The snapshot graph $\Delta y(x, t = 4\,s)$.
b. The snapshot graph $\Delta y(x, t = -2\,s)$.
c. The history graph $\Delta y(x = 0\,m, t)$ at $x = 0$ m.
d. The history graph $\Delta y(x = 4\,m, t)$ at $x = 4$m.

EXAMPLE 3: The figure shows a history graph made at $x = 0$.

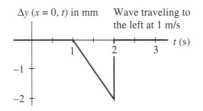

Draw:

a. The history graph $\Delta y(x = 1\,m, t)$.
b. The snapshot graph $\Delta y(x, t = 1\,s)$.

Expect *considerable* difficulty with these at first. However, doing several in class and assigning more for homework rapidly improves students' ability to visualize waves. If time permits, you might want do to a graphing/picture exercise for a longitudinal wave, such as the one shown in the Pedagogical Approach section.

Sinusoidal waves can be introduced in graphical form, emphasizing wavelength and wave fronts. Here's an example.

EXAMPLE 4: Consider the wave on a string shown here.

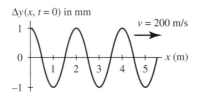

a. Find the wavelength and the period.

b. Draw graphs at $t = T/4$ and at $t = T/2$.

c. Draw a graph if the wavelength is halved.

d. Draw a graph if the frequency is halved but speed is unchanged.

e. Draw a graph if the phase constant increases by π.

f. Draw a graph if the tension is halved but the frequency is unchanged.

The beginning of day 2 is a good time to use videos or computer simulations of waves. Simulations that show a string as a series of dots are especially good because they help to clarify the distinction between motion of the wave and motion of particles in the medium. These demonstrations reinforce points made on day 1 and prepare students to discuss the mathematical representation of sinusoidal waves.

The essential idea, of course, is that $A \sin(kx - \omega t + \phi_0)$ describes the displacement at time t of a *particle* at position x in the medium (at least for string and sound waves) for a *wave* that is moving through the medium with speed $v = \lambda f$. The displacement might be either Δy or Δx, depending on the wave. For a light wave, the "displacement" is the change at (x, t) of an abstract quantity called the electric field. Students don't seem to have any difficulty talking about or using the "waviness" of light, even though they don't know what is waving. Perhaps this is not so surprising; after all, the wave theory of light was developed in the nineteenth century *before* physicists knew that light was an electromagnetic wave.

Sound waves and light waves can be used as the basis for example problems, giving students some familiarity with the order of magnitude of the parameters. You should insist that students immediately memorize the audible frequency range, the range of visible wavelengths, the speed of sound at room temperature, and the speed of light.

The superposition of waves turns out to be a difficult concept for many students, although it is helped if their visualization of waves has progressed to the point that they can picture two waves moving through each other. The biggest difficulty is wanting to add (or subtract) the *amplitudes* of the waves rather than adding the displacements. A number of computer simulations show wave pulses passing through each other, showing the individual waves and the net wave. It is important to have students look at several different points on a wave—not just at crests and troughs— and to confirm that the *displacements* are adding at each point.

An easy demonstration of superposition, and one of interest to students, is that of beats. You want them to recognize that two different sound waves are meeting, and superimposing, at their ear drum. Students who have played in bands or orchestras are usually familiar with beats, but they often don't know the cause. Nonmusician students have probably never heard beats. A nice demonstration is to use two audio oscillators, so that you can change the frequencies, and a microphone "ear" to display the beat pattern on an oscilloscope. The mathematical analysis is rather detailed and should be assigned for careful reading rather than derived in class. The only result you want for class use is that $f_{\text{beat}} = |f_1 - f_2|$.

There's no shortage of good demonstrations of standing waves, including the reflection of pulses on a string, different standing wave modes on a string or rope (noting the distinct frequencies), and the different harmonics of an overblown tube of air (organ pipe, whistle, musical instrument, or whatever you have that works). Another nice demonstration uses the corrugated plastic tubes—sold as toys—that you whirl over your head to get different tones. (These play the $n \geq 2$ harmonics of the tube, but the speed needed for the $n = 1$ fundamental doesn't draw air through the tube fast enough to set up oscillations over the corrugations. So $n = 1$ won't sound, and even $n = 2$ is often difficult.)

Such demonstrations lead naturally into a graphical description of standing waves, defining nodes and antinodes, and so on. You'll want to make sure that students understand what the "standard picture" of a standing wave shows and that they recognize why the distance between nodes is $\lambda/2$ rather than λ. Even without a mathematical proof, you can show graphically that $\lambda_n = (2L)/n$.

It seems natural that the frequencies would be $f_n = v/\lambda_n = n(v/2L)$. But this conclusion depends on the relationship $v = \lambda f$, which was found for *traveling* waves. Before using it, you have to convince students that what they see is the superposition of two traveling waves. Once they're convinced, *then* you can find the desired expression for the allowed frequencies.

Although all the important conclusions about standing waves can be obtained graphically, you do want to review the mathematical analysis. The two important ideas are the *amplitude function,* showing how the oscillation amplitude varies with position, and the *boundary conditions.* Students have met the idea of boundary conditions in calculus, in conjunction with the limits of integration, but this is likely the first time they've seen a real physical example where boundary conditions are important. Consequently, it is worth some time explaining what is meant by a boundary condition and showing how the boundary condition leads to the discrete frequencies of oscillation.

A nice demonstration that goes beyond the usual string and sound waves is to show a helium-neon laser with the cover removed so that students can see the discharge tube. Students are intrigued by lasers, but most have no idea how they work. You can discuss the laser as a standing wave cavity, emphasizing the smallness of the wavelength of light.

Standing wave resonances also make a nice demonstration. You need a long tube (2 m or so is a good length) with a loudspeaker at one end and a microphone at the other, connected to an oscilloscope. It's usually possible to see 8 or 10 resonances, although the lowest resonances may be too low in frequency for the typical small loudspeaker. You can quickly record a few frequencies with a frequency meter and ask students to identify the mode numbers. This is a useful exercise, because many students will automatically assume that the lowest recorded frequency must be $n = 1$. They're puzzled if you point out that their "$n = 1$" and "$n = 2$" frequencies don't have a 2-to-1 ratio!

Musical instruments, of course, provide many nice examples of standing waves if your class is musically inclined—and if the instructor is also musically inclined! These can lead to numerical exercises, such as measuring the length of a flute (open-open tube) or a clarinet (open-closed tube) and predicting the lowest note. Such predictions are close, but not as accurate (off by typically 10%) as students might expect. For a class that's interested, this can lead to a short discussion about whether real musical instruments are adequately modeled as a one-dimensional tube, about the actual locations of the antinodes (typically 1 cm or so beyond the physical ends of the tube), and so on.

Sample Reading Quiz or Discussion Questions

1. Define transverse wave and longitudinal wave.

2. Two sound waves of nearly equal frequencies are played simultaneously. What is the name of the acoustic phenomena you hear if you listen to these two waves?

3. When a wave pulse on a string reflects from a boundary, how is the reflected pulse related to the incident pulse?

4. There are some points on a standing wave that never move. What are these points called?

5. A wave front diagram shows
   a. the wavelengths of a wave.
   b. the crests of a wave at a particular instant of time.
   c. how the wave looks as it moves toward you.
   d. the forces acting on a string that's under tension.
   e. wave front diagrams were not discussed in this chapter.

6. The waves discussed in this chapter are
   a. sound waves.
   b. light waves.
   c. string waves.
   d. water waves.
   e. both a and b.
   f. both a and c.
   g. the three waves a, b, and c.
   h. all of a, b, c, and d.

7. The various possible standing waves on a string are called the
   a. antinodes.
   b. normal modes.
   c. resonant nodes.
   d. incident waves.

8. The frequency of the third harmonic of a string is
   a. one-third the frequency of the fundamental.
   b. equal to the frequency of the fundamental.
   c. three times the frequency of the fundamental.
   d. nine times the frequency of the fundamental.

Sample Exam Questions

1. The figure shows a snapshot graph $\Delta y(x, t = 2 \text{ s})$ taken at $t = 2$ s of a pulse traveling to the *left* along a string with a speed of 2 m/s. Draw the history graph $\Delta y(x = -2 \text{ m}, t)$ of the wave at $x = -2$ m.

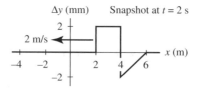

2. The figure shows a snapshot graph at $t = 0$ of a sinusoidal string wave traveling to the right at 50 m/s.

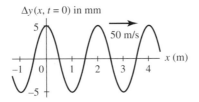

a. Write the equation that describes the displacement $\Delta y(x, t)$ of this wave. Your equation should have *numerical values,* including units, for all quantities except x and t.
b. What is the phase of the wave at $x = 3$ m?
c. Suppose the string's tension is increased by 20%. By what percentage does the wavelength change? Does the wavelength increase or decrease?

3. A tube, open at both ends, is filled with an unknown gas. The tube is 190 cm in length and 3 cm in diameter. By using different tuning forks, it is found that resonances can be excited at frequencies of 315 Hz, 420 Hz, and 525 Hz, and at no frequencies in between.

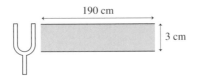

a. What is the speed of sound in this gas?
b. Draw a graph *on the picture* of the 315 Hz standing wave.
c. Can you determine the amplitude of the wave? If so, what is it? If not, why not?

4. Two strings with linear densities of 5 g/m are stretched over pulleys, adjusted to have vibrating lengths of 50 cm, and attached to hanging blocks. The block attached to string 1 has a mass of 20 kg and the block attached to string 2 has mass M. Listeners hear a beat frequency of 2 Hz when string 1 is excited at its fundamental frequency and string 2 at its third harmonic. What is *one* possible value for mass M?

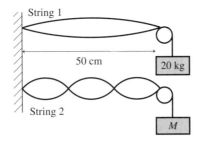

5. The picture below shows two pulses approaching each other on a stretched string at time $t = 0$. Both pulses have a speed of 1 m/s. Using the empty graph axes below the picture, draw a picture of the string at $t = 4$ s.

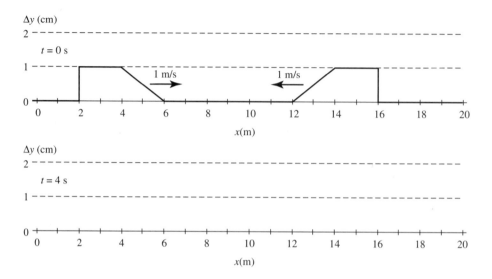

Thermal Physics

Background Information

Of all the major subjects in introductory physics, thermal physics looks most like a potpourri of unrelated topics and equations. Students—and even many instructors—find little logic or coherence in the conventional presentation of thermal physics. A major goal is thus to simplify and streamline the presentation so that the logic of thermal physics and thermodynamics is clear. The primary message to students has two main ideas:

- *Thermal physics* is concerned with understanding the properties of macroscopic systems. Macroscopic systems are characterized by a small number of "state variables."
- *Thermodynamics* is a more general study of how macroscopic systems transfer and transform energy.

There is relatively little research on student understanding of concepts in thermal physics and thermodynamics. I'll concentrate my remarks on a few topics known to cause student difficulties and on establishing a coherent view of thermal physics.

First, it has to be acknowledged that thermal physics is plagued with many inconsistent sets of units. As much as we might wish otherwise, all these many non-SI units are widely used throughout science and engineering. There's no point in trying to shield students from this harsh reality, and most texts present examples and problems in many different units of pressure, volume, temperature, mass, and so on. Not surprisingly, one of the most common student errors is failure to convert to the proper units for doing calculations. Instructors are urged to be very explicit about units and conversions while working example problems.

The measurement of volume is particularly telling. A significant fraction of students in most introductory physics classes cannot convert cm^2 to m^2 or cm^3 to m^3. Because $1 \, m = 100 \, cm$, you'll find many students using $1 \, m^2 = 100 \, cm^2$ and $1 \, m^3 = 100 \, cm^3$. Simply telling them the conversion factor has little effect, but you can quickly convince most of them with a simple exercise of computing the number of $1 \, cm \times 1 \, cm$ squares in a $1 \, m \times 1 \, m$ square and the number of $1 \, cm \times 1 \, cm \times 1 \, cm$ cubes in a $1 \, m \times 1 \, m \times 1 \, m$ cube.

Pressure can be introduced either in thermal physics or in a separate chapter on fluids. While all students have some sense of what pressure is, their understanding has many conceptual gaps. One difficulty is the idea that pressure is isotropic, pushing equally in all directions. We perhaps contribute unwittingly to this difficulty with the common explanation that pressure is due to the weight of the air or the weight of the water "pressing down." Consequently, students have a difficult time visualizing the sideways and upward pressure forces that are needed to understand buoyancy or to understand that the pressure forces are equal on all surfaces of a container of gas.

Students also have a difficult time recognizing the consequences of a pressure *difference*. An interesting way to confront these difficulties is with a vacuum cleaner. In common language, the vacuum cleaner "sucks" up dirt. If you place your hand over the tube, you feel it being "sucked" into the tube. Students have never thought about the mechanism by which this happens, and only a few immediately recognize the fact that air has no ability to *pull* on an object. A class discussion on this topic (and related topics, such as suction cups) is very revealing, and it prepares students to understand the usual presentation and demonstration of the forces on a pair of evacuated hemispheres.

The McDermott group at the University of Washington has done limited research on student understanding of pressure in gases. For example, they asked students to compare the gas pressure inside an inflated balloon with the gas pressure of the surrounding air. In a sophomore-level thermal physics class, of which more than 25% of students were physics majors, only 30% answered correctly. Nearly all the incorrect responses were that the two pressures are equal, overlooking the role of surface tension in increasing the pressure inside the balloon.

In the simpler situation shown below of a massless piston "floating" on a cylinder of gas while supporting a mass m, most students responded that the gas pressure inside the cylinder is equal to the air pressure above the piston. Few understood the role of the mass in establishing the pressure. Student understanding improves if they have an opportunity to analyze this situation qualitatively, drawing a free-body diagram for the piston and using the equilibrium conditions.

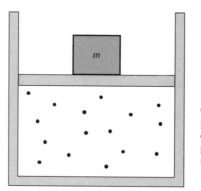

Compare the gas pressure inside the cylinder to the pressure of the surrounding air.

Instructors often assume that students understand the ideal gas law from chemistry classes. Most students have, indeed, seen the ideal gas law before, but very few have a working understanding. This is due, in part, to the difficulties students have with ratios. And although nearly all students have encountered moles in chemistry, few have any real sense of what is measured by moles or when two substances have equal or different numbers of moles.

The major new concept of thermal physics is *heat.* It's well known that students confuse heat with thermal energy. Most students think that heat is what a thermometer measures—not a surprising belief considering the everyday use of the word. It is important at every opportunity to emphasize that heat and work are energy that is *transferred* between the system and the environment. Neither heat nor work are properties *of* the system. Because temperature measures a property of the system, temperature must measure something other than heat.

Our own language can mislead students if we're not careful. I've caught myself saying, "I'm going to heat the gas in this cylinder by compressing it with a rapid push on the piston." There's clearly no *heat* involved in this process. Saying "I'm going to raise the temperature of the gas . . . " is awkward, but it's better to be awkward and correct than to convey the wrong message to students.

Student Learning Objectives

- To recognize and use the state variables—temperature, pressure, volume, moles—that characterize macroscopic phenomena.
- To understand pressure in static fluids and gases.
- To understand the ideal gas law.
- To understand and practice using pV-diagrams.
- To begin to understand heat and the process of heat transfer.
- To understand two important consequences of heat transfer—temperature change and phase change.

Pedagogical Approach

Unlike most other topics in the course, much of thermal physics is largely the presentation of facts and definitions. This is the basic knowledge of macroscopic phenomena on which thermodynamics is based, and students do need to know this material. This is a good chapter for an early quiz on basic facts, definitions, units, and the ability to carry out simple calculations with density, moles, pressure, and so on. Such a quiz requires students to consolidate this information quickly rather than waiting several weeks until a midterm exam.

Gases can be introduced from the very beginning as collections of rapidly moving atoms. It's interesting to ask students what they can conclude about gases from the fact that gases are compressible but liquids and solids (very nearly) aren't. Students have had little experience drawing conclusions from evidence, so this is a nice exercise in critical thinking.

pV diagrams will be very important in the chapter on thermodynamics, but these diagrams are rather hard for many students to understand. In keeping with a general emphasis on graphical representations of knowledge, I place more emphasis on *pV* diagrams than most textbooks, and I introduce them right along with the ideal gas law. This allows students to be familiar with their use before reaching heat engines. Because all of the processes considered in the introductory course have a constant number of moles, a point in the *pV* plane is a *complete* specification of the state of the gas. Many students need some help to see this, after which the usefulness of the *pV* diagram is more apparent.

Temperature needs to be associated from the very beginning with the thermal energy of molecular motion, not with heat. The explicit relationship between temperature and thermal energy will be found later, from kinetic theory, but students readily accept that "getting hotter" $(\Delta T > 0)$ is associated with higher molecular speeds and a higher thermal energy.

Because heat is an energy transfer, to be distinguished from thermal energy, an obvious question is, "What happens to a system when heat is added to it?" There are three possibilities:

■ A temperature increase,
■ A phase change, or
■ An isothermal expansion.

The most common result, and the only one students usually think of, is a temperature increase. Few texts explicitly call out the other possibilities when the subject of heat is introduced, thus starting students off with a misleading impression.

Specific heat and molar heat capacity can then be introduced to relate Q to ΔT when heat causes a temperature change. It's useful to write the relationship as $\Delta T = Q/Mc$. This has two effects. First, it makes clear that a certain amount of heat Q causes a well-defined temperature change ΔT. Second, it helps students see that, for a given amount of heat, an object with a small specific heat will have a larger temperature change than an object with a large specific heat. Or stated otherwise, it's harder to change the temperature of an object with a large specific heat. A discussion of these points helps students understand the significance of specific heat. You'll want to note that it's OK to compute ΔT in °C. After emphasizing during the ideal gas law the importance of converting T to kelvins, you'll find many students now doing needless—and error prone—conversions.

Calorimetry is not an essential subject, but it is a nice application of the ideas of heat and energy conservation. Calorimetry situations also help to establish ideas about how interacting systems reach thermal equilibrium. Although most students will admit to having seen calorimetry before, usually in chemistry, few can work a calorimetry problem. This is due, at least in part, to having memorized formulas in chemistry without any understanding of what it was all about. Another look at calorimetry, with the underlying principles now better explained, certainly won't hurt students.

The most common error in calorimetry problems is to ignore the sign convention for Q and to set $Q_1 = Q_2$, rather than $Q_1 + Q_2 = 0$. Some students can finesse the

ΔTs to make this work out, but far more end up with wrong answers due to sign errors. It's important to insist that they use $\Sigma Q = 0$ with all ΔTs being $T_f - T_i$. You can make this point by asking students to find the final temperature of a system of *three* interacting objects having different initial temperatures. One of the initial temperatures should be midway between the other two so that students can't tell whether ΔT for that object is going to be positive or negative.

Using Class Time

Thermal physics is a good place to reduce the encyclopedic scope of most texts by omitting all but the most essential topics. I cover fluid statics only, for the purpose of introducing pressure, but omit all fluid dynamics. I also omit thermal expansion, defer temperature scales to assigned reading, and omit the details of heat conduction, convection, and radiation. I stress the concept of heat through its consequences—temperature and phase changes. I also emphasize the ideal gas law and pV diagrams because thermodynamics in an intro course is based on ideal gas processes.

I like to begin with a brief introduction of what thermodynamics is all about and where the next few chapters are headed. A nice example is to draw a gas cylinder with a heat source and a piston, as shown below. Suppose this is the cylinder in a car engine. Gasoline is burned (the heat source), the piston pushes out and, through a series of gears, causes the drive wheels to exert a force on the road. The road, in reaction, exerts a force on the car and accelerates it forward. If 1000 J of energy are released by burning the gasoline, does the car acquire 1000 J of kinetic energy? If not, what happens to the rest of the energy? Are there any *fundamental limits* on what fraction of the gasoline's energy can be converted to the car's kinetic energy, or is it simply a matter of engineering to make cars ever more efficient?

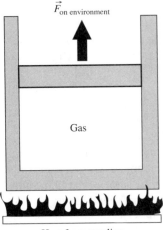

Heat from gasoline

These are the types of questions with which thermodynamics is concerned. They are both basic scientific questions—How is energy transferred and transformed?—as well as questions of great practical concern. Because energy is such an all-encompassing concept, thermodynamics has more widespread applications than any other area of physics. But before we can answer the above questions, we need more knowledge about *macroscopic systems*. Thus . . .

Day 1 should introduce the basic concepts of pressure. Textbooks usually start with hydrostatic pressure in fluids and only later shift to the pressure in a gas. That students don't always make the transition is seen in their occasional efforts to calculate the gas pressure in a cylinder of height h with $p = \rho gh$. To counter this, I prefer to start right in with demonstrations using both liquids and gases. The connection between liquids and gases is, of course, that both are fluids. They "deform" to fit their container, and this is how they exert pressure forces on all the surfaces.

The vacuum cleaner discussion described above is a good entry point that gets students' interest, followed by the typical demonstration of two hemispheres held together by pressure forces after being evacuated. (This is a good opportunity to call up the two largest men in the class and ask them to pull it apart.) A difficult concept for many students is that pressure forces push equally in all directions. A tall cylinder of water with holes down the side demonstrates that the water pressure pushes *sideways,* since the water is being pushed out of the holes. The cylinder also shows that the pressure increases with depth—important for liquids but *not* for gases because their density is so much less. This is a point worth emphasizing: The pressure of a liquid is determined by depth h, but the pressure of a gas (except for the full pressure of the atmosphere) is determined by its temperature and volume.

Derivation of the liquid pressure at depth h is standard, and you'll want to do a couple of examples that illustrate using and converting pressure units. Note that many texts first state $p = \rho gh$, without including the atmospheric pressure, as if it were a major equation. Even if the atmosphere is introduced a few paragraphs later, to give $p = p_{atm} + \rho gh$, many students will latch on to the first equation and use it—incorrectly—in all problems. Class discussion and example problems should always include the p_{atm} term.

A nice demonstration that can lead into a numerical example is to lift a colored fluid with a pipette or a clear drinking straw (see figure on next page). Students all know that they can lift soda by placing their finger tip over the straw, then release it by removing their finger. First ask them to *explain* why this works. Many explanations are quite weak, and you'll want to direct them to talk about forces on the liquid. Then ask them to compare the pressure of the trapped gas to the exterior air pressure. Nearly all students think they are equal. But by examining the pressure at the lower tip of the pipette, using an analysis similar to that of the barometer, you find that $p_{gas} + \rho gh = p_{atm}$. Thus the trapped gas pressure is *less* than atmospheric pressure by the amount ρgh. Apparently, some of the liquid flows out as you lift the pipette, increasing the trapped volume and lowering its pressure. If you ask for a numerical calculation, with typical values of ρ and h, the pressure reduction turns out to be $\approx 1\%$ of atmospheric pressure.

Fingertip

h

Pressure gauges, barometers, and manometers also make good demonstrations that can lead into numerical examples.

The ideal gas law should be introduced as an *empirical* discovery about gases. It can be interesting to ask students what the significance is of the fact that the gas constant R (the experimental slope of a pV versus nT graph) is the same for all gases. The ideal gas law is written two ways in most textbooks:

$$pV = nRT \qquad pV = NkT.$$

It's good to use both in examples. R can be characterized as "the gas constant per mole" while Boltzmann's constant k is "the gas constant per molecule."

Common student errors with the ideal gas law involve units—using pressures in atmospheres, volumes in cm^3, or temperatures in °C. A useful example is to have students compute the final pressure in a constant-volume process that raises the temperature of a gas from 5°C to 30°C. For those who get $p_f = 6p_i$, ask them if they really believe such a large pressure increase occurs for such a small temperature increase. Then ask what the final pressure would be if the starting temperature had been 0°C.

It's good to be able to demonstrate all three basic gas processes. Nearly all departments have a constant volume gas thermometer with an attached pressure gauge, and this can be used for a nice demonstration with a numerical calculation. A pump (with a decent seal) attached to a pressure gauge shows an isothermal process if you compress it fairly slowly. The difficult process to demonstrate, and the hardest for students to understand, is the isobaric process. Some departments will have a container with a tall glass tube and a precision-machined steel ball that just fits into the tube. The seal is sufficiently good that the ball "floats" in the tube for a minute or more before dropping significantly. This is adequate time to heat the container slowly with a hot-air gun, causing the ball to rise in an isobaric expansion.

As noted in the Background Information section, students do have some conceptual difficulties with how the pressure in a gas cylinder is established. A useful set of questions to ask of students is based on a cylinder with these characteristics:

- The piston can be locked or unlocked with a pin.
- Masses can be added or removed from the piston.
- The entire cylinder can be placed in a hot or cold liquid.

After establishing the experimental arrangement (see figure on next page), ask students:

- Can you decrease the volume without changing the pressure? If so, how?
- Can you decrease the volume without changing the temperature?

- Can you decrease the pressure without changing the temperature?
- Can you decrease the pressure without changing the volume?

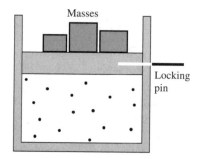

Typical problems simply state that "The pressure is doubled" or "The volume is halved." Students rarely get a chance to think about how such changes are brought about. (Note: Don't mention *heat* with this example. Placing the cylinder in hot liquid is simply a way to increase the temperature of the gas.)

Another thought question is to show two identical cylinders, both with the same mass on the piston and the same volume. One contains hydrogen, the other nitrogen, as shown below.
Then ask:

- Both gases have the same mass. Compare the temperature of hydrogen to that of nitrogen.
- Both gases are at the same temperature. Compare the number of moles of hydrogen to that of nitrogen.

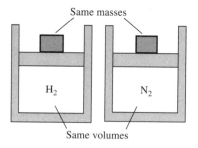

An extended analysis of a piston of mass M and cross section area A "floating" in a cylinder is worthwhile (see figure on next page). An equilibrium analysis of the piston shows that the gas pressure inside the cylinder is

$$p_{gas} = p_{atm} + Mg/A.$$

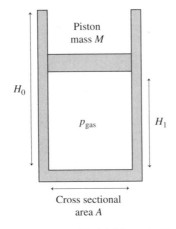

If the piston is "dropped" into a tube of initial length H_0 and initial pressure p_{atm}, it causes an isothermal compression. It is straightforward to show that the equilibrium height of the piston is

$$H_1 = \left(\frac{p_{atm}}{p_{atm} + Mg/A}\right)H_0.$$

A 2 kg piston dropped into a cylinder of diameter 4 cm comes to rest at 87% of H_0. This is a nice example to have students work through because it combines both a force analysis of the piston and the ideal gas law.

It's good to use pV diagrams with all examples of ideal gas processes so that students become used to seeing them. At least one example of interpreting a pV diagram is worthwhile. For example, the diagram shown below could be used to ask students

- What are the initial and final temperatures?
- What kind of process is this?
- What are some other ways to get from the initial to the final state?

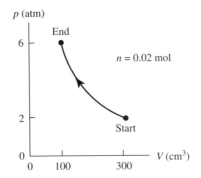

You can introduce thermal energy, heat, and specific heat by considering two blocks in contact with each other (see next page). Block A (initially 0°C) and block B (initially 100°C) are of equal size. Students will quickly agree that, left to themselves, the blocks will eventually reach a common temperature. This allows you to define thermal equilibrium. But the more important question to ask is: *How* does the system reach thermal equilibrium? Nothing moved, there were no visible changes, and yet clearly something happened.

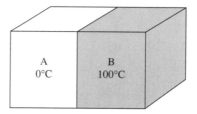

If you earlier associated temperature with the motion of atoms, you can now reap the benefits. Block A is gaining microscopic energy as its atoms speed up, while block B is losing energy. Students will even understand that this can happen as a result of collisions between A-atoms and B-atoms at the boundary, although few would have thought of this on their own. Now you can define both *thermal energy* as the "invisible" energy of the molecular motion and *heat* as the transfer of energy due to a temperature difference.

Next, ask students if $T_f = 50°C$. Most students, after some discussion, will agree that they cannot find T_f without knowing what the materials are. This leads naturally to *specific heat* as a quantity that determines how a material will respond to a heat transfer.

Further discussion of heat leads to the question "What happens to a system when heat is added to it?" There are three possibilities, as noted above:

- A temperature increase,
- A phase change, or
- An isothermal expansion.

You've already talked about temperature changes. Isothermal expansion should be acknowledged, but you'll defer the details to the chapter on thermodynamics where students will learn to calculate the work done. That leaves only phase changes and the need to introduce heats of transformation.

This discussion is easily generalized to the molar heat capacity of gases. The distinction between heat capacity at constant volume and constant pressure is not clear to many students. They need to see an example of heating a gas cylinder at constant volume (piston locked in place) and at constant pressure (allowing the piston to move out). Use a pV diagram to show that these are different processes with different end points, so it's not surprising that they yield different final temperatures. The more complete analysis, in which a computation of the work done at constant pressure process yields $C_p - C_V = R$, must be deferred until thermodynamics.

A series of simple exercises can highlight the main points:

- One mole of hydrogen gas and one mole of nitrogen gas are in equal-size rigid containers at 30°C. 1000 J of heat are added to each. Compare their final temperatures.

■ One mole of helium gas and one mole of nitrogen gas are in equal-size rigid containers at 30°C. 1000 J of heat are added to each. Compare their final temperatures.

■ 28 g of helium gas and 28 g of nitrogen gas are in equal-size rigid containers at 30°C. 1000 J of heat are added to each. Compare their final temperatures.

■ One mole of helium gas and one mole of nitrogen gas are in equal-size containers at 30°C. The nitrogen container is rigid, but the helium container has a piston that moves so as to keep the pressure constant. 1000 J of heat are added to each. Compare their final temperatures.

A useful extended problem that touches on many issues is as follows:
   A rigid 10 cm × 10 cm × 10 cm cube contains 0.4 g of helium at a pressure of 2 atm. The cube is held over a flame, and 250 J of heat are added to the gas.
   a. What is the final pressure?
   b. Show the process on a pV diagram.

Sample Reading Quiz or Discussion Questions

1. What is the SI unit of pressure?
2. What is a pV diagram and how is it used?
3. What was the original unit for measuring heat?
4. The ideal gas model is valid if
   a. the gas density is low.   e. both a and c.
   b. the gas density is high.   f. both a and d.
   c. the temperature is low.   g. both b and c.
   d. the temperature is high.   h. both b and d.
5. An ideal gas process in which the volume doesn't change is called
   a. isobaric.   c. isothermal.
   b. isochoric.   d. isentropic.
6. Heat is
   a. the amount of thermal energy in an object.   d. both a and b.
   b. the energy that moves from a hotter object to   e. both a and c.
      a colder object.
   c. an invisible fluid-like substance that flows   f. both b and c.
      from a hotter object to a colder object.
7. The thermal behavior of water is characterized by the value of its
   a. molar heat capacity.   c. heat constant.
   b. specific heat.   d. thermal index.

Sample Exam Questions

1. The cylinder of gas shown in the figure has a piston that can float up and down. You can:
      Lock or unlock the piston in place with a pin,
      Add or remove masses from the piston, or
      Place the entire cylinder in either a hot or cold liquid.

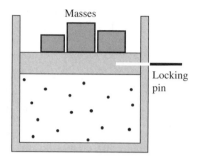

Masses

Locking pin

a. Can you increase the gas temperature without changing the pressure? If so, describe how you would do it. If not, explain why not.

b. Can you increase the gas pressure without changing the temperature? If so, describe how you would do it. If not, explain why not.

2. A U-shaped tube is open at one end and closed at the other, as shown. It is initially filled with air at 30°C and 1 atm pressure. Mercury ($\rho = 13{,}600$ kg/m^3) is poured slowly into the open end without letting any air escape, thus compressing the air. This is continued until the open side of the tube is completely filled with mercury. What is the length L of the column of compressed air?

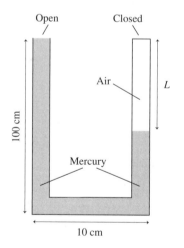

Open

Closed

Air

L

100 cm

Mercury

10 cm

3. A 1000 cm^3 box with perfectly insulating walls contains nitrogen at 1 atm pressure and 20°C. A 20 g block of copper at 500°C is placed in the box, then the perfectly insulating lid is quickly and tightly sealed. After a long time has passed, what is the pressure in the box? (Students will need a table of data. Include both C_V and C_P for nitrogen. Note that the copper excludes some of the gas from the box, so the gas volume is 998 cm^3.)

4. 0.4 g of helium gas is sealed in a rigid container, and the gas then undergoes the process shown in the pV diagram.

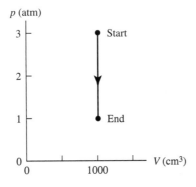

a. Is heat *added to* the gas or is heat *removed from* the gas during this process? Explain.

b. What is the quantity of heat that is added or removed?

Thermodynamics

Background Information

Thermodynamics, properly presented, is a grand subject. It's a sweeping, "big picture" view of energy. Unfortunately, this big picture is barely glimpsed in the standard presentation, which becomes focused on myriad details and tedious calculations. Even more unfortunate, the discussion of energy in thermodynamics is never connected with the earlier presentation of energy in mechanics.

In the energy chapter, Chapter 9, I recommended that you develop the law of conservation of energy in the form

$$\Delta K + \Delta U + \Delta E_{\text{therm}} = W_{\text{ext}},$$

where ΔE_{therm} arises from nonconservative forces within the system and W_{ext} is work done by external forces originating in the environment. From here, you need add just a few small steps to arrive at the first law of thermodynamics.

First, recognize that, for an extended system, K and U are the center-of-mass energies. With this recognition, the energy equation can be written

$$\Delta K_{\text{cm}} + \Delta U_{\text{cm}} + \Delta E_{\text{therm}} = W_{\text{ext}}.$$

Second, recognize that this statement can't be complete. If I put a pan of water over a flame, nothing happens to the center of mass. But $\Delta E_{\text{therm}} > 0$ while no external work is being done. Since the thermal physics chapters introduced heat as another means, a nonmechanical means, of energy transfer, it's a small step to place heat on an equal footing with work. Then

$$\Delta K_{\text{cm}} + \Delta U_{\text{cm}} + \Delta E_{\text{therm}} = W + Q,$$

where the subscript label on W has now been dropped.

This line of reasoning is summarized in what I call the *knowledge structure of energy*:

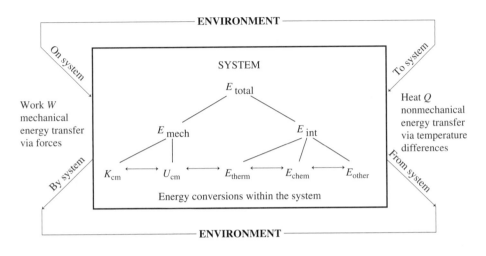

First Law of Thermodynamics

$$W + Q = \Delta E_{total} = \Delta K_{cm} + \Delta U_{cm} + \Delta E_{therm} + \Delta E_{chem} + \cdots$$
Net energy transferred = Total change in system energy

Strictly speaking, we should use E_{int}, which—as the figure suggests—includes not only thermal energy but also chemical, nuclear, and other forms of energy. But the only form of internal energy considered in the intro course is thermal energy. More significantly, it is E_{therm}, not the full E_{int}, that is related to the temperature.

Finally, the systems we want to study in thermodynamics are *stationary* containers of gases or liquids whose center-of-mass energy doesn't change. For such systems,

$$\Delta E_{therm} = W + Q.$$

This is, of course, the first law of thermodynamics. We have continued to modify and enlarge the concept of energy, but there has been no discontinuity of ideas in moving from mechanics to thermodynamics. You should encourage students to review the development that has taken us from kinetic energy to the first law, and you can ask students to articulate what new information and ideas were added at each step. The knowledge structure figure above, which students find very helpful, conveys the same ideas in pictorial form.

Here W is the work done *on* the system by the environment, as it was defined in mechanics, so W has the same sign convention as Q. It's customary in thermodynamics to change perspective and let W be the work done *by* the system. This changes the sign and gives the more common statement of the first law:

$$\Delta E_{therm} + W = Q.$$

Many textbooks use U as the symbol for thermal energy. This is a terrible notation, since U has been used previously to mean potential energy.

The introductory course invariably focuses on the thermodynamics of ideal gases, in order to avoid the additional complications of phase changes. Students are somewhat familiar with isochoric, isobaric, and isothermal processes, from their study of the ideal gas law in the thermal physics chapters, and they should be able to visualize how these processes take place. The thermodynamics chapter adds new information about how to calculate the heat and work associated with these processes. The entirely new process, and one that students do have difficulty understanding, is the adiabatic process. Because students associate temperature change with "heat," it is hard for them to understand a process in which $Q = 0$ but $\Delta T \neq 0$. Practical adiabatic process are compressions and expansions that take place "too quickly for heat to be exchanged with the environment." This is an idealization, and its good to point that out. It's also worth an explicit compare-and-contrast of an adiabatic compression and an isothermal compression.

As a word of caution, you should *not* use heat engine cycles with a downward-sloping straight line section such as shown in the figure below. As Dickerson and Mottman (*Amer. J. Phys.* **62**, 558 (1994)) have pointed out, the analysis of such a process is not as straightforward as is tacitly assumed by many texts. In a process with $\Delta T < 0$, which appears to be a cooling process, the heat Q is *not* purely Q_{out}. A more careful analysis, beyond what is appropriate for the introductory course, shows that heat flows *in* to the system until the straight line is tangent to an adiabat, then *out* of the system for the remainder of the process. The Q computed as $\Delta E_{therm} + W$ is the total $Q_{in} + Q_{out}$, but a simple analysis doesn't yield Q_{in} and Q_{out} separately.

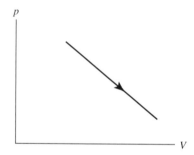

Student Learning Objectives

- To understand energy conservation as expressed in the first law of thermodynamics.
- To continue developing the concept of heat.
- To understand the thermodynamics of the four basic processes of an ideal gas.
- To understand the physics of simple heat engines.
- To recognize that thermodynamics has practical applications to real devices.
- To learn that there is a limit to the efficiency of a heat engine.

Pedagogical Approach

The main point to emphasize is that thermodynamics is about the transfer and transformation of energy. If you start with the energy conservation statement from mechanics, then add heat as an additional method of energy transfer, you find

$$\Delta E_{cm} + \Delta E_{therm} = Q + W_{ext}.$$

Although work *can* cause the entire system to move—and that's what we considered in mechanics—we're now interested only in systems whose center of mass remains at rest. Thus we can set $\Delta E_{mech} = 0$. With this restriction, and if we change W to be the work done *by* the system, the law of energy conservation becomes the first law of thermodynamics.

$$\Delta E_{therm} + W = Q.$$

Most textbooks never mention these simple points, leaving students unaware of any connection between thermodynamics and the previous discussion of energy. An explicit presentation of the origins of the first law is important for students to understand the logic behind why we call it a statement of energy conservation.

The heat engine embodies the central concepts of thermodynamics—transferring and transforming energy. A heat engine is a complex idea, a *model* of many different kinds of real engines. Most texts don't give a clear explanation of what a heat engine *is* or of the assumptions behind it. It's particularly difficult for students to understand

■ Why a heat engine must be a cyclical process.
■ That a heat engine requires both a heat input *and* a heat output.
■ That the common statement $W = Q$ means $W_{cycle} = Q_{in} - |Q_{out}|$.
■ Why W_{cycle} is the area inside the closed pV curve.
■ Why $\left(\Delta E_{therm} \right)_{cycle} = 0$.

It's important to look at these issues before getting into the details of specific heat engines.

I recommend avoiding the term "heat reservoir." This is a confusing and misleading term for students. First, it suggests only a *supply* (a heat source), and students don't recognize that a heat reservoir can also function as a heat *sink*. Second, the term suggests that heat is a *substance* that is "in" the reservoir—such as water in a lake. After endeavoring to convince students that heat is a transfer of energy, not a kind of energy, it's not good to use antiquated terminology that suggests otherwise. I simply refer to the "hot side" and "cold side" of a heat engine as being the source and sink for the transfer of heat energy.

The big practical difficulty with thermodynamics is keeping track of all the information. There are many different processes, many different parameters, and many different relationships between the parameters—some general, others applicable only to specific processes. It takes quite a bit of practice and experience to avoid drowning in this sea of details.

There is, however, a systematic *procedure* to follow when analyzing heat engines:

Analyzing Heat Engines

- Identify each process in the cycle and, by implication, the equations and relationships that are specific to that process.
- For each process, determine whether Q and W are positive, negative, or zero.
- Recognize relationships, such as the first law, that are valid for *any* process.
- Analyze each process in the cycle.
- Combine the processes to find Q_{in}, Q_{out}, and W_{cycle}. Verify that $(\Delta E_{therm})_{cycle} = 0$.

By being highly systematic in your examples, you can help students distinguish the big picture from the details.

A final issue is what to do about introducing the concept of entropy. The standard approach is to use the Clausius definition $dS = dQ/T$. The difficulty is that it requires a fairly sophisticated understanding of thermodynamics to see any rationale behind this. It's the rare student for whom the thermodynamic sense of entropy has any real meaning. Students may be able to calculate the entropy change of heating water, but it's an exercise with little physical significance. What have they really learned?

The more interesting sense of entropy is that associated with statistical mechanics and information. But a one page section on "probability and disorder" isn't going to get this idea across to students. It *is* possible for introductory physics students to understand some elementary statistical mechanics and to arrive at probabilistic definition of entropy, but such an approach needs an entire chapter. This is the path I follow in my textbook, as do Chabay and Sherwood in *Matter and Interactions* (1999) and Moore in *Six Ideas That Shaped Physics* (1998).

My preference is to omit entropy altogether unless there's time to develop the probabilistic approach and thus give the concept some real significance. This doesn't mean having to give up the second law. The Kelvin-Planck statement of the second law, that it's impossible for a cyclical heat engine to convert heat to work at 100% efficiency, can be introduced as an empirical discovery. Similarly, the Carnot statement of the second law, that a reversible engine working between two temperatures has the maximum possible efficiency, is easily shown with a simple definition of "reversible." This approach ties the second law to the types of thermodynamic processes that students have been studying, and it answers important questions about the limits of thermodynamic efficiencies. This is a new idea to nearly all students, and one that most find quite interesting. Suddenly defining entropy at the end of the chapter doesn't add anything to their understanding. It's a topic that is easily deferred until students take a full course on thermodynamics.

Using Class Time

This is a good chapter to start with a mini-lecture before getting into student activities. Remind students of where you left off with the development of the concept of energy back in mechanics, then follow the procedure described in the Background Information section until—*Voila!*—you've linked the new first law of thermodynamics back to subject matter that is (mostly) familiar. This leads naturally into a discussion of heat as an energy transfer and of the sign convention for heat and work.

A nice exercise is to describe a situation and ask students if W, Q, ΔE_{cm}, ΔE_{therm}, and ΔT are each positive, negative, or zero. Situations might include:

- A steel block slides to a halt on a rough surface.
- A steel block is held in a flame.
- An expanding spring pushes a rigid cylinder of gas across a frictionless surface.
- A rigid cylinder of gas is held in a flame.
- An expanding spring pushes in a piston on a well-insulated cylinder of gas.
- A copper gas cylinder is placed in a tub of ice water and left for a long time. Then a piston is pushed in slowly so that the gas temperature does not change. Consider just the gas to be the system.

The goal is for students to feel comfortable with the first law and to begin seeing it as "common sense."

Demonstrations of the first law are worthwhile, but initially you want to use single-step processes and stay away from heat engines. A well-insulated bike pump with a thermocouple inside can be used to show a temperature increase when work is done but $Q = 0$.

A dramatic and memorable demonstration, but one that's hard to analyze well, is the imploding soda can. Take an empty aluminum soda can and pour just a few millimeters of water in the bottom. Hold the can with tongs over a flame for few seconds to bring the water to a boil and fill the can with steam. Then quickly invert the can into a large container of cold water. You get a dramatic implosion the instant the top of the can enters the cold water, due to the large pressure drop as the steam condenses. The cold water removes heat from the gas $(Q < 0)$ while the environment does work on the can to crush it $(W < 0)$. A complicating factor—and one you may want to pass over silently—is the release of latent heat as the steam condenses. Even though the analysis may be a little shaky, this demonstration of the roles of heat and work is a big hit with students.

After introducing the first law, the next tasks are to associate heat and work with each of the basic ideal gas processes, to establish the relationship between C_P and C_V, and to introduce adiabatic processes. Students initially have difficulty knowing whether a process as seen in a pV diagram corresponds to work done on or by the system, or whether heat is added or removed from the system. It's good to draw a process, such as an adiabatic expansion, and ask students to articulate what is happening to the gas. (Does the piston move? Do you need to burn fuel? and so on.)

Another useful exercise is to start with a point in the pV plane and ask students to draw a pV trajectory that meets certain criteria (see figure on next page). For example, you could ask them to draw

- An isobaric process in which heat is removed from the gas.
- A process in which the gas is compressed without any transfer of heat.
- A process in which W is negative.

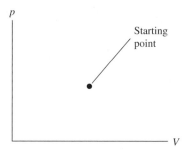

Some questions may have specific answers, others a range of possible answers. The latter can generate class discussion by asking a student who answers A if another student, who answers B, is also right. Qualitative exercises such as these can easily lead into numerical examples.

The issue of C_P versus C_V can be approached with a qualitative exercise in which you ask students to rank order the heat transfers Q_1 through Q_4 associated with each of the four processes shown in the figure below. (You first want to establish that all students recognize that each process has the same ΔT.) Many students will have difficulty with this, so after they've struggled for a couple of minutes you can ask them to start by comparing ΔE_{therm} for each process. Once they realize ΔE_{therm} is the same for each, you can ask them "What is the basic principle about energy conservation that is true for all processes?" This should be enough of a hint, but if students are still having difficulty you can ask "How is the work done related to the geometry of the pV trajectory?"

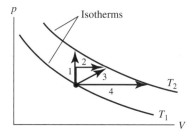

The point, which you should get students to articulate in their own words, is that it takes more heat to raise the temperature *and* do work than merely to raise the temperature. Consequently, it must be the case that $C_P > C_V$. *How much* larger requires a computation, but students should understand qualitatively *why* C_P is larger.

A useful example is to draw the diagram shown on the next page and ask for the value of the heat Q associated with this process. Many students will begin by seeking a formula to calculate Q, and it's worth letting them find that there is no formula. They must find W as the area under the curve, the temperatures from the ideal gas law, ΔE_{therm} from $nC_V\Delta T$, and finally use the first law $Q = W + \Delta E_{therm}$. Nearly every important idea is used in this example, and it helps convince students that—despite the numerous equations—there really is a *logic* to thermodynamics.

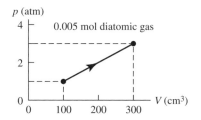

After these preliminaries, you're ready to introduce heat engines and the idea of the efficiency of an engine. This is all pretty straightforward, but you do want to emphasize that a heat engine has both a heat input Q_{in} *and* a heat output Q_{out}, with Q_{out} being a negative number.

The previous example can be extended to a three-sided heat engine that you and the students can analyze together. The hardest part of the analysis is already done. Putting the results in a table, to show Q, W, and ΔE_{therm} for each stage of the cycle, encourages students to track all the information carefully. It's easy to lose sight of information in these computationally intense problems.

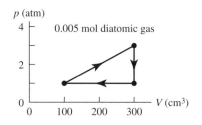

A somewhat more sophisticated example, but still with only three sides, involves one adiabatic process. This is a good example to have students analyze qualitatively before getting into the mathematics. In particular, ask them to

- Determine whether W is positive, negative, or zero for each process.
- Determine whether Q is positive, negative, or zero for each process.
- Describe in words and pictures how the cycle is accomplished.

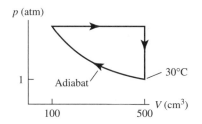

Then they can proceed to determine

- p, V, and T at each corner.
- W, Q, and ΔE_{therm} for each process.
- W_{cycle}, Q_{cycle}, and $(\Delta E_{\text{therm}})_{\text{cycle}}$.
- The efficiency of the heat engine.

As a follow-up, ask for the power output of the engine if it runs at 1200 rpm.

After this, the students are ready to look at the Carnot engine and move into the issue of maximum efficiency. This is a topic worth emphasizing. No one is surprised that an engine can't violate conservation of energy—you can't get something for nothing. But it is truly surprising that not only is there a limit to the efficiency of a heat engine, but you can state quite clearly what the limit is. And for most practical devices, the limiting efficiency is not very high.

It's worth giving a few simple exercises about engine efficiency. The figure below shows an example, for which you can ask "Is there some engine for which this is a physically possible behavior?" In this case, the efficiency of 30%, although low, exceeds the Carnot efficiency of 27%. So *no* engine can work this efficiently between temperatures of 100°C and 0°C.

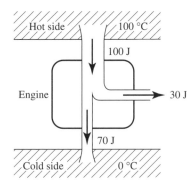

This discussion leads right into the second law of thermodynamics, a satisfying conclusion to this "big picture" look at energy.

Sample Reading Quiz and Discussion Questions

1. What is the name of an ideal gas process in which no heat is transferred $(Q = 0)$?
2. What is the generic name for a cyclical device that converts heat energy into work?
3. The maximum possible efficiency of a heat engine is determined by
   a. its design.
   b. the maximum and minimum pressure.
   c. the maximum and minimum temperature.
   d. the amount of heat that flows.
   e. the compression ratio.

4. The engine with the largest possible efficiency uses a
   a. Brayton cycle.   d. Joule cycle.
   b. Carnot cycle.   e. Otto cycle.
   c. Diesel cycle.   f. Stirling cycle.

5. $(\Delta E_{therm})_{cycle}$ for a heat engine is
   a. always positive.   d. equal to Q.
   b. always negative.   e. equal to W.
   c. always zero.

Sample Exam Questions

1. a. *Why* is C_P larger than C_V? Write a short explanation in which you make explicit use of a pV diagram.
   b. *Why* is C_V for a diatomic gas larger than C_V for a monatomic gas?

2. A monatomic gas is adiabatically compressed to 1/8 of its original volume. Do each of the following quantities change? If so, does the quantity increase or decrease, and by what factor? If not, why not?
   a. The rms velocity v_{rms}.   c. The collision frequency f_{coll}.
   b. The mean free path λ.   d. The molar heat capacity C_v.

3. Two gas cylinders, A and B, each hold 100 cm^3 of nitrogen gas at a temperature of 30°C and a pressure of 1 atm. Cylinder A undergoes an isobaric expansion to 300 cm^3. Cylinder B undergoes an isothermal expansion that does the same work as cylinder A.
   a. What are the final volume, temperature, and pressure of each cylinder?
   b. Show both processes on the pV diagram. Provide an appropriate numerical scale on each axis.
   (Note: It's good to provide an empty set of pV axes.)

4. A heat engine uses 120 mg of helium gas $(A = 4)$ as its working substance. The gas follows the thermodynamic cycle shown.

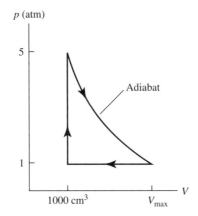

   a. What is the thermal efficiency of this engine?
   b. What is the maximum possible thermal efficiency of an engine that operates between T_{max} and T_{min}?

5. Three students submit their solutions to a design project. Each had been asked to design a heat engine that operates between temperatures of 300 K and 500 K. The heat input/output and the work done by their designs are shown in the following table:

Student	Q_{in}	Q_{out}	W
1	250 J	−140 J	110 J
2	250 J	−170 J	90 J
3	250 J	−160 J	90 J

Write an assessment of each student's design. Can it work? If not, why not?

14

Electrostatics

Background Information

Electricity and magnetism occupy a major portion of the introductory physics course—nearly one-third of most texts. Although electricity and magnetism are undeniably important components of science and technology, the level of detail and the level of mathematics often reflect more the interests of physicists than the needs of students. As noted earlier in this guidebook, the majority of engineering and science disciplines utilize only static electric and magnetic fields. Their interests are much more with understanding basic electric and magnetic phenomena, including the electric and magnetic properties of materials, than with mathematical sophistication.

Essentially all current textbooks present electricity and magnetism as a highly abstract, mathematical theory—the most sophisticated mathematics of the introductory physics course. Basic information about electric and magnetic phenomena is introduced quickly, with minimal discussion, as the text moves rapidly into the definitions and properties of fields and potentials. The emphasis is on the formal properties of fields and potentials, and these ideas are only loosely connected with electric and magnetic phenomena.

The "mathematization" of electricity and magnetism in the introductory course—the use of vector integrals, the inclusion of Gauss's law, Ampere's law, and Maxwell's equations—is a fairly recent entry into introductory physics. Unfortunately, only a small minority of students have yet reached vector integrals in their calculus course. For the large majority, this is new and frightening math that obscures the physical concepts of electric and magnetic fields. As an example, the tutorials designed by Lillian McDermott's group at the University of Washington have been highly successful at improving success rates, to typically >80%, on topics known to cause student difficulties. But on the topic of Gauss's law, the success rate on exam questions was still <20% *after* a Gauss's law tutorial. The evidence suggests that a vector integral approach to fields is simply beyond the capability of nearly all students in introductory physics.

Is this an appropriate pedagogical approach to the subject? Is it effective? Are students learning what we want them to about electricity and magnetism? Anecdotal information suggests that nearly all instructors are disappointed by their students' ability to learn and use the concepts of fields and potentials. And instructors who teach the upper-division electricity and magnetism course generally report that students seem to remember essentially nothing about fields from their introductory course.

There is an increasing body of research that we can turn to for guidance. This research has probed student understanding of both basic concepts and theoretical constructs in electricity and magnetism. Unfortunately, much of it is not yet published. Many of the results described in this chapter are based on research reports presented at meetings and from unpublished Ph.D. dissertations.

Students have a high level of familiarity with the basic phenomena of mechanics—namely, forces and motion. Although they harbor many misconceptions, they have little difficulty realizing that the Newtonian theory of mechanics is seeking to provide a quantitative explanation of the phenomena of motion.

By contrast, the large majority of students have essentially *no* familiarity with the basic phenomena of electricity and magnetism. For most, their experience is limited to changing batteries, turning on light switches, and using refrigerator magnets—and they hold major misconceptions about all three of these. Surveys of students show that only a small fraction of the men and essentially none of the women have ever engaged in hands-on activities with circuits, meters, motors, or magnets. If you hand each student a battery, a flashlight bulb, and a piece of copper wire and ask them to make the bulb light, only about half of a typical class can do this readily. Quite a few of the students are afraid of being shocked by 1.5 V batteries. Without basic knowledge of the phenomena, students don't know what the *theory* of electric and magnetic fields is attempting to explain.

Students have grown up in a technological society, and it's true that they know many of the *words* of electricity and magnetism. They immediately begin referring to voltage, current, and electrons—giving the impression of knowledge. But it takes only the slightest probing to find that there is no physics understanding associated with the terms. Research has found that student misconceptions and alternative conceptions in electricity and magnetism are at least as widespread and significant, perhaps more so, than in mechanics.

As an example, students at the University of Washington were asked:

> Consider two objects A and B. Object A has a net charge, object B is uncharged. Based on this information, can you tell whether or not either object is a conductor or an insulator? Explain your reasoning.

When given as a pretest in the electricity and magnetism quarter of the calculus-based course, only one-third of the responses were correct. Over 50% said that charged object A *had* to be a conductor. 20% said that uncharged object B *had* to be an insulator.

Follow-up questions found that a significant fraction of students believe that "an insulator cannot be charged." Many of these students, after seeing that glass and plastic rods can be charged, then conclude that glass and plastic are conductors! Part of the difficulty is that students do not differentiate between *charge* and *motion of charge* (i.e., a current). Because current will not flow through an insulator (no *motion* of charge), students erroneously conclude that the insulator cannot be charged.

An especially important phenomenon in electrostatics is the observation that charged objects of either sign attract neutral objects. A pretest at the University of Washington showed a charged rod held near an uncharged metal ball, and students were asked to predict the ball's response.

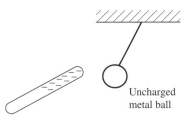

Approximately two-thirds predicted an attraction to the rod—seemingly a good response. But when asked for an explanation, 85% of those who predicted *attraction* gave a wrong answer. The three main categories of explanations were:

■ Charge moves back and forth between the rod and the ball. As one student said, "The rod has the electrons and the ball does not, so the ball tries to get the electrons."

■ The attraction is somehow magnetic.

■ "Neutral" is a third type of charge. Thus "negative" and "neutral" are "opposite charges" and "opposite charges attract."

In a related class demonstration, below, students were shown the attraction between balls A and B after A was charged. When asked to explain, ≈70% stated that ball B somehow acquires a net charge from A.

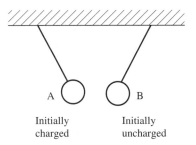

Thus there is a widespread misconception that an *isolated* conductor can become charged due to the nearby presence of a charged object. Many students think the deflection of the leaves of an electroscope implies a *net* charge on the electroscope.

Because the leaves deflect when a charged rod is held near, but not touching, they infer that charge moves from the rod to the electroscope. The charge moves back to the rod as it is withdrawn.

In general, research has revealed the following:

- Many students believe that insulators cannot be charged.
- Students do not distinguish between an object (insulator/conductor) and its charge state (charged/neutral).
- Students do not distinguish between charge and current (motion of charge).
- Students think of charge as an *object* rather than as a property of matter. Charge is a substance that can be "painted on" to matter. We unintentionally reinforce this thinking if we refer to "two charges . . . rather than "two charged objects . . . " or "two charged particles . . . "
- Students think that "neutral" is a third type of charge. The attraction of a neutral object to a charged object is "because opposites attract."
- Students don't recognize charge conservation.
- Most students think there is a fundamental reason that electrons *have* to be negative.
- Many students think that a positively charged object has received an excess of protons. Protons can be as mobile as electrons in ordinary materials.
- Students have little or no understanding of the structure or atomic properties of solids. They don't know what *neutral, non-neutral,* or *charged* mean at the atomic level.
- A large majority of students think that batteries are a source of constant current, delivering the same current to any circuit.
- Many students think that current is "used up" as it flows through a circuit.

These are widely held views, not minority views. Most instructors are not aware of them because they never ask their students to give qualitative explanations of basic phenomena.

The use of pre/post tests shows that these beliefs are little changed by conventional instruction; nearly the same percentages hold these beliefs *after* completing introductory physics. Although conventional laboratories produce little or no change in these beliefs, tutorials and labs that have been carefully designed to confront student misconceptions *do* significantly improve student conceptual understanding.

Moving on to forces and Coulomb's law, second-year physics students at the University of Washington were asked about the forces between two unequal charges. Only one-third correctly identified the forces as an action/reaction pair with equal magnitudes. There is a strong tendency, as shown in part a) of the figure on page 195, to think that the force of the large charge on the small charge will be stronger than the reverse situation. Further studies of students' perception of electric forces found major difficulties with the principle of superposition. Some students employ a "dominance principle" in which the largest of several charges "wins" and the forces due to other charges are neglected. Others, as shown in b) of the figure, believe that one charge can block or shield the force of another charge. In addition to these conceptual difficulties, a significant fraction of students still have trouble with the basic vector addition of superposition.

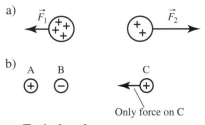

Typical student responses.

Most students don't notice that Coulomb's law applies only to *point* charges, and they will use it indiscriminately for extended objects. Thus they will use Coulomb's law to predict *no* force between a charged sphere and a neutral sphere because $Q = 0$ for the neutral sphere. Many students will make this prediction *after* having seen demonstrations of the attractive polarization force because of their strong belief in the correctness of equations. However, students who have an adequate mental model for charges are less likely to make this error.

Students are comfortable with the idea of force, even if many still have difficulties with Newton's third law. Force is a tangible quantity, and the effects of forces "make sense." The electric field, however, is an abstract, intangible concept. Although we will later claim "reality" for the electric field, at the beginning it is an arbitrary definition. It shouldn't be surprising that students have a very hard time with the field concept. Specific difficulties at the beginning include:

- Students are bothered by the term "field." It seems like it ought to have some significance, but they don't know what.
- Students don't understand what it means to "assign a vector to each point in space."
- Students don't understand the role of the test charge. They don't recognize that the field exists independently of the test charge or that the test charge is merely measuring the field.

If asked to describe the electric field, many students say it is "the area around a charge." Specifically, the area in which the charge has "influence," an area that is usually considered bounded and finite. This is a vague and amorphous idea, not at all the well-defined vector field that a physicist has in mind. In addition, many students believe that the electric field exists *only* at the point of the test charge or *only* at the points along a field line. The ability to draw a more-or-less correct field diagram is not an indication that a student understands what a field *is*.

Students find it particularly hard to understand that a field vector diagram represents a non-spatial vector quantity at discrete *points* in space. There is a strong desire to interpret the vector as a *spatial* quantity that "stretches" on the next page along the length of the vector. Thus the single vector shown in the figure on the next page could represent the field "along" the line of the vector.

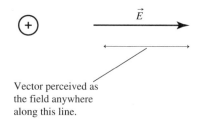

Vector perceived as
the field anywhere
along this line.

If students are asked whether the two figures below represent possible or impossible electric fields, a large majority state that a) is possible but b) impossible. When asked to explain why, they refer to the point where the vectors cross as being a point at which there would be "two fields."

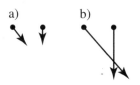

Although electric field vectors are hard for students to interpret, field-line diagrams are even harder. When asked what the field lines represent, a significant fraction of students say it is the "flow of charge." This response is likely connected to the belief that charge moves back and forth between charged objects.

What conclusions can be drawn from these observations?

- Students have little familiarity with, and many misconceptions about, the most basic phenomena of electricity and magnetism—the very phenomena that the *theory* of electricity and magnetism purports to explain.
- Students have no *mental model* for thinking about the behavior of charged objects.
- From the students' perspective, there is no rationale or motive for a theory of fields and potentials, nor are they able to judge whether the theory has any connection to reality.
- At the end of instruction, students cannot use the concepts of field and potential to *reason* about a situation. They may be able to manipulate them mathematically, but it is an exercise in pure mathematics that has no connection to physical reality.
- Exposure to standard lecture demonstrations and to conventional laboratories has little positive effect.

A theory of electric fields is not likely to make any sense if erected upon such flimsy foundations.

Returning to the earlier idea of an educational transformer, physics education research strongly suggests that the elaborate formal theory of electricity and magnetism that we present to students is highly mismatched to where they are in their learning. Without understanding concepts or knowing about basic phenomena, the theory becomes "just math." It is worth noting that the *ideas* of field and potential were introduced by Ampere, Faraday, and others in close conjunction with specific phenomena they were trying to understand. Only after the utility of the ideas was established, and scientists were comfortable using fields and potentials to reason about physical events, did Maxwell establish their formal mathematical properties.

Fields and potentials *are* difficult, but that doesn't mean they shouldn't be taught. The field model of interactions is an essential and important part of our modern understanding of physics, and students should gain some familiarity with it. Research and practical experience suggest that the goals for teaching electricity and magnetism should be:

- To acquire, through hands-on experience and interactive demonstrations, familiarity with the basic phenomena of electricity and magnetism. This requires not just passive observation but active discussion, explanation, and reasoning.
- To develop and use a mental model of charges. The *charge model* is built on empirical evidence about the properties of charges, the electric properties of materials, charging processes, and an atomic-level description of charge. This is a model with explanatory power, and students should be able to use the charge model to *reason* about and explain electric phenomena.
- To understand what fields and potentials are and how they are used. This is the *field model*. In developing the field model, the emphasis needs to be on using superposition, rather than formal symmetry-based statements about the field. The mathematics needs to stay within the comfort level of students, and theoretical constructs need to be connected to experiments and observations at every opportunity.

The pedagogy to implement these goals is threefold:

- Place a strong emphasis on observation and experiment.
- Close the loop: does the theory that's developed *explain* the phenomena?
- Require students to *use* the charge and field models to reason about phenomena and to predict new phenomena.

This approach will leave students well prepared to use electric and magnetic concepts in their careers. Those students who will meet Maxwell's equations in a later course will be fully prepared and will have a firm *physical* foundation on which to build mathematical sophistication.

Student Learning Objectives

- To become familiar with basic electric phenomena.
- To learn the *charge model* and apply it to conductors and insulators.
- To understand polarization and the attraction between neutral and charged objects.
- To understand and use Coulomb's law for point charges.
- To recognize and use the principle of superposition for electric forces.
- To begin the process of understanding the *field model* and the concept of a field.
- To learn the electric field of a point charge.

Pedagogical Approach

A major goal of the initial chapter on electrostatics is to develop, through observations and experiments, a *charge model*. The purpose of such a model is to allow students to reason about electric phenomena in terms of charges, forces, and mobilities before starting to learn the field model. The charge model should explain:

- What charge is.
- How we know about charges and their properties.
- How we know there are two and only two kinds of charge.
- Forces between charges and the superposition of forces.
- What we mean by "neutral," and that neutral does not mean "no charges."
- The charging process and the transfer of charge.
- Insulators and conductors and their basic charge-related properties.
- Charge conservation.
- The relation between charges and current.
- The micro/macro link between atomic-level charges and macroscopic phenomena.

Students cannot *see* charge. What we know about charge, and what we want students to learn about charge, are inferences drawn from a variety of different experiments. As is well known, electrostatic demonstrations are quirky, weather-sensitive, and hard to reproduce. It requires *many* such demonstrations for any patterns to become clear.

Unfortunately, many instructors rush through a small number of demonstrations with inadequate explanation and discussion of *what is happening*. The basic charging of an electroscope reveals nothing to a student who thinks that insulators can't be charged, that charge moves between the rod and electroscope *before* they touch, and who is unaware how charge gets from the knob to the leaves. To make matters more challenging, the most common and familiar electrostatic phenomenon is the attraction of a *neutral* object to a charged object via polarization forces. This exceedingly basic observation is never mentioned in most textbooks or presentations, thus giving the erroneous perception that most of the world consists of oppositely charged objects.

Drawing conclusions about charge from electrostatic observations is not at all straightforward. If students are to understand the basic foundations of the subject of electricity, they need an opportunity to see and think about many different phenomena. These can be presented as

- Interactive lecture demonstrations in which students have to *predict* the outcome of an experiment, then discuss it with neighbors and the instructor, and finally arrive at an understanding and explanation before going on to the next demonstration. These give the instructor an opportunity to confront student misconceptions.
- Carefully structured laboratory experiments or tutorials. Although electrostatic labs are common, most are ineffective because students are just following a list of "things to do." Labs *can* be effective if the activities are well chosen, properly sequenced, and—most important—if they are used to confront student misconceptions and to require students to *explain* their observations using the charge model.

■ Do-at-home experiments using the charging properties of transparent tape, as has been described by Arons (1997) and by Chabay and Sherwood (1999). These can be very effective if given in a worksheet format where students do a few prescribed observations then have to answer questions and write explanations. A drawback of do-at-home experiments is that some students won't take them seriously and others, in the absence of instructor feedback, will go astray and draw the wrong conclusions. Even so, this is a worthwhile activity because it quickly lets students gain practice and familiarity with electric phenomena. Instructors are urged to consult the references for more details.

My approach is to place significant emphasis on understanding how a charged object attracts a neutral object. This is a "puzzle" that catches students' attention when first demonstrated. A related puzzle is charging a conductor by induction. If students can understand these phenomena and give a coherent description, they will have made major progress toward understanding the charge model. In addition, the idea of polarization will reappear when we want to know how a refrigerator magnet sticks to the refrigerator.

Coulomb's law needs to be stated very carefully. A traditional way of writing it has been

$$F = k\frac{q_1 q_2}{r^2}.$$

Unfortunately, this version leads students to think that positive F means a repulsive force and that negative F is an attractive force. In actual use, of course, F is the component of a vector, and the sign of F depends on which way the vector points. This is analogous to the student error of thinking that positive acceleration always means "speeding up."

A common way around this difficulty is to write Coulomb's law as

$$\vec{F}_{12} = k\frac{q_1 q_2}{r^2}\hat{r}.$$

Although this is now a vector equation, there are still major drawbacks. Is this the force of charge 1 on charge 2, or vice versa? Does $\hat{r}$ point from 1 to 2, or vice versa? More significantly, the equation has introduced an asymmetry, suggesting to students that one charge causes the force while the other experiences it. In addition, most students have not yet encountered the unit vector $\hat{r}$ in math, and they find it to be a difficult idea. I've found the use of $\hat{r}$ hinders, rather than helps, students' understanding of Coulomb's law and electric fields.

I prefer to state Coulomb's law as follows:

1. If two point charges q_1 and q_2 are a distance r apart, the charges exert forces on each other of magnitude

$$|\vec{F}_{1\,\text{on}\,2}| = |\vec{F}_{2\,\text{on}\,1}| = k\frac{|q_1||q_2|}{r^2}.$$

These forces are an action/reaction pair, equal in magnitude but opposite in direction.

2. The forces are along the line joining the two charges. The forces are repulsive for two like charges, attractive for two opposite charges.

Although this version is wordier, it has distinct advantages:

- The magnitudes and the directions of the forces are considered separately. The charges are given with absolute value signs so that the magnitude is positive.
- Using $\vec{F}_{1 \text{ on } 2}$ and $\vec{F}_{2 \text{ on } 1}$ makes it explicit that Coulomb's law describes *two* forces, an action/reaction pair of forces between the interacting charges.
- The restriction to point charges is emphasized.

These will help students avoid many of the more common errors associated with Coulomb's law.

The field model is then introduced to provide a mechanism for action at a distance. It's important to let students know we are *defining* the electric field, and we define it in a way that, through experience, has been found to be most useful. In developing the field concept, you want to emphasize that

- We are distinguishing between the *source charges* that cause the field and the *test charge* that experiences or moves in the field of the source charges.
- The field of a point charge exists at *all* points in *all* of space.
- The field is independent of the test charge. The test charge measures or samples the field, but it doesn't cause or create the field. The field is present at a point whether there is a charge there or not.
- The vectors in a field-vector diagram illustrate the electric field at discrete points in space. Each vector on the page is in the direction of $\vec{E}$ *at that point*, and it's length is *proportional* to the magnitude of $\vec{E}$. The vector is associated with just that one point, it does *not* have a spatial length or extent.

I stopped using field-line diagrams many years ago. Instead, for both electric and magnetic fields, I use field-vector diagrams that show field vectors at representative points in space. As was mentioned above, most students don't understand the information in a field-line diagram. The vectors in a field-vector diagram are more tangible and help students to visualize this abstract thing called the electric field. There's evidence from the University of Washington both that students learn electric field concepts better with field-vector diagrams and that students familiar with field-vector diagrams can easily acquire an understanding of field lines if and when the need arises.

Using Class Time

Electrostatics is a very important chapter that establishes the foundations of electricity. It needs a minimum of three days, with the first day on basic phenomena, the second on polarization and Coulomb's law, and the third on superposition and the field model. If available, an extra day is well spent elaborating the charge model and having students use the charge model to reason about different demonstrations and phenomena.

Day 1: Day 1 should be a demonstration-oriented day intended to establish the basic ideas of the charge model. To be effective, demonstrations need to:

- Start with the most basic of issues, such as how you know if an object is charged.
- Build the ideas in a systematic way.
- *Involve the students* actively by asking them to make predictions, give explanations, or draw conclusions.

The goal is both to confront student misconceptions (e.g., insulators can't be charged) and to provide students with evidence for a charge model.

Although there are many ways in which electrostatic demonstrations can be used, an interactive sequence that I have found effective is as follows.

1. Have a roll of plastic wrap or transparent tape. Remind students that quickly pulling a piece of plastic or tape causes it to cling to their hand—an attractive force. What is the nature of this force? Is it a familiar or a new force? Note that this is a long-range force, not a contact force. Unlike gravity, this is a long-range force that is amenable to exploration.
2. Have a student run a comb through his or her hair and then pick up small pieces of paper. (Surprisingly, many students have never seen this demonstration.) You could have everyone in the class try this. Explicitly note that a comb that has not been recently used does *not* pick up the paper, so the "rubbing" is somehow altering the comb. Show that a "rubbed comb" attracts a hanging pith ball but an "unrubbed comb" does not. Don't yet talk about *charges* or about *insulators* or *conductors*.
3. Rub the usual rubber/plastic rods with fur and glass rods with silk to show that they also pick up paper and attract a pith ball. (A plastic rod rubbed with wool acquires the same negative charge as a rubber rod rubbed with fur, but wool on plastic often produces a larger charge, especially on humid days.) You can now *define* the rubbing process as "charging" the objects. Note that this is merely a name, it tells nothing about what is actually happening.
4. Have students, in their own words, answer the question "How can you test whether or not an object is charged?"
5. Now charge two pith balls with the rubber/plastic rod. Test each ball against a neutral ball to prove that each is now charged, then show the repulsive force between them. Repeat using the glass rod. All students are familiar with "static cling," but many have never before seen a repulsive electrostatic force.
6. Touching two balls with the same rod seems to cause a repulsive force between the balls, so what do students predict will happen if you touch one ball with a rubber/plastic rod and the other with a glass rod? Although many will know or suspect attraction, challenge them to give a *reason* for their prediction. Charge the balls and test them to show that they *both* are charged. (This is an important step because you've already shown that a charged object attracts a neutral ball.) Then show the attractive force.
7. Summarize the findings thus far by introducing the idea of *charge* and the existence of positive and negative charge. Comment that "negative" is an arbitrary definition (a rubber rod rubbed with fur is, by definition, negative) and that the

names "positive" and "negative" are arbitrary names. This is surprising to many students who think there is an absolute significance to the terms. Note that simple electrostatic experiments are *consistent* with a two-charge model but do not prove it. (Ben Franklin's single-charge excess/deficit model is also consistent.) This is worth pointing out because students may be puzzled. The subsequent discovery of atomic particles with inherent positive and negative charge is the "missing link."

Have students recognize that *naming* the charges does nothing to reveal what charges *are*. They will assume that a negatively charged object has acquired electrons, perhaps by stripping them from the other object. In fact, frictional charging is mostly due to breaking molecular bonds in the large organic molecules of rubber, fur, and silk. The bonds break in such a way as to create positive and negative molecular ions.

8. The conclusions thus far are:
    i. Rubbing causes some objects to become charged.
    ii. A charged object attracts small neutral objects. This is the "test" for charge.
    iii. Charge can be transferred by contact.
    iv. There are *at least* two different kinds of charge. Like charges repel, opposite charges attract.

These are important pieces of the charge model. It is most beneficial to students if you ask them to draw these conclusions, in discussions with neighbors, rather than you doing it. Remind them that they are "doing science."

9. Before proceeding, solidify their understanding thus far with two questions:
    i. They are handed an object. How can they determine if it is neutral, positive, or negative? (You want to make sure they don't see "attraction to a positive ball" as being an indication for "negative object." Repulsion is the only sure test.)
    ii. Suppose there is a third type of charge. What test would reveal that an object has this third charge? (Two-step test: First verify that it *is* charged by testing against a neutral ball. Then see if it is attracted to *both* positive and negative charges. The second test alone, which some students will propose, doesn't rule out a neutral object.) Few, if any, students will recognize that the existence of *only* two kinds of charge is an experimental question. A third charge type is not logically impossible, but no object has ever been found that passes the two-step test.

10. The next demonstrations should distinguish between conductors and insulators. While there are various possibilities, a straightforward demonstration is to place a rod (metal, glass, or plastic) in contact with a insulated metal sphere, then touch the opposite end of the rod with a charged rod (see figure on next page). You can then show, using pith balls, that the sphere is charged (same as the charged rod) if the intervening rod is metal but not if it is glass or plastic. You can now define a conductor as an object that charge moves through and an insulator as one where charge does not *move*. Since you can *charge* the glass and plastic, an insulator *can* be charged but the charge is "stuck" on the surface and cannot move. You can go ahead and define the flow of charge through the metal rod as a *current*.

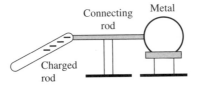

As a follow up, ask where the charge *is* on the metal sphere. You want students to recognize that

   i. Charge was transferred to the sphere by contact.

   ii. Charge is free to move through a metal.

   iii. Like charges repel each other.

Consequently, charge placed on a metal will quickly spread as far apart as possible and will charge the entire metal surface. This is in contrast to the charge placed on an insulator. The time for the charge to spread out on a conductor is "instantaneous" as far as simple experiments are concerned.

11. This is a good point at which to demonstrate polarization of a metal. Place a neutral pith ball very near one end of the metal rod and hold a charged rod near (not touching) the other. Because the pith ball is attracted to the metal, the end of the metal rod near the pith ball must be charged.

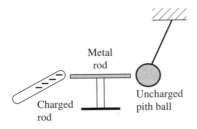

Students need two ideas to understand this. First, that charge is free to move in a conductor. Second, and vitally important, is that a conductor must contain *charge carriers* that are free to move. A neutral rod is thus not a situation of "no charges," but of balanced positive and negative charges, one (or both) of which is free to move. Students should be led to see how these two ideas lead to the conclusion that the metal rod is polarized by the charged rod. You can also show that a plastic rod is not polarized (or at least not sufficiently to affect the pith ball), which is evidence that an insulator does *not* have charge carriers free to move around.

12. *Finally* you can proceed to standard demonstrations with a foil-leaf electroscope. Although typical classes start with an electroscope, it is only after all the preceding demonstrations have helped to clarify many ideas about charges, the charging process, and the properties of insulators and conductors that students can understand an electroscope. You'll want to:

i. Go over the construction of the electroscope, noting particularly which pieces are conductors and which are insulators.

ii. Give the electroscope a net charge by touching it with a charged rod. Ask *why* the leaves repel. What would happen if the post of the electroscope were made of glass?

iii. Hold a charged rod near but not touching, causing the leaves to deflect because of polarization. Ask students, while holding the rod near, if the electroscope has a net charge. Prior to class, most students would have answered "yes"—a documented misconception of students—which means that most would not have understood what they were seeing. Now, however, most will be able to recognize the effects of polarization.

iv. Charge the electroscope negative, then bring both positive and negative rods near and ask students to explain the observations. Bring a positive rod so close that the leaves totally collapse and then spring open again, then ask for an explanation. By now, most students should be able to use the charge model to *reason* about events.

v. Call students' attention to the fact that you've been "discharging" the electroscope by touching it. Ask if they can explain this, then lead into a simple discussion of *grounding*. Students have very strange ideas, nearly all of which are wrong, about what it means to be "grounded."

13. As a final demonstration for day 1, place two metal spheres in contact, hold a charged rod near one (not touching), then separate the spheres. (You might want to test the spheres first to show they are neutral.) After separating the spheres, ask students to predict their charge states, then test them. You've demonstrated charging by induction—a *new* way to charge objects—and students should now be able to give a pretty good explanation. A useful homework problem is to explain how an electroscope is charged by induction while touching it. (You should demonstrate this first in class.)

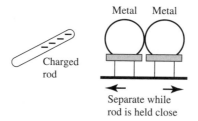

14. Now is a good time to have students finish articulating the basic ideas and concepts of the charge model. Make sure they can tie each idea to a specific piece of evidence. Two good follow-up questions are:

i. How could you give two identical metal spheres *exactly* equal charges?

ii. How could you give two identical metal spheres charges of opposite sign but *exactly* equal magnitude?

Day 2: Day 2 emphasizes forces, both the forces between point charges and the still unexplained force of a charged object on a neutral object. A nice start is to draw a "+1 charge" and a "+5 charge" (no units yet) and ask students to draw the force vectors, with the length proportional to the size of the force. A significant fraction will draw unequal lengths. Rather than answering, appeal to a demonstration. Charge two pith balls unequally, then bring them close together. Have students compare the angles at which they hang, then ask what they can conclude about the two forces. Only after they've seen the evidence and thought about it do you want to bring Newton's third law into the discussion. (A useful review question is to ask them to draw free-body diagrams of the pith balls on the strings. Many will need some review of forces before getting into the issues of charged-particle motion.)

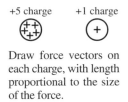

+5 charge   +1 charge

Draw force vectors on
each charge, with length
proportional to the size
of the force.

Then charge two pith balls and have students observe the angles as you change the distance. Although not quantitative, they can conclude that "the electric force weakens as the distance between the charges increases." This is sufficient to understand polarization forces.

Use a charged rod to pick up a small piece of aluminum foil. Ask them to discuss this observation with their neighbors and try to *explain* it based on what they've already learned about charges. The main ideas, of course, are to combine "force decreases with distance" with their day 1 conclusions about how a metal is polarized. Because this is a flat foil rather than a long rod, many will not think about polarization and may need some hints.

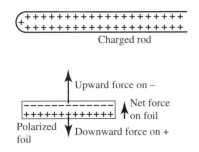

A difficulty for many students is an unstated and unexamined assumption that the top layer of charges "shields" the lower layer of charges from the force of the charged rod. Even after recognizing the polarization, they will show only the force on the upper charge layer. You'll need to convince them that the rod exerts forces on *all* the charges. As a follow-up question, ask what the foil will do if the charged rod has the opposite sign.

The attraction of a metal foil can be understood on the basis of macroscopic polarization, but you finally have to get into atomic issues to understand the dipole polarization of an insulator. At this time, a "simple atom" picture of an induced dipole moment is sufficient to explain how polarization forces act on an insulator. Students have made significant progress if they can give a reasonably coherent explanation of how a charged object attracts a neutral object, and they are now ready for the quantitative theory of electrical interactions.

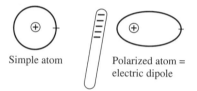

Simple atom   Polarized atom =
                     electric dipole

Coulomb's law is straightforward if stated correctly, although you will want to stress that it applies only to *point charges*. The charges of typical electrostatic experiments are in the 1 to 100 nC range, so these are appropriate charges to use in examples. A simple but useful exercise is for students to compare the electric force between two 10 nC pith balls 3 cm apart with the gravitational force on a pith ball with 10 mg mass. You want students to realize that electric forces are usually larger than gravitational forces.

Students have difficulty with the superposition of electric forces, both qualitatively and especially quantitatively. A useful exercise is to draw a group of charges and ask students to draw both individual force vectors and the net force vector on a specified charge. For example, the figure below shows four situations in which students are to draw the individual and net force vectors on charge B. After doing these qualitatively, it's good practice to find the net force $\vec{F}_B$ quantitatively for at least one arrangement. Ask students to find the magnitude and the direction of $\vec{F}_B$. Most instructors are surprised, at this point in the course, how poorly most students do on this task.

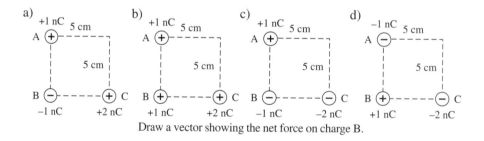

Draw a vector showing the net force on charge B.

If time permits, have students find $\vec{F}_B$ when four charges are arranged at the corners of a rectangle. Because students still find vector addition hard, and because superposition is such a critical idea for finding the electric field of continuous charge distributions, it's worth spending time in recitation, lab, or class to go over several of these examples.

Day 3: The field concept is introduced as the means by which two charges interact without contact. The initial idea has two aspects:

- The *source charges* somehow alter the space around them. This is the electric field.
- *Test charges* in the field experience a force *from the field.*

This is essentially Faraday's idea of fields. It is important to tell students that this is a *hypothesis* about electrical interactions. We have no evidence at the moment that the electric field actually exists. We will develop evidence as we go along.

The field is an abstract idea, and even a simple demonstration can help it seem more real to students. Charge a pith ball, then clamp the charged rod in place (the source charge) while you move the pith ball (the test charge) around the rod. Point out that the test charge experiences a force at *every* point in space, although in practice the force may get too weak to detect if the distance is too large. While holding the pith ball close, so that the force is fairly large, reiterate the idea that the pith ball is simply responding to the electric field of the rod. Remove the pith ball, then point to the location where the pith ball had been. Is there now anything at that point in space? You want students to understand that

- The field exists at *all* points in space, even though diagrams may show only a few illustrative vectors.
- The field is present whether the test charge is there or not. The test charge measures the field, but it does not cause the field.
- The field at each point in space is a *vector* because it causes a test charge to experience a force in a particular direction.

As an exercise, tell students that a positive charge $+q$ experiences the three forces shown if placed at points 1, 2, and 3 (see figure below).

- Have them draw the three points, then draw the electric field vectors $\vec{E}_1$, $\vec{E}_2$, and $\vec{E}_3$ at the points.
- After drawing the field vectors on the board, point to a position in the middle of one of the vectors (right on the vector) and ask, "Is there an electric field at this point? If so, can you tell what direction it points?" Because many students think that the field vector is a *spatial* quantity stretching through space, they will respond that the direction is the same as the vector you are pointing to. In this case, of course, there's not enough information to determine the field at any other point.
- Ask students to draw the force vectors for a charge $-q$ at point 1 and $+2q$ at point 3.

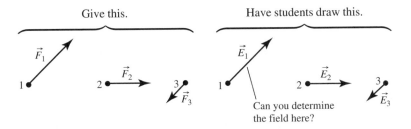

As a follow-up question on the issue of what the picture of an electric field vector means, draw the two pictures shown below. Ask if either or both of these is a *possible* electric field. If not, what is wrong with it? Many students will answer that the picture on the right is *not* possible because there would be two fields at the point where the vectors cross. It takes quite a few specific examples for them to understand that the vectors, which are drawn as spatial intervals, describe the fields only at the *points* A and B.

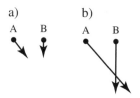

If time permits, it's useful to return to the issue of how a charged rod exerts an attractive force on a neutral piece of metal, this time using the field perspective. First note that the rod creates an electric field in the metal, thus exerting forces on the electrons and causing them to move in the direction *opposite* the field. (The amount of motion and the fact that $\vec{E}_{net} = 0$ in equilibrium will be taken up later, but you can introduce these ideas now if there's time.) Once the metal is polarized, the electric field of the charged rod exerts forces of slightly different strengths, and opposite directions, on the positive charges and the negative charges. Thus the attractive force.

Sample Reading Quiz or Discussion Questions

1. What is the SI unit of charge?

2. A charge alters the space around it. What is this alteration of space called?

3. An electroscope has been given a negative charge, and the leaves are standing apart. A positively charged rod is brought close to the electroscope knob, but it doesn't touch. Describe what happens to the leaves. Your description should include both i) an explanation in words and ii) pictures that show the charges.

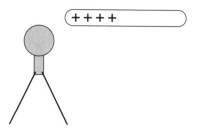

4. Could this picture represent electric field vectors? If so, show how you could arrange source charges to create them. If not, explain why not.

5. If a negatively charged rod is held near a neutral metal ball, the ball is attracted to the rod. This happens
   a. because of magnetic effects.
   b. because the ball tries to pull the rod's electrons over to it.
   c. because the rod polarizes the metal.
   d. because the rod and the ball have opposite charges.

6. The electric field of a charge is defined by the force on a(n)
   a. electron.   c. source charge.
   b. proton.   d. test charge.

Sample Exam Questions

1. Two neutral metal spheres on insulating stands are placed in contact. A negatively charged rod is then brought directly over the top of the left sphere, as shown, but does not touch either sphere. While the rod is held still, the right sphere is moved so that the spheres no longer touch. Then the charged rod is withdrawn. After the rod is withdrawn, what is the charge state of each sphere? Use both words *and* pictures to explain your answer.

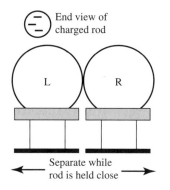

2. A positive point charge is brought near a permanent electric dipole that is oriented as shown. Describe *and explain* how the dipole responds.

3. A positively charged glass rod is held over a thin, uncharged piece of metal foil. How does the foil respond? Give a step-by-step explanation, using both words *and* pictures.

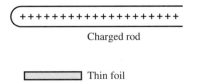

Charged rod

Thin foil

4. What is the force on the +1 nC charge? Give your answer as a magnitude and direction.

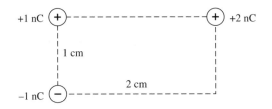

Electric Fields

Background Information

The electrostatics chapter introduced the basic ideas of the field model. This chapter will now develop the quantitative analysis of electric fields and of the motion of charged particles in the field. The goal is not for students to become experts at calculating fields, although they should be able to do calculations for simple charge distributions, but for them to recognize and use common electric fields, such as the field due to a line of charge or the field of a plane of charge.

Most textbooks do a couple of simple field calculations, then introduce Gauss's law. As the previous chapter noted, research has found that only a very small fraction of students are able to understand and use Gauss's law. This seems to be due to several factors:

- Few students have reached vector integrals in calculus. Even though we may do only "simple cases," they are far from simple to the students.
- Gauss's law requires reasoning on the basis of symmetry. As important as this is, it requires a high level of conceptual ability. Few students have ever used this kind of reasoning, and few will master it in the short time available.
- Students have misconceptions about electrostatics and difficulty understanding the basic ideas of the field model. These difficulties are only compounded when they are overlaid with new and very abstract mathematics.

For these reasons, I prefer to base all electric field calculations directly upon the principle of superposition. Superposition is a straightforward and readily understood procedure. It helps students with their conceptual difficulties by explicitly linking all field calculations to Coulomb's law. And with one exception, all the major fields of interest can easily be found using superposition. The exception is the field of a sphere of charge, for which the integrations—although doable—are too complex for many introductory classes. However, it is quite believable, without a rigorous proof, that the exterior field of a sphere of charge is the same as that of a point charge. We routinely make this same assertion without a proof for Newton's law of gravity.

Although line and surface integrals may be unfamiliar, students have been doing calculus long enough that nearly all can carry out a basic integration. However, students are still very inexperienced at *using* calculus to analyze a problem. Their difficulty with electric fields—and it is a major difficulty for most students—is not with doing integrals but with knowing *what to integrate*. Students are completely unfamiliar with the procedure of dividing a charge distribution into small pieces, writing the superposition of fields as a sum, then converting the sum to an integral by using the charge density. The emphasis in this chapter needs to be on the analysis procedure that leads up to an integration.

Students also have a hard time visualizing fields in three dimensions. The fields of interest in this course usually have an axis of symmetry, so you need to encourage students to imagine the two-dimensional drawings of the textbook, or that you draw on the board, as being rotated about that axis. Several software packages give three-dimensional renderings of fields, including tilts and rotations, and these can be useful classroom demonstrations.

It is important to explain clearly what the standard two-dimensional drawings are meant to represent. For example, a parallel-plate capacitor is usually drawn as shown in the figure. Many students don't recognize, unless you point it out, that this drawing represents two *planes* of charge that extend above and below the plane of the drawing. These students will interpret the drawing literally as a two-dimensional field due to two short line segments of charge.

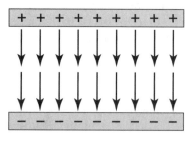

The parallel-plate capacitor encounters another common student misconception: that one layer of charge "blocks" the electric field due to the other layer. The conclusion that the field outside the capacitor is zero relies on the superposition of the fields of two planes of charge, but many students don't recognize that the field of the negative plane is present in the space above the positive plane.

Student Learning Objectives

- To use the principle of superposition to calculate the electric field of multiple point charges and of continuous distributions of charge.
- To learn the electric fields of common charge distributions.
- To study the motion of charged particles and dipoles in simple electric fields.

Pedagogical Approach

The electric field chapter typically derives the fields of several important and common charge distributions:

- A dipole.
- A line segment of charge.
- An infinitely long charged wire.
- A plane of charge.
- A parallel-plate capacitor.
- A sphere of charge.

With the exception of the sphere these are all easily derived by application of the principle of superposition, thus emphasizing an important property of electric fields.

For instructors who feel it is important to introduce Gauss's law, the following approach will give students the best chance of success with this topic:

- First calculate the fields of a line of charge and a plane of charge via superposition.
- Discuss symmetry. Note how the symmetry of the charges gives you information about the field.
- Introduce the concept of flux and the mathematics of surface integrals very slowly and carefully, keeping in mind that this is quite new to most students.
- Finally, rederive the field of lines and planes with Gauss's law. Be very explicit about where and how symmetry arguments are being used.

This slower introduction of Gauss's law will give students more confidence in their ability to understand and use it.

Electric field calculations are more lecture oriented than demonstration oriented. Even so, it is to the students' benefit if you guide them through a few field calculations, letting them do much of the work, rather than doing the entire calculation yourself. The difficult part for students is not carrying out the integrations, but knowing *what* to integrate and setting up the integrals. The following explicit strategy can guide them through this process. Be sure to stop after each step to verify that the students are with you.

Strategy for Calculating the Electric Field of a Continuous Charge Distribution
- Draw a picture and establish a coordinate system.
- Divide the total charge Q into small regularly shaped pieces of charge Δq, using shapes for which you already know how to determine $\vec{E}$. This is often, but not always, a division into small pieces that can be treated as point charges.
- Look for any possible symmetry of the charge distribution. A properly chosen coordinate system or a clever "pairing" of the Δq may allow you to conclude that some components of $\vec{E}$ are zero. Setting up and evaluating integrals can be a lot of work, with many chances for error, so you want to avoid doing all that work only to get a result of zero.
- Use superposition to form an algebraic expression for each of the three components of $\vec{E}$ (unless you are sure one or more is zero) at a point in the field. Recall that $\vec{E}_{net} = \sum \vec{E}_i$ is really three separate equations, one for each component.

- Replace the small charge Δq with an equivalent expression involving a charge density and the small geometric quantity that describes the shape of charge Δq. This is an essential step in making the transition from a sum to an integral because you will need a coordinate to serve as the integration variable.
- Let the sum become an integral. The integration will be over the coordinate variable that is related to Δq by the charge density. All angles and distances must be expressed in terms of the integration variable. Think carefully about the integration limits for this variable; they will depend on the coordinate system you have chosen to use. Then carry out the integration and any subsequent simplification of the result.
- Check that your result is consistent with any limits (such as $z \to \infty$) for which you know what the field should be.

You'll need to be explicit with many of the steps that seem obvious to you. For example, relating the charge Δq to a segment length Δx and charge density λ via $\Delta q = \lambda \Delta x$ is a troublesome step for students, and you'll see them trying to integrate with dq. Another difficulty is expressing all quantities in terms of the integration variable, such as replacing $\cos \theta$ with $y/(y^2 + r^2)^{1/2}$. These are the kinds of "details" that students get hung up on.

Most textbooks use the binomial approximation to look at solutions in various limits. Students who started calculus concurrently with physics may not have seen the binomial expansion in calculus. Even if they have, students may not recognize how to apply the binomial approximation in practice. They are further confused because calculus uses simple expressions such as $(1 + x)^{-2}$ whereas we want to apply the approximation to expressions like $(y + \frac{1}{2}s)^{-2}$ when $y \gg s$. Students need to see *all* the steps in the reasoning before they can use these techniques in their own work.

This is also the first time for many students to see the idea of *limiting cases,* such as checking to see if the field of a charged rod looks like that of a point charge as $r \to \infty$. This turns out to be a difficult idea for many, so you need to explain not only *what* you are doing but also *why.*

The motion of charged particles is a good opportunity to spiral back to mechanics. Many students have forgotten what they previously learned about projectile and circular motion, but this is a good opportunity for further practice. At this point you want to consider only projectile motion in uniform fields and circular motion around point charges or charged wires. The motion of charges in more complex fields will be considered from an energy conservation perspective after the electric potential is introduced. Be sure to include examples with negative charges, as many students don't automatically distinguish between positive and negative.

To further the discussion of polarization forces, you want students to recognize that a dipole will experience a torque and will rotate until it aligns with the field, but it has a net force only in a non-uniform field. These ideas will be utilized again in magnetism.

Using Class Time

Two days are marginally adequate for calculating electric fields via superposition, although three days are preferable. Gauss's law, if you include it, takes at least another full day.

It's important to introduce superposition with discrete point charges, appealing to what students know about the superposition of forces and to the definition of the electric field. Charges arranged on the corners of rectangles make nice examples, as shown in the figure below. Although this is straightforward vector addition, students have *far* more difficulty with these calculations than you might expect. You may want to break it into several steps, asking them to draw the field of each charge, then graphically determine the net field, then decide which vector components are positive and which negative, and only then start numerical computations.

Find the electric field at A.
Find the acceleration of an electron at point A.

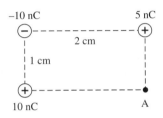

Students can usually read about the dipole field in the text, but it's worth a summary in class of what the dipole field looks like. This is a good initial opportunity to start talking about symmetry. You may want to use class time to discuss how the binomial approximation is used to simplify the field expression when the dipole spacing is very small.

Continuous charge distributions introduce the concepts of linear charge density λ and surface charge density σ. (Note that the symbol σ is used both for surface charge density and, in a couple of chapters, to represent conductivity. I prefer to use η for surface charge density, to avoid confusion.) Many students have a hard time understanding linear and surface densities. A useful exercise is:

Charge Q is spread uniformly to make a rectangle of sides a and b.
a. What is the surface charge density?
b. The original rectangle, identified as number 1, is broken into the smaller rectangles 2 and 3. Compare the surface charge densities σ_1, σ_2, and σ_3.

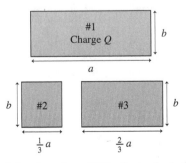

McDermott's group found that less than 50% of a typical class correctly recognized that $\sigma_1 = \sigma_2 = \sigma_3$.

The field in the midplane of a finite line charge is so important that it is worth going through the derivation in class—an unusual recommendation for this guidebook! You should do this on an interactive basis with the students, using the strategy given above and asking students to contribute the major ideas at each step. At the end, when you want the field of an infinite wire, many students are puzzled by the idea of letting the line become infinitely long without changing the linear charge density, so this needs extra emphasis.

Alternatively, if you think that most of your class understood the line-of-charge derivation in the text, use class time to find the field a distance r from one tip of the wire. This is a more challenging calculation since E_x is not zero.

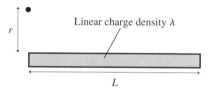

A nice motion problem to end this calculation is to ask what speed an electron would need to orbit at a radius of a thin 1 cm wire having a linear charge density of 10 nC/m.

Most textbooks use Gauss's law, rather than superposition, to obtain the field of a plane of charge. You'll need to supply the derivation in class if you're following the approach recommended here. Although not technically difficult once you've found the field of an infinite wire, it does require care to portray the three-dimensional aspects. Unless you have superior artistic skills, you may want to make a transparency in advance that shows the coordinate system, the division of the plane into parallel wires, etc.

Many students are unwilling to believe that the field of a plane can be independent of distance from the plane. You need to remind them that we're considering an *infinite* plane, so it spans your entire field of view no matter how far away you get. No real plane is infinite, so the field of a real plane *will* decrease with distance. Even so, the infinite plane is a good approximation to a real plane at points that are much closer to the surface than to an edge.

Our primary interest in the field of a plane is to be able to put two planes together to form a parallel-plate capacitor. As noted above, many students think that one plane of charge will somehow "block" the field of the other plane and prevent the field from passing through. You'll need to convince students that the fields of both planes extend throughout space. Then ask them to do the superposition that leads to the conclusion $E_{outside} = 0$ and $E_{inside} = \sigma/\varepsilon_0$. You'll want to emphasize the importance of capacitors as being a practical way to generate a *uniform* electric field—the electrical equivalent of flat-earth gravity.

Because many students aren't exactly sure what a uniform field is, a good exercise is to ask them to rank order the electric field strengths at the four points labeled 1–4 in the following figure.

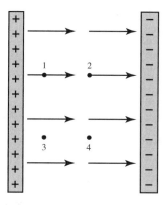

There are three possible misconceptions that lead to mistakes:

- The field gets weaker moving away from the positive plate, so $E_1 > E_2$.
- The field is stronger at 2 than at 1 because there's a full arrow leaving point 2 but only a half arrow leaving point 1.
- There are no field at points 3 and 4 because they aren't on field vectors.

Students need to realize that the field is the *same* at *all* points inside the capacitor.

Capacitors lead naturally into problems about charged particle motion. As a warm-up, draw a positive charge in the center of a capacitor, then ask students to sketch the trajectory if the particle's initial velocity is zero, is straight down, or is to the right.

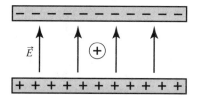

For a numerical example, consider two 2 cm × 2 cm square plates that are charged with ±0.03 nC. The plates are balanced on an insulating table and are 6 mm apart. A 0.1-mm-diameter charged water droplet that is missing 5×10^5 electrons is released from rest 2 cm above the table and halfway between the plates. Where does the droplet hit the table?

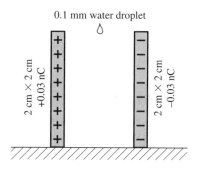

As yet another example, consider three (or more) dipoles between the plates of a capacitor. For each dipole, have students say whether it

a. does not move,
b. moves to the right,
c. moves to the left,

d. rotates clockwise,
e. rotates counterclockwise, or
f. some combination of these.

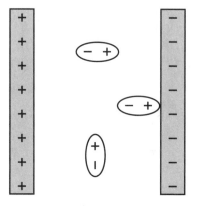

You'll want to end with the exterior field of a sphere of charge. Unless you're using Gauss's law, you'll simply need to present the result. Unlike the field of wires and planes, which couldn't have been guessed in advance, students have no trouble accepting that the field of a sphere is the same as the field of a point charge. Without Gauss's law you can't find the interior field of a uniformly charged sphere. However, that result has no practical value in the introductory course.

Sample Reading Quiz or Discussion Questions

1. What device provides a practical way to produce a uniform electric field?
2. For charged particles, what is the quantity q/m called?
3. Which of these charge distributions did *not* have its electric field calculated in this chapter?
   a. A line of charge.
   b. A ring of charge.
   c. A plane of charge.

   d. A parallel-plate capacitor.
   e. A sphere of charge.
   f. They were *all* calculated.
4. The worked examples of charged-particle motion are relevant to
   a. a transistor.
   b. magnetic resonance imaging.
   c. holography.

   d. a cathode ray tube.
   e. cosmic rays.
   f. lasers.

Sample Exam Questions

1. Two unequal negatively charged rings are placed on the *x*-axis. Draw *on the figure* the electric field vectors at the three points identified by dots. The lengths of your vectors

should indicate the relative field strengths at each of these points. (Assume that you've derived the on-axis field of a ring either in class or as homework.)

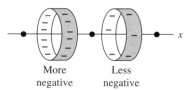

More negative   Less negative

2. Charged particles of -2 nC and $+8$ nC are located at $x = -2$ cm and $x = +2$ cm, respectively. At what point or points on the x-axis is the electric field $\vec{E} = 0$?

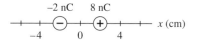

-2 nC   8 nC

x (cm)

3. a. What is the electric field at the point indicated by a dot? Give your answer in component form.

   b. What is the acceleration of an electron at that point? Give your answer as a magnitude and direction.

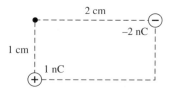

2 cm

-2 nC

1 cm

1 nC

4. A 100 mg pith ball hangs by a thread in the center of a parallel plate capacitor. The capacitor plates are 5 cm × 5 cm and are spaced 1 cm apart. When the ball is charged to $+10$ nC and the capacitor plates are charge to $\pm Q$, the ball hangs at a 3° angle. What is Q?

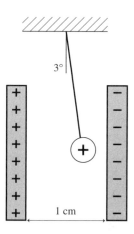

3°

1 cm

5. A parallel plate capacitor is formed of two 10 cm × 10 cm plates spaced 1 cm apart. The plates are charged to ±1 nC. An electron is shot through a very small hole in the positive plate. What is the slowest speed the electron can have if it is to reach the negative plate?

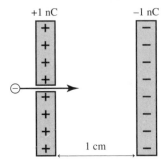

6. A very long sheet of charge has width L and surface charge density σ. The sheet lies in the xy-plane between $x = -L/2$ and $x = L/2$. What is the electric field $\vec{E}$ at a point in the xy-plane that is outside the sheet of charge? That is, at position x, where $x > L/2$. Give your answer in component form.

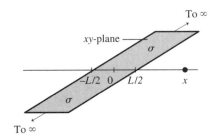

16

The Electric Potential

Background Information

Of all the concepts in introductory physics, the electric potential is the hardest to grasp for most students. It is a very abstract idea, far removed from most phenomena with which they are familiar. There's good evidence that most students leave introductory physics with almost no knowledge of the electric potential or how it is used.

Students do recognize the term *voltage* because of its common use. However, they associate voltage with vague and imprecise notions about the "flow of electricity." Most students are unfamiliar with the terms *potential* and *potential difference*, and even those who recognize the words have essentially no knowledge of the concepts to which they refer.

There is a strong tendency of students to confuse the electric potential with either the electric field or with the electric potential energy—they are liable to use the three interchangeably. After stressing the importance of vectors for electric field calculations, and giving many practice examples using vector superposition, I've been surprised to find students attempting to find the "components" of the electric potential. The fact that the expressions for the field and the potential of a point charge look so similar likely contributes to students' misunderstanding.

The primary investigation of student understanding of the basic concept of electric potential (apart from the application of the concept to circuits) is unpublished work at the University of Washington. Students there received a conventional introduction to the electric potential. They also had a laboratory where they used a voltmeter to measure potential differences and find electric fields on weakly conducting carbon paper with painted-on conducting electrodes. Student understanding was examined during the laboratory, a week after the laboratory and lectures, and on the final exam.

There was strong evidence that most students did not understand the principles of the laboratory and gained little or no knowledge of electric potential from the experience. The majority of students didn't know what the voltmeter measured. When questioned, they stated that it measured "the flow of charge" or "the difference of charge" between two points. When asked questions about the electric field, nearly all students answered by referring to the electrodes on the carbon paper and to the positive and negative terminals of the power supplies, rather than to any measurements they were making with the voltmeter. A week after the lab, only 10% of students could describe how to use a voltmeter to find the electric field at a point.

On the final exam, students were shown equipotential drawings, such as the one below, and asked to compare the electric field strengths at points A and B and at C and D. They were also asked about the field direction at different points. Only about 25% of students were generally successful with these questions. Roughly two-thirds of students thought that the field strengths would be the same at two points on an equipotential, such as points A and B. Only 35% correctly found $E_C < E_D$, with 30% stating explicitly that the field is zero at point D on the 0 V equipotential. Both responses show a confusion between field and potential. Only 50% could determine the field direction at a point such as B.

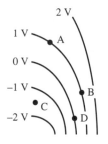

In general, research has found that students

- Have great difficulty using energy conservation in problems involving the motion of charged particles.
- Don't acquire a conceptual model of potential or potential difference.
- Are unable to relate the electric potential to the electric field.
- Rarely invoke potential or potential difference when asked to explain a phenomenon.
- Are later unable to use the idea of potential difference in circuits. In particular, students view batteries as constant current sources rather than as sources of potential difference.

Standard numerical problems that use the electric potential do little to build a conceptual understanding of what potential is all about. Students need ample opportunities to work with potential graphs and equipotential surfaces if they are to understand the links between the electric field and the electric potential.

Student Learning Objectives

- To introduce electric potential energy and use it in conservation of energy problems.
- To define the electric potential.

- To find and use the electric potential of point charges, charged spheres, and parallel-plate capacitors.
- To find the electric potential of a continuous distribution of charge.
- To establish the relationship between $\vec{E}$ and V.
- To introduce and use potential graphs and equipotential surfaces.
- To learn the properties of a conductor in electrostatic equilibrium.
- To find the connection between charge and potential difference for a capacitor.
- To introduce batteries as a practical source of potential difference.

Pedagogical Approach

There are several alternatives for introducing the electric potential: through its connection with the electric field; as the work per unit charge when the charge moves in a field; or as the electric potential energy per unit charge. I prefer the third alternative, since I've already invested significant effort to teach the concept of energy. Electric potential energy is introduced first, and students get to practice using conservation of energy for charged particles. Then the electric potential difference is defined as $\Delta V = \Delta U/q_{\text{test}}$, in analogy with the earlier $\vec{E} = \vec{F}/q_{\text{test}}$. Thus the potential and the field, which are properties of the source charges, are distinguished from the potential energy of and force on a test charge *in* the field.

Using energy conservation is still difficult for many students. You'll want to work through several examples of charged particle motion near point charges or in parallel-plate capacitors, having the students assist with some steps. Students need to identify the "before" and "after" situations, as they learned previously, and to articulate whether kinetic/potential energy is increasing or decreasing as the particle moves.

Textbooks usually introduce the *electron volt* at this time as a new unit of energy. Many instructors are aware that students don't recognize any difference between *volts* and *electron volts*, and are likely to use them interchangeably. There is already enough confusion between potential and potential difference, no need to add more. Consequently, I defer introducing electron volts until we get to modern physics.

As you move from potential energy to potential, it is important to emphasize that

- the field and the potential exist *throughout* space, regardless of whether a test charge is present or not. They are properties of the source charges.
- the field and the potential are not separate, independent ideas. The link between them will be established later in the chapter, but students should recognize early that the two "go hand in hand" just like force and potential energy—two different perspectives of a *single* physical situation.

Because of the nearly identical names and symbols, students are easily confused between potential and potential energy—and many use V and U interchangeably. It is worth devoting several class exercises to having students articulate, in the context of a specific example, the distinction between potential and potential energy.

Another source of confusion for many students is the unfortunate fact that we use the symbol V to represent the electric potential and, at the same time, use V as an abbreviation for the unit "volt." It's not uncommon to hear a student say "The volts at this point . . ." While we may overlook such statements as merely being imprecise, they often reveal conceptual difficulties. Careful use of language can help to clarify the situation.

It is, of course, critically important to distinguish between *potential* and *potential difference*. Unfortunately, all too many textbooks use the symbol V when they actually mean a potential difference ΔV. Examples are Ohm's law $(V = IR)$ and the capacitor equation $(Q = CV)$. Writing equations this way reinforces students' misunderstanding and encourages mistakes.

The electric potentials of most interest are those of a point charge, a charged sphere, and a parallel-plate capacitor. Finding the electric potential of an arbitrary charge distribution is a lower priority. Most textbooks do solve for the potential of a charged rod and a charged disk, but the results are not utilized. These derivations could easily be omitted. It is preferable to focus on the basic concept of the electric potential and its use in common situations involving point charges and capacitors.

An important task is to establish the general relationship between the electric potential and the electric field. Students need help and practice with

- Reading two-dimensional and three-dimensional graphs of the potential. The analogy with topographic maps is helpful, but a surprising number of students are not familiar with and cannot read a topographic map.
- Understanding equipotential lines and surfaces.
- Recognizing that $\vec{E}$ points "downhill" on the potential graph and that the field strength E is the slope of the "potential hill."
- Recognizing that $\vec{E}$ is always perpendicular to equipotential surfaces.
- Understanding that $\Delta V_{\text{closed loop}} = 0$.

These are best approached with a wide variety of graphical and qualitative exercises.

If students at your school start calculus concurrently with physics, there's a good chance that they will not have reached multivariable calculus and partial derivatives. (It's worth getting a calculus syllabus from the math department to find out.) If your students don't yet know partial derivatives, there's little harm in stating the relationship of potential to field as $E_x = -dV/dx$ rather than $-\partial V/\partial x$. None of the applications in introductory physics depend on understanding the distinction, and students who encounter the potential in later courses will immediately recognize that $\vec{E} = -\nabla V$ is the proper statement.

The reverse relationship $\Delta V = -\int E_s ds$ is a line integral. However, students never have to evaluate the integral, so this "detail" can be glossed over. They only need to understand two ideas based on this integral:

- If $\vec{E} = 0$ at all points, such as inside a conductor in electrostatic equilibrium, then the potential difference between any two points is $\Delta V = 0$.
- If two points are on an equipotential, so that $\Delta V = 0$, then either regions of positive and negative E_s happen to cancel when doing the integral, or E_s, the component of $\vec{E}$

along the equipotential, is zero at all points. Because this is true for *any* two points, it must be that the component E_s along the line is zero. Thus $\vec{E}$ is perpendicular to the equipotential surface. This is another subtle argument, but an important one.

Many schools have students perform a laboratory experiment to find equipotentials and measure electric fields by using weakly conducting carbon paper and voltmeters. The University of Washington research cited above finds that this experiment, as commonly used, does not help students to understand the relationship between field and potential, and it can even reinforce misconceptions. The experiment assumes that the students understand

- what a voltmeter is and what it measures, and
- what ΔV is and how it can be used to determine $\vec{E}$.

Both assumptions turn out to be false for a majority of students.

The University of Washington has found *some* improvement in student outcomes when this laboratory is preceded by exercises in which

- Students relate voltmeter readings to the work needed to move a charge.
- Students use two voltmeter probes with a *fixed* 1 cm spacing to measure ΔV in small increments along multiple paths between two fixed points. They compare these to each other and to the overall ΔV as measured with two independent probes.
- Students measure ΔV for different points on a 1-cm-radius circle about a fixed point and answer questions about the implications of their measurements.

The main experiment is then more effective if the power supply leads and the conducting electrodes are hidden under cardboard. Otherwise, students answer questions about the field based on observing the electrode positions and their connections to the positive/negative terminals of the power supply rather than by referring to the voltmeter readings.

The relationship between the electric field and potential can be used to establish that a conductor in electrostatic equilibrium is an equipotential, that $\vec{E}_{in} = 0$, and that $\vec{E}_{out}$ is perpendicular to the surface. The argument that $\vec{E}_{in} = 0$ is subtle and difficult for many students to follow because it is based on what *doesn't* happen—because the charge carriers are free to move but *aren't* moving (no currents), there must not be an electric field. It will be necessary to repeat many times that these results apply only to conductors in electrostatic equilibrium, *not* to current-carrying conductors.

An important idea to establish is that a potential difference is *created* when positive and negative charge are separated. This is what happens in a capacitor, for which I prefer to emphasize the potential *difference* by writing $Q = C\Delta V$ instead of the more traditional $Q = CV$. But the charge separation (and potential difference) of a capacitor can't be maintained if the capacitor is "used." If the capacitor plates are connected by a wire, charges from the positive plate "fall downhill" through the wire to the negative plate. This discharges the capacitor. If we wanted to maintain the potential difference and the current, we would need some means to "lift" the charges back to the positive plate, giving them a new supply of potential energy.

Thus batteries can be introduced as being a "charge escalator" that moves charge from the negative terminal to the positive terminal. This is a *non-electric* means of creating a charge separation. Whether the charge escalator functions due to chemical reactions (batteries) or due to mechanical forces on the charges (generators), they do *work* to "lift" the charge back to the positive terminal. The battery's emf $\mathcal{E}$ can be defined as "the work per unit charge to create a charge separation and thus cause a potential difference."

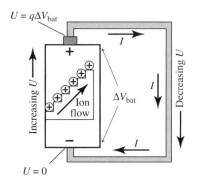

A battery is a "charge escalator" doing work to lift charges to the opposite terminal, thus creating a potential difference.

Students find it easy to visualize the charge-escalator model. It can be used to establish three major ideas about batteries:

■ Batteries are a source of potential difference because they have an internal energy supply that is used to separate charge, thus giving potential energy to the charges.
■ The potential difference is a *fixed* quantity because the escalator lifts all charges the same height.
■ The battery can supply *varying* amounts of current by running the escalator faster or slower, thus lifting more or less charge per second. Regardless of *how much* charge is lifted, it is all lifted the same height. Thus the potential difference doesn't change as the current varies.

This is particularly important for countering the strong student tendency to think that batteries are sources of constant current.

Using Class Time

The electric potential requires a minimum of three days, preferably four.

Many students will need to start with a review of energy conservation. It remains a difficult idea for them, despite all the earlier attention, but this is a good opportunity to reinforce the main ideas. Examples should explicitly define "before" and "after," articulate and interpret the energy transformations, and use potential energy graphs. Unlike gravitational energy problems, where the potential energy always increases "up," charge problems can have arbitrary orientations as well as charges of both signs. Examples should be as different as possible.

Before getting into numerical examples, it's good to ask students to determine whether the change in potential energy ΔU is positive, negative, or zero as a particle moves from an initial position i to a final position f. The figure below shows some possible situations. Your examples should ask about ΔU for protons in some cases, electrons in others, and for "hydrogen atoms" in yet others. It's not obvious to many students that the electric potential energy of a neutral object, like a hydrogen atom, is zero.

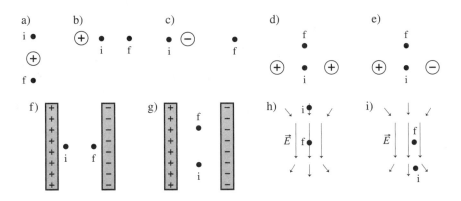

Is ΔU of the particle positive, negative, or zero as the particle moves from i to f?

A nice example that can lead from a qualitative explanation into a numerical example is to consider a small charged particle shot toward a larger stationary charge.

First ask students to decide whether the particle speeds up or slows down, basing an explanation first on the concept of force, then on the concept of electric field, and finally on the concept of energy. You can ask them to graph the potential energy (and maybe also the force) as a function of separation r. Also ask whether or not the particle has constant acceleration. Repeat these questions if the large charge is negative rather than positive. Finally, pose a problem such as:

A 2-mm-diameter plastic bead is charged to -1 nC.
a. An alpha particle (remind them what an alpha particle is) is fired at the bead from far away with a speed of 1×10^6 m/s, and it collides head-on. What is the impact speed?

b. An electron is fired at the bead from far away. It "reflects," with a turning point 0.1 mm from the surface of the bead. What was the electron's initial speed?

Conventional examples of a charged particle moving inside a capacitor are appropriate. These problems should be explored graphically as well as numerically. It's good to have at least one example that involves a turning point.

The electric potential energy of charged particles leads easily into the idea of an *electric potential*. My approach is to make a strong analogy with $\vec{E} = \vec{F}/q$ by defining $V = U/q$, emphasizing that both $\vec{E}$ and V are properties of the source charges and that $\vec{E}$ and V exist throughout space whether a test charge is present or not. This also allows me to emphasize immediately that the field and the potential are two perspectives of the same situation and that one important goal of the chapter is to find the connection between $\vec{E}$ and V.

Qualitative and conceptual examples are worthwhile before doing any computations. For example, ask students to rank order the potentials at the points indicated by dots in each of the figures below. For the capacitor, you might want to first identify the negative plate as $V = 0$, then ask if their answer would change if the positive plate were chosen as $V = 0$. The capacitor and the field vector situations offer an opportunity to emphasize that the potential always decreases along the direction of the electric field.

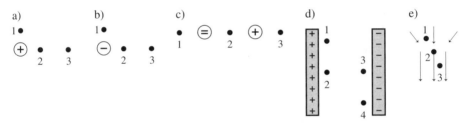

Rank order the electric potentials at the numbered points.

If you used the figure on the previous page to ask students about ΔU for a charged particle, you can return to the same figure and now ask about ΔV. This is a good exercise to emphasize that the potential exists because of the source charges and is *independent* of the test charge. It's easy to improvise more such exercises if students are having difficulty.

A simple numerical exercise makes important points:

For the figure shown, find

a. The potential at points a and b.
b. The potential difference between a and b.

c. The potential energy of a proton at a and b.

d. The speed at point b of a proton that was moving to the right at point a with a speed of 4×10^5 m/s.

e. The speed at point a of a proton that was moving to the left at point b with a speed of 4×10^5 m/s.

Also ask students if they can answer the question, "What is the *potential* of a proton 2 cm from the charge?" Many will say "Yes" and will compute the *potential energy*.

As a follow up, show the following figure and say that a proton moves along a trajectory that passes through points c and d. The proton's speed at point c is 4×10^5 m/s. What is its speed at point d? The issue, of course, is whether students recognize that they've already solved this problem.

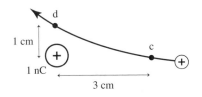

The graph below shows a final exercise that can both reinforce earlier ideas about energy diagrams below and prepare students to think about the graphical relationship between potential and field. First ask them to draw the potential energy diagram for a -20 nC charge that moves in this potential. Then tell them that the charged particle is shot from the right with an initial kinetic energy of 1 μJ.

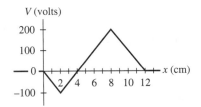

■ Where is the point of maximum speed?

■ What is the particle's kinetic energy at that point?

■ Where is the turning point?

■ What is the force on the particle at the turning point?

If time permits, you might want to calculate V for a continuous charge distribution— for example, calculating the on-axis potential of a charged disk. But this is a lower priority than point charges and capacitors.

The next task is to establish the connection between the electric potential and the electric field. The initial examples should be primarily graphical, with minimal computation. Starting in one dimension, you can provide a graph of V-versus-x, such as the example shown below, and ask for a graph of E_x-versus-x. Then give a graph of E_x-versus-x and ask for a graph of V-versus-x. These exercises are, of course, reminiscent of kinematic graph exercises as well as graphical exercises using potential energy curves. You want students to understand that E_x is the negative of the slope of a potential graph and that V is the negative of the area under a field graph.

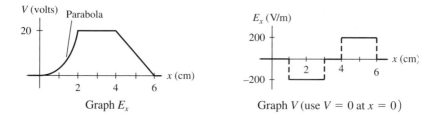

Graph E_x Graph V (use $V = 0$ at $x = 0$)

An interesting follow-up question is to ask students how they could *create* the electric field shown in the graph on the right. (Back-to-back capacitors.)

Equipotential surfaces can be introduced by considering a point charge of value $q = 5/9$ nC. This value conveniently gives a potential of 500 V at $r = 1$ cm. Ask the students to

- Determine the values of r at which the potential is 500, 400, 300, 200, and 100 volts.
- Graph V-versus-x along an x-axis passing through the charge (including both positive and negative values of x) and note the hyperbolic shapes.
- Use the slope to determine if E_x points toward or away from the charge and how the field strength changes with distance.
- Draw a potential map showing the five equipotential lines. Call their attention to the fact that each ΔV is the same, just like the contours on a contour map.
- Draw electric field vectors on the map. Make sure they have the directions right and that the vectors get shorter at larger distances.
- Use the information *on the map* to estimate the field strength at a point where $V = 400$ V. Don't tell them how to go about it, but you can provide hints about $-\Delta V/\Delta x$ as needed.
- Finally, ask them to compare their estimate to a calculated field strength.

As a follow up, draw a set of elliptic equipotentials with several points labeled.

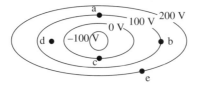

Ask them to:

- Describe the shape of the potential surface in a three-dimensional graph (*V*-versus-*xy*).
- Compare the field strengths E_a and E_b. Are they equal, or is one larger than the other? Explain.
- Compare the field strengths E_c and E_d. Are they equal, or is one larger than the other? Explain.
- Draw the electric field vectors at points a-e.

As noted in the last chapter, students have a tendency to think that field strengths are equal at two points on the same equipotential and that $\vec{E} = 0$ at points on a 0 V contour.

To go the reverse direction, from field to potential, draw a *uniform* electric field in an arbitrary tilted direction and label it 200 N/C (note the units, *not* V/m). First ask whether point 1 or point 2 is at the higher potential. Then ask for a *value* of ΔV_{12}. (They'll have to recognize that 200 N/C = 200 V/m.) Finally, ask them to draw a series of equipotential surfaces spaced every 5 V. This example has obvious implications for the upcoming discussion of field and potentials in current-carrying wires.

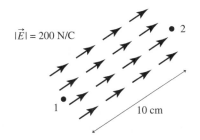

As a final example, draw a square (conveniently chosen to have a diagonal of 1 cm) and give the value of the potential on each of the corners. Ask students to estimate the electric field $\vec{E}$, both magnitude and direction, at the center of the square.

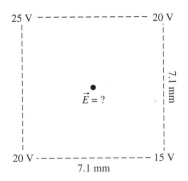

An important implication of these ideas is that a conductor in electrostatic equilibrium is an equipotential. It has zero interior electric field and an external field perpendicular to the surface. An interesting exercise is to ask students to consider two metal spheres of different diameters connected by a long, thin wire. If $Q_1 = 2$ nC, what is Q_2? Most students find this difficult and don't know how to proceed. If you point out the wire, their first inclination is to think that the charge spreads out until $Q_1 = Q_2$. You can lead them through a series of questions: Is there a current flowing, or is this electrostatic equilibrium? Is this a single conductor or two independent conductors? What do you know about conductors in electrostatic equilibrium? What do you know about the potential at the surface of a sphere?

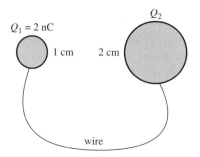

A final task for the chapter is to learn about *sources* of a potential difference — capacitors and batteries. Capacitors are a "chicken and egg" issue. The standard approach is to define the capacitance of two conductors as $C = Q/\Delta V$, where ΔV is a potential difference between them. (Actually, most textbooks use just Q/V, maintaining the confusion between potential and potential difference.) But where did the potential difference come from? To students, this seems an arbitrary definition.

I think a better approach is to imagine moving charge from one conductor to another. This will create an electric field between the conductors and thus, based on what students have learned in this chapter, a potential difference. It's the separation of charge that causes the potential difference, not vice versa. It's not hard to show that ΔV is proportional to the amount of charge moved: $\Delta V = kQ$. If, for practical reasons, we define the capacitance as $1/k$, then $\Delta V = Q/C$. Capacitance is not an arbitrary definition; it has a physical significance in terms of the proportionality between charge separation Q and potential difference ΔV. It's then straightforward to note that capacitance depends only on the *geometry* of the conductors and to derive the standard expressions for planar or spherical capacitors.

The concept of capacitance is important, and simple examples with parallel-plate capacitors are worthwhile. Series and parallel combinations of capacitors and dielectrics are of little significance in an introductory course and can easily be omitted if you're pressed for time.

A capacitor is a source of potential difference only as long as it remains charged. It's easy to show that a capacitor "runs down" as it is used. If you charge a high-capacitance $(C > 1$ F$)$ capacitor to a potential difference of a few volts, you can run a light bulb or motor for a few seconds as the capacitor discharges. A voltmeter can show the decreasing potential difference.

A practical source of potential difference needs some way to *maintain* the charge separation. And this, of course, is a battery. The "charge escalator" described earlier provides students with a concrete *model* of how a battery works. More important, it is a model that shows why a battery is a source of constant potential difference—all charges are lifted the same "height"—rather than a source of constant current.

The charge escalator also provides a *mechanism* for charging a capacitor. Textbooks consider it obvious that a capacitor attached to a battery has the same potential difference, but this isn't at all obvious to many students. The reasoning steps you want them to understand are

- The battery's charge escalator moves charge from one capacitor plate to the other, creating a charge separation and thus a potential difference between the capacitor plates.
- The motion of charge from one plate *through the battery* to the other plate is a current. But this current is very brief—less than a nanosecond. The capacitor plates and the connecting wires very quickly reach a state of electrostatic equilibrium.
- A battery terminal, the connecting wire, and one capacitor plate form a single conductor in electrostatic equilibrium. Consequently, they are an equipotential. Stated another way, each capacitor plate quickly reaches the same potential as the battery terminal it is connected to.
- Thus the potential difference across the capacitor is the same as that of the battery.

You can explore these issues by showing two "rods" that have been connected to a battery for "a long time." First ask students the values of ΔV_{12}, ΔV_{34}, and ΔV_{23}. This is a nontrivial question for many students. Be sure to point out that $\Delta V_{\text{loop}} = 0$. (This is good preparation for upcoming circuit questions that students have difficulty with.) Then ask them to sketch equipotential surfaces (spaced every 1 V) and field vectors in the region between the conductors.

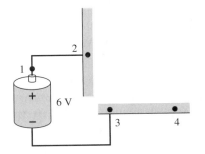

Finally, consider two capacitor plates connected to a battery (see figure on next page). First ask about ΔV_{12}, ΔV_{23}, and ΔV_{34} after the battery has been attached for a long time. Then ask for the *value* of the electric field strength E_5 at point 5. Not all students will yet recognize that E_5 is simply $(3 \text{ V})/(1 \text{ mm}) = 3000 \text{ V/m}$. Next, say that insulating handles will be used to pull the plates slightly apart while the battery remains attached. Do ΔV_{cap}, Q, σ, C, and E_5 increase, decrease, or stay the same? Can they justify their responses? Repeat, but this time with the battery disconnected before the plates are separated.

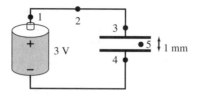

Sample Reading Quiz or Discussion Questions

1. What are the units of *potential difference?*
2. What quantity is represented by the symbol $\mathcal{E}$?
3. What is the SI unit of capacitance?
4. A parallel-plate capacitor has the negative plate at $x = 0$ and the positive plate at $x = 1$ mm. The capacitor is charged to 10 V. Draw a V-versus-x graph of the potential inside the capacitor.
5. New units of the electric field were introduced in this chapter. They are:
   a. V/C.   d. J/m^2.
   b. N/C.   e. Ω/m.
   c. V/m.   f. J/C.
6. The electric potential inside a capacitor
   a. is constant.
   b. increases linearly from the negative to the positive plate.
   c. decreases linearly from the negative to the positive plate.
   d. decreases inversely with distance from the negative plate.
   e. decreases inversely with the square of the distance from the negative plate.
7. The electric field
   a. is always perpendicular to an equipotential surface.
   b. is always tangent to an equipotential surface.
   c. always bisects an equipotential surface.
   d. makes an angle to an equipotential surface that depends on the amount of charge.

Sample Exam Questions

1. A parallel-plate capacitor consisting of two 2 cm × 2 cm square plates is connected to a battery. There is a vacuum between the plates. An electron released from rest at the surface of the negative plate crosses the capacitor and strikes the positive plate with a speed of 1.78×10^6 m/s.
   a. What is the emf of the battery?
   b. Can you determine the electric field inside the capacitor? If so, what is it? If not, why not?

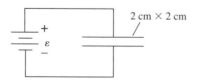

2. a. What is the electric potential at the center of the 1 cm square?
   b. What is the potential energy of an electron at that point?

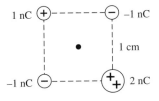

3. The figure shows the electric potential at the corners of a 1 cm square. What is the electric field—both magnitude and direction—at the center of the square?

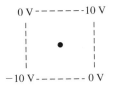

4. The graph below shows the x-component of the electric field as a function of position on the x-axis. If the electric potential at the origin is 3.0 V, what is the electric potential at $x = 3$ cm?

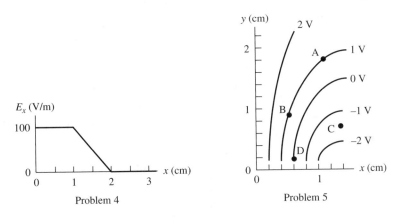

Problem 4   Problem 5

5. The figure shows a series of equipotential curves.
   a. Is the electric field strength at point A larger, smaller, or equal to the field strength at point B? Explain your reasoning.
   b. Is the electric field strength at point C larger, smaller, or equal to the field strength at point D? Explain your reasoning.
   c. From the information on the figure, determine the electric field $\vec{E}_D$ at point D.

6. An alpha particle ($q = +2e$, $m = 4$ u) is shot toward a uranium nucleus ($q = +92e$) from far away. The alpha particle reaches a closest distance of 10 fm before being pushed back. What was the initial speed of the alpha particle? (The uranium nucleus can be assumed not to move since it is so much more massive than the alpha particle.)

Current and Conductivity

Background Information

The electric field and electric potential chapters contained a high density of new ideas and new mathematical techniques. That information can now be put to use. The goal of this chapter, which is less technical, is to apply the charge model and field model to an important topic—current flow—before introducing the more practical issues of circuits.

Students all know the *word* "current," but most don't really know what a current is.

- Many students think that current is "used up" when passing through devices such as light bulbs. In this instance, they are confusing current flow with energy transfer.
- Students know that current has something to do with electrons, but most students:
  Think that electrons are the charge carriers for *any and all* kinds of currents.
  Think it is somehow "obvious" that electrons flow through wires.
  Don't distinguish the conventional current I from the electron flow rate.
- Most students do not understand the conditions under which current flows.

Consequently, much of this chapter is focused on developing a concrete model of current flow.

Student Learning Objectives

- To use the charge model and the field model to develop a concrete model of current flow through conductors.
- To examine the evidence by which we know that current in a metal is due to the motion of electrons.

- To develop a micro/macro connection between the motion of charge carriers and the conventional macroscopic current.
- To introduce conductivity and resistivity as important parameters describing the electrical properties of materials.
- To introduce resistance and Ohm's law.

Pedagogical Approach

I like to introduce current as an application of the charge and field models. The potential isn't needed until it's time for Ohm's law. This approach keeps the analysis of current tied closely to the tangible concepts of charges and forces.

An important issue in electrostatics was the *evidence* by which we know about charges and their properties. The emphasis on evidence needs to continue here with the questions

- How do we know that something "flows" when we say that there is a current?
- How do we know *what* it is that flows?

As noted above, students think there's some obvious reason (although they're not sure what it is) why we say that current is due to the motion of electrons. If you light a bulb with a battery and ask students if there's any *evidence* that something flows from the battery to the bulb, many will answer "Yes" and will become quite defensive if you challenge them to justify their response.

Consequently, it is important to establish the evidence for our model of current. The discharge of a capacitor is a good starting point because it can be understood on the basis of the charge model. Essentially all students will now understand that you can charge a capacitor by moving charge from one plate to the other. They have also seen evidence that charge can move through a conductor. If you connect two capacitor plates together with a wire, the Coulomb force will cause the charges to move through the wire until the capacitor is discharged. (As simple as this idea seems, it is important to have the students explicate each step in detail.)

If you charge a high-capacitance capacitor ($C > 1$ farad) with a few volts, you can run a light bulb or a small motor for several seconds as the capacitor discharges. Discharging it through a coil of wire will deflect a compass needle. If you then switch to a battery, you can again light a bulb, run a motor, and deflect a compass needle. Only this time the process is continuous, rather than running down after a few seconds. The logic you want to emphasize is

- The wire that discharges a capacitor can light a bulb, run a motor, and deflect a compass.
- The wire attached to a battery can do the same.
- Thus there is evidence that the *same physical process* is happening inside the wire in both cases.

- The wire that discharges the capacitor has charge moving through it—a current.
- Therefore, the wires that make up a "circuit" have charge flowing through them, and this is what we *mean* when we say that "a current is flowing in the wire."

This may seem needlessly elementary, but it turns out to be quite important. Without the demonstrations and the logic, many students see no connection at all between the current in a circuit and the charges they've studied for the last few chapters. "Current" simply becomes a term, and students form no mental image of what a current *is* or how it flows.

These demonstrations provide evidence that current is a flow of charge, but they give no hint about the nature of the charge carriers. Of course, this was the situation throughout the nineteenth century, when current was defined as the flow of positive charge. Although J. J. Thompson's discovery of the electron was suggestive of electrons as charge carriers, the first direct evidence about the charge carriers in metals didn't appear until the Tolman-Stewart experiment of 1916. Tolman and Stewart gave a sudden and large acceleration to a metal rod. They anticipated that inertia would shift the charge carriers to the rear of the rod and create a potential difference between the ends. Tolman and Stewart found that the rear surface of the accelerating rod was negative, indicating negative charge carriers. It's worth relating this ingenious experiment to your class as you discuss the issue of, "How do we know that electrons are the charge carriers?"

Once the evidence is in place that electrons are the charge carriers in metals, it's easy to develop a microscopic Drude model of the electron current and to link the rate of electron flow to the conventional definition of current. There's no need to dwell on the details, but there are three major points you want students to understand:

- Electron-ion collisions cause the electrons to experience "friction" as they move through the metal. This is rather like pushing a block across a table at constant speed. The electrons need a steady force from an electric field in order to move at constant average speed through the metal.
- The drift velocity is proportional to the field strength. There is no current if $E = 0$, and a larger field causes a larger drift velocity and thus a larger current.
- The drift velocity is surprisingly small. Large currents can flow through wires not because of the high average speed of the electrons but because the number of electrons is so large.

You'll need to emphasize that a current-carrying wire is *not* in electrostatic equilibrium. An electric field in the wire is not only allowed, it is essential.

The Drude model leads directly to $J = \sigma E$. I like to call this the "physicist's version of Ohm's law." It is true for any conductor, regardless of its shape or dimensions. An electric field of 10 V/m will cause the same current density in any piece of copper in the universe. Note that current density, like densities in general, does pose conceptual problems for some students. You'll want several examples relating current density J to actual current I.

Using Class Time

A very important demonstration, described in the Pedagogical Approach section, is to show that current really is the flow of charge through a wire. The large-capacitance ($C > 1$ farad) capacitors that are now readily available make this approach worthwhile. You can tell students that this is a high-tech version of a parallel plate capacitor with two "plates" that can be charged to $+Q$ and $-Q$. Since few students know about units of capacitance or about RC time constants, the exceptionally long time constant means nothing to them at this point and so there's no need to call attention to it.

You can use the capacitor to light a bulb, run a motor, or deflect a compass needle for a few seconds. From the charge model, students should accept that charge flows through the wire until the capacitor is discharged, then stops. Thus you can use "lights a bulb, runs a motor, and deflects a compass needle" as an *operational definition* of "current due to the motion (flow) of charge." If you now do the same things with a battery-powered circuit, you have established convincing evidence that charge is moving through the wires. This is what we *mean* by "a current." A battery is apparently a device that can *sustain* the flow of charge.

The flow of water through pipes is a useful and easily visualized analogy for current. A demonstration with a pump and turbine is nice, but otherwise just draw a picture and say that a battery is to charge what a pump is to water. Note that simply seeing the turbine spin, like seeing a motor spin, is no *proof* that anything is flowing through the pipes. They would have to conduct experiments to ascertain that a material substance is flowing.

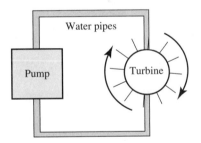

This is also a good time to note the micro/macro connection for water flow. While it is usually easiest to think of water as a liquid flowing through the pipes, they know that the flow of water is "really" the motion of vast numbers of individual water molecules. Likewise, it is usually easiest to think of current as the liquid-like flow of charge—and this is how current was understood in the nineteenth century. But students know that charge is associated with atomic-level charged particles, so it must be that current is "really" the motion of charged particles.

What are the charged particles—the *charge carriers*—that form a current? Students are sure that they are electrons, so it's worth a class discussion on two questions:

- Are electrons *always* the charge carriers in all currents?
- *How do we know* what the charge carriers are? What's the evidence?

You want students to recognize that these are not questions with easy or obvious answers. Although this chapter focuses on currents in metals, students should be aware that currents in liquids or in gas discharges may have ions as the charge carriers.

An important idea for students to understand is that $\vec{E}_{net} = 0$ inside a conductor in electrostatic equilibrium but *not* in a current-carrying conductor. A potential difference between the ends of a wire, from a capacitor or a battery, creates a *net* electric field inside the wire, and this field is the *cause* of the charged particle motion that we call a current. This is an important aspect of the mental model we want students to acquire about current.

I like to define the "electron current" $\Delta N_{elec}/\Delta t$ before introducing the conventional current I because the electron current provides a more concrete image. A good demonstration is simply water flowing out of a rubber tube. You can ask the question, "How many water molecules flow out of the tube every second?" It is worth going through the short derivation that $\Delta N_{water}/\Delta t = nvA$. If water is incompressible, then water must be flowing *through* the tube at the same rate. Most students find this easy to grasp, so it is a good way to establish the electron current $\Delta N_{elec}/\Delta t = nv_dA$ as the rate at which electrons flow through a wire with drift velocity v_d. It is then a simple step to conventional current I as the rate at which *charge* flows, but now students have a better image of what that means.

A nice demonstration to start day 2 is to have three light bulbs hooked up in a row, but with the switch open so that the bulbs are off. First examine a separate battery and light bulb, noting that the bulb has two terminals and that the current flows *through* the bulb to make it light. (A surprising number of students *don't* know this!) Don't say anything about the current being conserved, just note that it goes through. Also note that the brightness of a bulb depends on the *amount* of current, and perhaps demonstrate this.

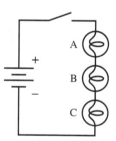

Now turn to the three bulbs. First, show the wiring and draw a picture, so that students clearly understand how the bulbs are connected. Then point out that, when the switch is closed, bulb A will be connected to the positive side of the battery and bulb C to the negative side. Ask students to *predict* the brightness of the bulbs, from most bright to least bright. Students who think that current is "used up" by the bulbs will predict A > B > C. A few, now that they "know" that current is really the flow of negative electrons, will predict C > B > A. Only 50–60% of students in a typical class correctly predict A = B = C, although the fraction will be higher if you let them discuss it with their neighbors before making a prediction.

Conservation of current is such an important idea that its worth a follow-up demonstration to show that the current measured on either side of a light bulb is the same. This demonstration also shows the size of typical currents—a few tenths of an amp for a 3 V battery and a flashlight bulb. (Note that batteries are preferred to power supplies for these demonstrations. Students know that batteries have a positive and negative terminal, even if they understand little else about batteries, whereas a power supply is truly a "magic black box.")

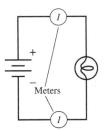

Current density, like density ideas in general, turns out to be a fairly hard concept for most students to grasp. They tend to use current and current density interchangeably. Once again, water flow can provide a concrete analogy. A possible exercise is:

Pipe A has a 1 cm² cross section and carries 2 gal/min.
Pipe B has a 3 cm³ cross section and carries 6 gal/min.
Pipe C has a 5 cm² cross section and carries 8 gal/min.
Compare the currents of these three pipes. Compare the current densities of the three pipes.

You can follow up with a similar exercise using wires and amperes. Two examples are shown in the figure.

Wires 1 and 2 are made of the same materials. Wire 2 has twice the radius and one-third the electric field of wire 1. Compare I_1 and I_2. Compare J_1 and J_2.

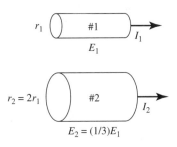

A wire consist of two segments, a and b, of different diameters. Compare I_a and I_b. Compare J_a and J_b.

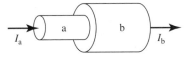

A nice demonstration of conductivity is to have several wires of different materials but the same dimensions. You can show that different currents flow when each of the wires is connected to the same battery. That is, different metals under *identical* circumstances have different currents.

It is then an easy step to introduce resistance and Ohm's law by considering the current flow I through a particular wire, of known dimensions, when it is attached to a battery of potential difference ΔV. I prefer to write Ohm's law as $I = \Delta V/R$ because this, like $a = F/m$, conveys a better sense of cause and effect. Regardless of whether you use this version or the conventional $\Delta V = IR$, I encourage you to use an explicit ΔV for potential difference, not V.

You want students to start reasoning on the basis of Ohm's law before starting into the chapter on circuits. Here's a couple of useful examples:

Two wires of equal length and identical metals but different diameters are connected to identical batteries. Compare currents I_1 and I_2.

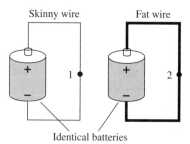

Here's where students' tendency to see batteries as constant-current sources will enter, with many wanting to claim $I_1 = I_2$. You can give an analysis based on resistance, you can talk about the charge escalator running at different speeds but always to the same height, and you can use the analogy of more water flowing through larger pipes. The more analogies you can use, the more likely students are to begin seeing batteries as voltage sources.

A wire connected to a battery has sections of different diameters.

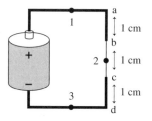

a. Compare the currents I_1, I_2, and I_3 at points 1, 2, and 3.
b. Compare the current densities J_1, J_2, and J_3 at points 1, 2, and 3.
c. Compare potential differences ΔV_{ab}, ΔV_{bc}, and ΔV_{cd}.

You need not dwell on these exercises. Students will see similar issues in the chapter on circuits, so for now you just want to introduce the basic ideas.

The major points you want students to leave this chapter with are

- A battery is a source of emf.
- A battery causes a potential difference along a wire and an electric field inside the wire.
- The charge carriers in a metal wire are electrons. The electric field causes the electrons to move through the wire with a steady drift speed.
- The current density is $J = \sigma E$. It depends on both the strength of the field and on the electrical properties of the material.
- The current itself is $I = JA = \Delta V/R$.
- Current is conserved.

Sample Reading Quiz or Discussion Questions

1. What quantity is represented by the symbol J?
2. What is meant by the *drift speed* of an electron?
3. The electron drift speed in a typical current-carrying wire is
   a. extremely slow ($\approx 10^{-4}$ m/s). c. very fast ($\approx 10^4$ m/s).
   b. moderate (≈ 1 m/s). d. could be any of a, b, or c.
4. Current will be larger in a wire that has a larger value of
   a. conductivity. d. net charge.
   b. resistivity. e. potential.
   c. the coefficient of current.

Sample Exam Questions

1. How do we *know* that a current in a metal wire is really the flow of electrons? Summarize the experimental evidence by which we know that i) *something* is moving through the wire, and ii) the *something* is electrons.

2. A conducting wire is connected between the terminals of a battery. Write a short essay explaining *how* and *why* a current flows in the wire. Your explanation should discuss the role of
   - The emf of the battery.
   - The conductivity of the wire.
   - The electric field inside the wire.
   - The relationship between current and potential difference.
   - Any other ideas that are necessary for understanding how current flows in a circuit.

3. A 2-mm-diameter copper wire and a 1-mm-diameter aluminum wire are joined into a single wire. A current flows through the wire. The electric field strength in the copper section of the wire is $E_{Cu} = 0.010$ V/m. What is the electric field strength E_{Al} in the aluminum section of the wire?
   Data: $\sigma_{Cu} = 6.0 \times 10^7 \ \Omega^{-1}\text{m}^{-1}$ $\sigma_{Al} = 3.5 \times 10^7 \ \Omega^{-1}\text{m}^{-1}$

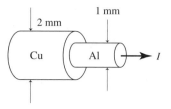

4. The figure shows the equipotentials along a wire that has conductivity $\sigma = 1.0 \times 10^5 \ \Omega^{-1}\text{m}^{-1}$.

  a. What is the electric field $\vec{E}$ at the point indicated with a dot?

  b. How much current is flowing in the wire? In which direction?

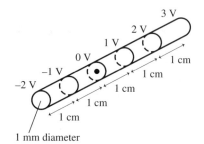

DC Circuits

Background Information

The Physics Education Research chapter of this guidebook gave an extensive summary of research findings about student understanding of circuits. From that research, we learn that

- Students do not differentiate between the concepts of current, voltage, energy, and power. To them, it's all just "electricity." The situation is analogous to students' use of the term "motion" to describe either velocity or acceleration.
- Students think almost exclusively about current, rarely or never about potential difference. Thus the majority of students, even after seeing experimental evidence that suggests otherwise, continue to believe that batteries are constant-current sources.
- Students cannot reason with the concept of potential difference, and they rarely invoke potential difference spontaneously. When potential differences arise, it is almost invariably in the context of Ohm's law: $\Delta V = IR$. In these instances, students see a potential difference as an *effect* rather than a cause. Students do not use $\Delta V_{\text{loop}} = 0$ to reason about circuits, and many have trouble following an argument that does so. Their understanding of potential difference is not so much a *misconception* as *no* conception.
- Students reason locally, not globally. In most circumstances, they do not see that changing one circuit component will affect the voltage and current at other points in the circuit.
- Students have no micro/macro understanding of circuits. They do not see any connection between macroscopic quantities, such as current or resistance, and their previous study of charges, forces, and fields. To students, circuits are a subject entirely independent of electrostatics.

These conclusions are perhaps not surprising. Surveys at the University of Washington have found that less than 20% of students report even rudimentary experience with battery-and-bulb types of circuits. I've confirmed this in surveys of my own students, and also found that the percentage of women students with any previous circuit experience is very close to zero. What *is* surprising is that students show essentially no improvement in their conceptual understanding of circuits following conventional instruction.

Many of these students *can* successfully apply Kirchhoff's laws to the analysis of circuits. This is an algorithmic procedure, and the ability to follow this procedure apparently does not imply that students understand the physical concepts or that they can *reason* about circuits. Indeed, studies have found that students, when asked a question such as "What happens if I remove this light bulb?" immediately begin trying to apply formulas rather than reasoning.

In addition to the findings described in the Physics Education Research chapter, other research has found the following:

- Students generally have no trouble with series and parallel resistance when just two resistors are involved. However, unusual circuit arrangements can cause confusion. For example, some students fail to recognize that R_1 and R_2 in the circuit below are in parallel. Apparently they consider the definition to refer to a geometric rather than an electric relationship. More serious difficulties appear when there are more than two resistors. For example, many students consider R_3 and R_4 to be in series.

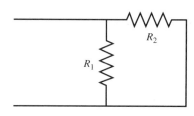
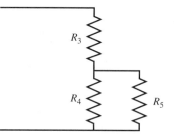

- Students interpret schematic diagrams literally. They find it difficult to relate a schematic diagram to the layout of an actual circuit. Even simple rearrangements of the diagram can cause confusion. For example, many students perceive the two diagrams below to represent different circuits.

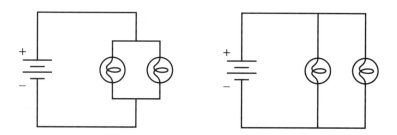

- For the most part, students do not understand how meters function, what their properties are, or how they're used in a circuit.
- Most students do not know what it means for a light bulb to be a "100 watt bulb." The majority interpret this to mean that the bulb *always* dissipates 100 W of power.
- Even though they use the term freely, almost no students know what *grounding* a circuit means or why it is done.

Student Learning Objectives

- To develop and use a conceptual model of simple DC circuits.
- To understand series and parallel resistances.
- To apply Kirchhoff's laws to the analysis of circuits.
- To understand energy transfer and power dissipation in circuits.
- To understand how and why circuits are grounded.

Pedagogical Approach

The primary goal is to establish the *physics* of circuits. Hence I place more emphasis on conceptual understanding and physical processes, less on applications or on the detailed analysis of circuits. Most science and engineering students will take a later circuits class where they can learn the details of circuit analysis. Their most urgent need is a conceptual model for *thinking* about basic circuit properties.

It is difficult with a classroom presentation to overcome students' lack of familiarity with simple circuits and to convey the basic ideas of a conceptual model. Hands-on experience at a student-controlled pace is essential. Although most schools have one or two "circuits laboratories," research has found that conventional measurement-oriented labs provide essentially no improvement in a student's conceptual understanding of circuits. Standard laboratories, despite how simple they appear to the instructor, are mismatched to the conceptual level where most students are starting.

It is far more useful to devote two or three lab periods to hands-on experiences and questions of the type described in Evans (1978) or Shaffer and McDermott (1992), preferably starting *before* circuits are introduced in lecture. These are "guided discovery" experiments designed to help students acquire a mental model of basic circuit processes.

If such lab experiences are not possible, an alternative is to spend an entire class period working through similar demonstrations in an interactive mode, asking students to predict, explain, and reason. Such an approach adds an extra day to the chapter. Whether it is hands-on experience or interactive demonstrations, it is vital that students have an opportunity to see and to reason about circuits qualitatively, thus building conceptual understanding, *prior* to beginning any quantitative analysis.

To compensate for students' lack of experience and familiarity, I start my presentation at a more basic level than most textbooks. I call out explicitly many ideas that usually go unsaid in textbooks because they seem too "obvious."

I place a strong emphasis on conservation of current and on Kirchhoff's loop law. These are the primary reasoning tools for quantitative analysis, and you want to reinforce wherever possible the fact that you're really using conservation of charge and conservation of energy. The issue of signs when using the loop law is very confusing to students, so you need to discuss this very explicitly when working examples. They need to learn that

- The potential *decreases* in the direction that current I flows through a resistor. You can relate this to the fact that current flows in the direction of $\vec{E}$ and that $\vec{E}$ points in the direction of decreasing potential.
- The ΔV used in the loop law is $V_{downstream} - V_{upstream}$. Thus ΔV of a resistor is *negative* when the loop law is applied in the direction of current flow.

Most textbooks don't mention power dissipation until late in the chapter, but there's a strong argument for introducing power early. Students will more readily abandon the common "current is used up" misconception if they begin to see that it is *energy* that is transferred from the battery to the resistor and "used up" by being converted to thermal energy. Note that an earlier definition of power as the rate of energy transfer, rather than the rate of doing work, now makes it much easier to discuss energy transfer and power dissipation in circuits.

Class examples of series and parallel resistors need to include unusual geometries since, as noted above, students tend to think of the geometric relationship between the resistors rather than the electric relationship. Short circuits present a major difficulty, with most students failing to see that a wire is a 0 Ω resistor. For example, the figure shows a series/parallel reduction problem with a shorting wire. Most students ignore the shorting wire, as if it weren't there.

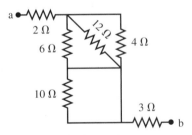

Find the equivalent resistance between a and b.

If students are pressed for a qualitative explanation of a circuit's behavior, many will respond with "Because it's grounded." Almost no students know what it means to be grounded or how it affects the behavior of a circuit. Because grounding introduces a 0 V reference point, it becomes possible to talk about *the* potential at a point in the circuit. This provides an opportunity to reinforce ideas about the electric potential. Determining the value of the potential at different points in a circuit, such as in the figure, is *very* hard for most students. Several class exercises are well worthwhile.

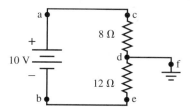

Find the potential at the points a–f.

Using Class Time

Single loop circuits can be covered in two days *if* students have good opportunities for hands-on experience in laboratory. If not, a preparatory class day devoted to interactive demonstrations, as described above, is recommended. I omit multiloop circuits; they introduce technical and mathematical details, but no new physics. Allow an extra day if you plan to cover them.

Day 1: As elementary as it appears, it is worth spending up to half of day 1 on the most basic circuit of a battery and a single resistor. To make the example concrete, you can have students compute the resistance of a 1-mm-diameter, 3-cm-long carbon resistor (1.33 Ω, conveniently giving 3.0 A from a 4 V battery.)

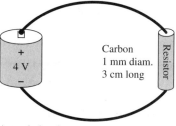

Actual circuit.

Issues to be considered during this example:

- Relating the physical circuit to the schematic diagram.

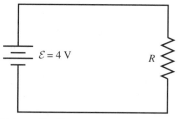

Schematic diagram.

- Calculating resistance.
- Conservation of current.
- The role of the wires. The conductivity of copper is 2000 times that of carbon, so $R_{wires} \ll R_{carbon}$. This leads to the concept of an *ideal wire*.
- Using the loop law $\Delta V_{loop} = \Delta V_{batt} + \Delta V_{wires} + \Delta V_{resistor} = 0$.
- Using Ohm's law to determine the current.
- Discussing and graphing the potential as it changes throughout the circuit, as shown in the figure below.

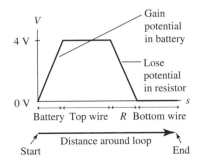

- Power supplied by the battery, energy transfer mechanisms, and power dissipated by the resistor.

It's worth using a voltmeter and an ammeter in a class demonstration to show that

- The current is the same before and after the resistor.
- The potential difference along the wire is ≈ 0 V.
- The potential difference across the resistor equals the potential difference of the battery.
- Changing the resistor causes the *current* to change but *not* the potential difference.

Such demonstrations should use resistances large enough that the battery's internal resistance is negligible. Note that batteries (large dry cells can be used) are much preferable to power supplies, which are magic black boxes to students.

You need not discuss how meters actually work in order to use them in demonstrations. Simply note that an ammeter measures the current flowing through it, that a voltmeter (with standard red and black leads) measures the potential *difference* $\Delta V = V_{red} - V_{black}$, and that both are designed to measure the circuit without affecting the circuit.

You can use the voltmeter to demonstrate the loop law, including the relevant signs. By going "around" the loop with the red lead always "ahead" (clockwise) of the black lead, you can show that $\Delta V_{battery}$ is positive but $\Delta V_{resistor}$ is *negative*. Doing this measurement in conjunction with the graph of the potential "around" the circuit reinforces the idea that potential increases in the battery and decreases in the resistor. On a microscopic level, charges gain potential *energy* in the battery (via the charge escalator) and lose potential energy in the resistor (transfer to kinetic energy and then, via collisions, to thermal energy).

Once you've made all the points in this simplest of circuits, you can have students try several single-loop circuits with multiple resistors and/or multiple batteries. It's good for at least one such example to have a demonstration set up where you can measure all the voltages around the loop, including proper signs, and again reinforce the idea that $\Delta V_{loop} = 0$.

As a challenge, show students a simple two-bulb circuit with identical bulbs as shown below. Ask for the potential difference ΔV_{ab} "across" bulb B. Tell them they cannot use any formulas, because they don't know the resistances of the bulbs, but they must justify their answer by *reasoning* about the circuit. Most students will be able to do this. It's good to have a demo set up where you can confirm their answer. Then ask, "What will the potential difference ΔV_{ab} be between points a and b if I remove bulb B from the circuit?" Students overwhelmingly predict that the potential difference will become zero, based on two possible lines of reasoning:

- There is no current between a and b, so $\Delta V = IR = 0$.
- There is no resistance between a and b, so $\Delta V = IR = 0$.

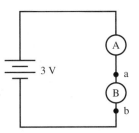

Note the confusion between "no resistor" $(R = \infty)$ and "no resistance" $(R = 0)$. Few students spontaneously consider using the loop law to answer this question. Before discussing it, it's good to let them "see" the answer with a demonstration— unscrew the bulb while a voltmeter is attached.

As a final demonstration and discussion for day 1, have a 60 watt and 100 watt light bulb and a circuit board where you can place the two either in series or parallel with standard 120 V line voltage. Show the 100 W bulb alone, then ask students to predict whether the 60 W alone will be brighter, dimmer, or the same. Congratulate them for all correctly predicting that the 60 W is dimmer. Then draw a picture showing how you're going to place them in series. (You may want to avoid using the term "series"

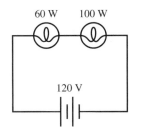

until you talk about series and parallel resistors on day 2.) Ask students to predict which will be brighter. Nearly all will predict that the 100 W will be brighter, and they are amazed that the 100 W bulb barely glows at all when you turn on the power.

This demonstration points out that most students have no idea what it means for bulbs to be labeled as "60 W" or "100 W." Many think that the bulb will *always* dissipate this much power, and essentially none will recognize that this has anything to do with the bulb's resistance. After the demonstration has captured their attention, you can talk about power dissipation in resistors and have them compute the resistance of each bulb. You can then have them analyze the series circuit to find the current, voltage, and power dissipation of each bulb (23 W for the 60 W bulb, 14 W for the 100 W bulb). It's best to "overlook" the resistance change with temperature, although you might want to discuss this qualitatively at the end if there's time.

Day 2: The second day is a good time to get into series and parallel resistors. Students have a hard time with the concept of an *equivalent resistance*, so it's worth a demonstration where you measure the current to and voltage across a resistor combination, have students compute the equivalent resistance, replace the combination with a single resistor of value R_{eq}, then show that the current to and voltage across the single resistor are identical to that of the combination. You might also want to use an ohmmeter to measure the net resistance of a combination.

Students also find it hard to believe that the equivalent resistance of parallel resistors is *less* than any resistor in the group. Common sense suggests that "more resistors" should give "more resistance." Getting more water through parallel water pipes is a useful analogy, but nothing beats an actual measurement to convince students of the results.

After several examples of series and parallel resistors, show the three-bulb circuit of the figure with the switch open. Ask students to predict what will happen to the brightness of bulbs A and B when you close the switch. Most will predict that B will dim and that A will not change. Their prediction about A is based on a continuing belief that the battery is a constant current source as well as on their tendency to think locally, rather than globally. Because the switch is "downstream" from A, they don't consider that its closure might affect A. After they're surprised by the results, this is a good circuit to explain by reasoning on the basis of parallel and series resistance.

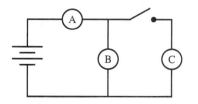

You'll want to give a detailed example of the analysis of at least one complex circuit, such as the one shown in the figure below. You can ask for the voltage across and current through each resistor, as well as for the power supplied by the battery. It's best to follow a systematic procedure, being very explicit at each step and confirming that the loop and junction laws are satisfied at each step. Even though the procedure is straightforward, students need to see and practice several of these in great detail before they really get the hang of it. As I mentioned earlier, I consider only circuits that can be reduced to a single loop, not multiloop circuits that require simultaneous equations.

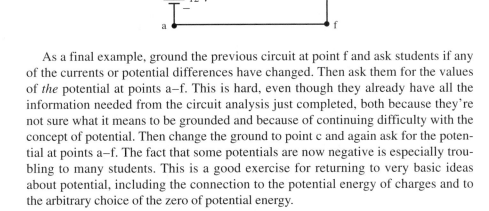

As a final example, ground the previous circuit at point f and ask students if any of the currents or potential differences have changed. Then ask them for the values of *the* potential at points a–f. This is hard, even though they already have all the information needed from the circuit analysis just completed, both because they're not sure what it means to be grounded and because of continuing difficulty with the concept of potential. Then change the ground to point c and again ask for the potential at points a–f. The fact that some potentials are now negative is especially troubling to many students. This is a good exercise for returning to very basic ideas about potential, including the connection to the potential energy of charges and to the arbitrary choice of the zero of potential energy.

Sample Reading Quiz or Discussion Questions

1. How many laws are named after Kirchhoff?
2. What property of a real battery makes its potential difference slightly different from that of an ideal battery?
3. Which of the following are *ohmic* materials?
   a. Batteries.   f. Materials a–d.
   b. Wires.   g. Materials b and c.
   c. Resistors.   h. Materials a and b.
   d. Light bulb filaments.   i. All of a–e.
   e. Semiconductors.
4. The equivalent resistance for a group of parallel resistors is
   a. less than any resistor in the group.
   b. equal to the smallest resistance in the group.
   c. equal to the average resistance of the group.
   d. equal to the largest resistance in the group.
   e. larger than any resistor in the group.

Sample Exam Questions

1. In the circuit shown on next page:
   a. Rank order the brightness of bulbs A–D, from most to least bright.

b. Describe what, if anything, happens to the brightness of bulbs A, B, and D if bulb C is removed from its socket.

For both parts a and b, you must *explain your reasoning*.

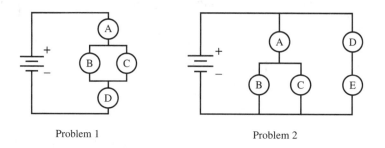

Problem 1   Problem 2

2. In the circuit shown above, rank order the brightness of bulbs A–E, from most to least bright. Explain your reasoning.

3. In the circuit shown below, describe what happens to the brightness of bulbs A and B when the switch is closed. Explain your reasoning.

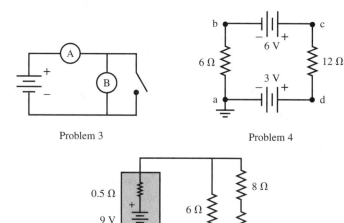

Problem 3   Problem 4

Problem 5

4. In the circuit shown above:
   a. How much power is dissipated by the 12 Ω resistor?
   b. What is the value of the potential at points a, b, c, and d?

5. The battery in the circuit shown above has an internal resistance of 0.5 Ω. What is the terminal voltage of the battery?

Magnetic Fields

Background Information

Research has found that more than 50% of a typical class of students begins their study of magnetism believing that

- A positively charged rod held near the center of a pivoted bar magnet causes the magnet to rotate, with the positive charge repelling the north pole and attracting the south pole.
- If one end of a bar magnet attracts a paper clip, the opposite end will repel the paper clip. (A number of students giving this response were asked about refrigerator magnets. Several replied that the decoration is pasted on the "repulsive side" of the magnet!)

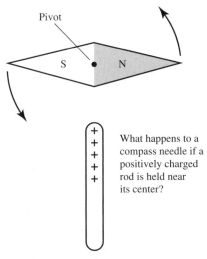

Common student response.

In addition to treating electric charges and magnet poles as more-or-less equivalent, students tend to use electric and magnetic fields interchangeably. These beliefs are little changed by conventional instruction.

For most students, their experience of magnetism extends little beyond the use of refrigerator magnets. Virtually all school children have used a magnet to pick up small steel objects, but the high percentage that expect one end of a magnet to repel the object suggests that these experiences have been infrequent. Informal surveys find that many students have never seen or experienced the *repulsive* force between two magnets. And, not surprisingly, very few students are familiar with electromagnets.

The presentation of the subject usually begins by reminding students that they are familiar with refrigerator magnets and perhaps other permanent magnets, then switches immediately to the development of a theory of *electro*magnetism. This is a classic case of bait-and-switch, with few texts ever making any connection between electromagnetism and permanent magnets. Few, if any, conventional textbooks ever answer the most obvious question that a student might have, namely "How does the magnet stick to the refrigerator?" This is the magnetic equivalent of "How does a charged comb pick up pieces of paper?"

Student Learning Objectives

- To acquire familiarity with basic magnetic phenomena.
- To develop a dipole model of magnetism, analogous to the charge model of electricity, that allows students to understand and reason about basic magnetic phenomena.
- To learn the magnetic fields due to currents in wires, loops, and solenoids.
- To study the motion of charged particles in magnetic fields.
- To understand the magnetic forces and torques on wires and current loops.
- To learn a simple atomic-level model of ferromagnetism.
- To connect the theory of electromagnetism to the phenomena of permanent magnets.

Pedagogical Approach

Many students are unfamiliar with the most basic of magnetic phenomena. Without such knowledge and awareness, the theory that is to be developed is not grounded in physical reality. Consequently, my approach to magnetism emphasizes

- The *phenomena* of magnetism. Although basic characteristics are described in textbooks, this cannot substitute for demonstrations and hands-on experience.
- A *dipole model* of magnetism that allows students to reason qualitatively about magnetic phenomena in terms of forces and torques on permanent and induced dipoles. This model is analogous to the charge model of electricity.
- The *connection* between the electromagnetism of moving charges and the macroscopic phenomena of permanent magnets. An important goal is to answer the question "How does the magnet stick to the refrigerator?"

A troublesome issue is how to introduce and define the magnetic field. The electric field was introduced by starting with the electric force between two charges and then using the electric force on a test charge to *define* the field. If only there were magnetic monopoles, we could follow an analogous procedure to define $\vec{B}$. Alas, we have to find an alternative. Unfortunately, the magnetic force on a moving charge does not uniquely define the magnetic field.

From a pedagogical perspective, the best analog of a test charge is a compass needle—a "test dipole." Students generally know, and it is easily demonstrated, that a compass responds to a magnet. They readily accept that the compass orientation *defines* the magnetic field direction. Furthermore, students' experience in earlier chapters with electric dipoles should allow them to recognize that the magnetic field is exerting equal but opposite forces on the two inseparable poles of the dipole, causing a torque until the dipole aligns with the field. Although the test dipole gives only the shape of the field, not the strength, that is still a big step forward toward understanding magnetism.

Oersted's discovery that a current deflects a compass implies that an electric current is another way to make a magnetic field. But a current is just moving charges, so students can quickly see and accept that at least one way to make a magnetic field is with a *moving* charge. The Biot-Savart law can then be introduced as an *empirical* finding that is logically equivalent to the empirical Coulomb's law of electricity. This is my preferred starting point. I then develop the magnetic field of common current distributions *before* considering the effects of magnetic fields on charges.

A simple current loop is an important link between electromagnets and permanent magnets. It is easily demonstrated that

- A current loop orients in a magnetic field.
- One side of a current loop attracts the north pole of a bar magnet and the other side of the loop repels the north pole.
- Two facing current loops will either repel or attract each other.

This, of course, is the same behavior that students have seen with a bar magnet.

These observations can be understood by attributing north and south magnetic poles to the two faces of the current loop, with the north pole being the side *out* of which the magnetic field points. They also suggest a *mechanism* for understanding magnetic forces in terms of the interactions between moving charged particles.

Much of this presentation is rather qualitative. It is the *reasoning* about force and torques, and their relationship to charged particles, that helps students to develop an understanding of what magnetism is all about.

I do not use Ampère's law, for much the same reasons that I don't use Gauss's law: the mathematical sophistication of vector integrals is beyond where nearly all students are in calculus, and superposition is a much more tangible way for students to first learn about fields. It's not difficult to find the magnetic field of a wire and the on-axis field of a current loop from the Biot-Savart law. The latter can then be used to find the field of an infinite solenoid.

The motion of charged particles in magnetic fields is difficult for two reasons: the need to visualize the situation in three dimensions, and the use of the vector cross product. Students do not learn the cross product in high school, and the usual calculus syllabus delays vectors until late in the course. You'll want to find out whether or not your students have met cross products in calculus. If not, allow extra time on the mathematics.

The magnetic properties of materials are often treated in a separate chapter, and often omitted. While I'm fully in favor of omitting paramagnetism and diamagnetism, students need at least an introduction to ferromagnetism if you're going to close the loop and make the connection between electromagnets and permanent magnets. A full treatment of ferromagnetism isn't needed, but you do need a simple model of magnetic domains based on the electron's magnetic moment.

Point out that electrons have an *inherent electric property*—namely, their charge. Students then find it believable that electrons also have an *inherent magnetic property* and act—for this presentation—like tiny bar magnets. Magnetic domains provide a qualitative explanation of both permanent magnets and of the magnetic force on the induced moment in nonmagnetized steel. Thus we close the loop between electromagnetism and permanent magnets, arriving at the answer to the question "How does the magnet stick to the refrigerator?"

Using Class Time

The magnetism chapter or chapters have a very high density of information, and essentially all of it is new to students. Although the mathematical level of my approach is somewhat less sophisticated than in most textbooks, the physical reasoning is perhaps *more* demanding. Magnetism can be covered in three days if more than usual reliance is placed on student reading. Four days are much preferred, however, to allow adequate time for demonstrations and for exercises about the right-hand rule, forces, torques, and so on. One laboratory period devoted to hands-on experience with *basic* magnetic phenomena is highly recommended.

Day 1: Magnetism demonstrations are both entertaining and instructive, and most of the main ideas are best illustrated with demonstrations. Examples include:

- Attractive and repulsive forces between magnets.
- The response of demonstration-size compasses to various magnets.
- Moving both a charged rod and a magnet toward the center of a compass to see if the compass pivots.
- Showing that *both* ends of a magnet attract steel objects.
- Cutting a magnet in half (if you have a cheap one to spare).

You want students to recognize from the beginning that

- Magnet poles are *not* the same as electric charges, and magnetic forces are distinct from electric forces.
- A compass needle is itself a small bar magnet.
- A compass's orientation can be used to define the idea of a *magnetic field*. The field is defined to point in the direction of the compass needle's north pole.

■ Using a compass as a probe reveals that the magnetic field comes *out* of the north pole of a bar magnet and goes *in* to the south pole. For a horseshoe magnet, this creates a magnet field *between* the pole tips.

To establish the big picture, let students know that the goal of this chapter is to understand how and why these magnetic phenomena occur. The starting point, however, will seem to have little to do with magnets. They will have to be especially attentive to the logic that leads from moving charged particles to refrigerator magnets.

The primary demonstration that begins the development of magnetism is, of course, Oersted's discovery that a current deflects a compass. Although this can be demonstrated with a horizontal wire on the lecture bench, a more convincing demonstration can be done with a copper rod cemented into a clear plastic plate. With legs or spacers to hold it up, you can place this on the overhead projector and set a transparent compass at different points around the wire. (You'll need a power supply that can deliver about 10 A.) You can also sprinkle iron filings on the plastic and project the circular field patterns onto the viewing screen.

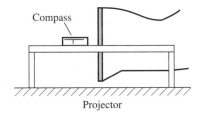

For most students, this is their first experience with a right-hand rule and with the
• and **x** notation for vectors and currents perpendicular to the page. Students will need many opportunities to practice before they are comfortable with these ideas. A good starting example is simply to draw a wire with a current and ask students to show the magnetic field, as deduced from the right hand rule, both in the plane of the wire and as seen with the current coming toward them. I use only magnetic field vectors, not magnetic field lines, for the same reasons that I use electric field vectors

Draw the magnetic field as seen in this plane. Then draw the magnetic field as seen if the bottom of the wire is coming out of the page toward you.

rather than field lines. But students should be getting more comfortable with the field concept, so this is not a bad time to switch over to using field lines.

Oersted's discovery implies that moving charges create a magnetic field, and you can introduce the law of Biot and Savart as an empirical discovery. Once again, students need several opportunities to practice using the right hand rule. For example, ask them to draw magnetic field vectors at each of the dots for the two moving charges shown in the figure. In the case of the negative charge, make sure the vectors at the corners are shorter than the vectors on the sides.

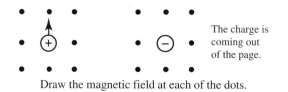

The charge is coming out of the page.

Draw the magnetic field at each of the dots.

This is enough for day 1 if you have four days for this chapter, but in a three-day presentation you'll want to go ahead and introduce current loops. All textbooks derive the on-axis field of a circular loop, so you need not repeat this in class. Focus on a qualitative understanding of the *shape* of the loop's magnetic field.

I like to use a square loop for class discussion. We've just finished talking about the magnetic field of a straight wire, so a square loop can be considered as the superposition of four straight current-carrying segments. Note that the field of a *finite* straight wire encircles the wire, but it's weaker at the ends than in the center. That's all you need for a qualitative determination of the magnetic field of a square loop via superposition.

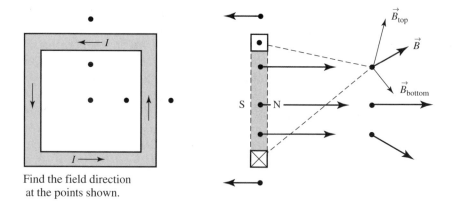

Find the field direction at the points shown.

First have students deduce the field direction at the center. Then have them consider off-center points *inside* the loop. They should be able to conclude that the field at all points inside the loop is (for the current shown) pointing straight out of the page. The field-current relationship is given by the right-hand rule for current loops. Then consider points outside the loop. This one is harder, but you want them to realize that the superposition is dominated by the nearest segment and thus points *into* the page.

Next, draw the loop as seen from the edge. If you consider just the top and bottom segments of the loop (current out of and into the page), students can use superposition to see that the field looks as shown in the figure above. (The side segments—current up and down—do contribute, but their fields at the "corners" of this diagram are weaker and nearly cancel. Neglecting them does not affect the conclusions about the shape of the field.) You may want to do one or two points to illustrate the reasoning, then have students do one or two others.

You want students to leave this exercise with a good sense that

- The field of a current loop converges into one side of the loop, passes through, and diverges from the other side.
- The field passes through the loop in a direction given by the right-hand rule.
- This field is the superposition of the fields of individual moving charges.

A nice demonstration is to have a heavy current loop cemented into plastic, like the straight rod shown earlier, so that it can be projected onto the screen as an edge view. Iron filings can then be used to illustrate the field pattern and to confirm the superposition reasoning. Alternatively, use a large current loop and a compass to confirm the basic shape of the field.

Day 2: You may want to summarize how the fields of a wire or a current loop are calculated from the Biot-Savart law, but in a three-day presentation you'll have to let students gather most of this information by reading the text. The emphasis is more on *using* these results than on doing field calculations.

A good demonstration is to mount a vertical current loop on a low-friction pivot or suspend one by a thread. You can show that the current loop axis orients along a north-south line, just like a compass needle. Also, use current loops to show that

- A loop exerts attractive and repulsive forces on bar magnets and compass needles.
- Two current loops attract or repel each other, depending on their orientation.
- A current loop attracts small steel objects.

In other words, a current loop is a magnet, and the face from which the magnet field emerges is the north pole. These are novel phenomena to nearly all students, so the demonstrations are especially important. The important idea to stress is that a current loop exhibits the same magnetic behavior as a permanent magnet.

Draw two facing current loops and ask students to use superposition and their knowledge of the shape of a current loop field to predict the field at three points in

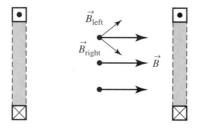

the mid plane. If you have two large current loops, such as a Helmholtz coil, you can use a compass to verify that the field vectors in the mid plane are all parallel.

Once students understand the idea for two loops, they'll readily believe that a stack of closely spaced current loops will produce a uniform field in the interior. Thus you can introduce the solenoid as a practical way to generate a uniform magnetic field—the magnetic equivalent of a parallel-plate capacitor. Demonstrations with solenoids should show that they are equivalent to a bar magnet. It is straightforward to derive the field by superposition from the on-axis field of a current loop.

Students need several opportunities to practice associating the current direction with the field direction, using the right-hand rule. Good exercises are to draw solenoids in random orientations. Either show the current and ask for the field direction, or show the field and ask for the current direction. This reasoning will be important in the next chapter on Faraday's law.

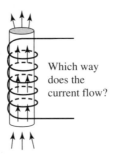

At this point, students have seen how to establish a magnetic field by the controlled motion of charged particles—although it's still an unanswered question how the field of a permanent magnet is established. Now it's time to switch over to the *effects* of magnetic fields. Because a current exerts a torque (a pair of forces) on a compass (a magnet), does a magnet exert forces on a current? Textbooks usually describe Ampère's experiment with two parallel wires, but an impressive demonstration is to run a wire through the pole tips of a strong laboratory magnet. Closing a switch to send a few amps of current through the wire will cause it to jump completely out of the magnet!

With this as a motivation, you can then introduce the idea of a magnetic force on a moving charged particle. The idea of a force perpendicular to the motion (and to the

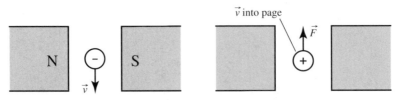

field) is very hard for most students to grasp. A Crookes tube or CRT that can be deflected with a magnet is an important demonstration. To make effective use of such demonstrations, you want students to *predict* how the electron beam will behave if you bring a specified end of a bar magnet in from a specified direction—for example, if you bring the south pole in from the bottom of the tube. Directional exercises such as those shown in the figure are also important. Students find these very hard to do at first and need quite a bit of practice.

It's reasonable to end day 2 with a quick derivation and simple example of the cyclotron frequency. A three-day presentation will leave no time for class discussion of examples of cyclotron motion, but students can read about this in the text.

Day 3: Day 2 demonstrated the magnetic force on a current-carrying wire. After having discussed the force on a charged particle, you can now go back to *explain* the force on a wire and to find the relationship between force and current.

It is interesting and important to extend this idea to the torque on a current loop in a magnetic field. First ask students to *explain* the torque in terms of forces on the currents. Then ask them to identify the north and south poles of the current loop. You want them to realize that the rotation of the current loop can also be understood in terms of the attraction/repulsion of opposite/like poles, but that this is just a short-hand way of characterizing the interactions of moving charged particles.

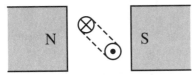
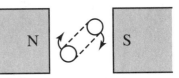

Show force on top and bottom of current loop. What is the equilibrium position?

Which way is current flowing in the loop?

Torque on a current loop leads naturally to the demonstration and discussion of a DC motor. It's nice to start with a motor using permanent magnets. This allows you to focus on the torque and on the split-ring commutator. Then switch to a motor that uses coils to generate the magnetic field. Now you have a nice example of using a current both to create the field and to experience a torque in the field. Only a tiny fraction of students have any idea how a motor operates, so this is a demonstration and explanation that most find to be especially interesting.

If you have a current balance, you may want to do a quick demonstration of the force between two parallel current-carrying wires. Stress that parallel currents *attract* each other. If this were merely an electrostatic effect, the wires would be equally charged and *repel* each other. A good exercise is simply to have students identify the field direction at each wire due to the current in the other wire, then use the right-hand rule to identify the force direction on each.

An especially important exercise is to have students consider two facing square current loops. The figure on the next page shows antiparallel currents. Have students first identify the field direction at the top and bottom of each loop due to the other

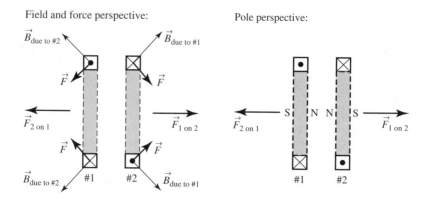

Field and force perspective:

Pole perspective:

loop, then the forces. They can find for themselves that two loops with antiparallel currents repel each other. Then have them identify the north and south poles of the loops (few will think to do this on their own). Now they see two north poles repelling each other. The importance of this exercise lies in discovering a *mechanism*—in terms of interacting charged particles—for *why* like poles repel. The attractive/repulsive force between parallel/antiparallel current loops will be important in the next chapter on Faraday's law.

A three-day presentation will leave little time to talk about ferromagnetism, so you'll have to leave this mostly for students to read. It's worthwhile to summarize the main ideas in a short mini-lecture. The main points are

■ Electrons are inherent magnets, with a north and south pole.
■ In some elements, the atomic-level magnets throughout a small piece of the material all align to create a macroscopic magnetic moment. The reasons *why* are quantum mechanical. We'll just accept it as an empirical finding. These small regions are magnetic domains, and the materials themselves are ferromagnetic.
■ A chunk of material in which most domains are oriented in the same direction is a permanent magnet. Thus permanent magnets trace their magnetism back to the electrons.
■ An external magnetic field exerts torques on the moments of the magnetic domains (and moves domain boundaries), thus creating an *induced* magnetic moment. This is analogous to polarization in electricity. The orientation of the induced magnetic moment is such that the piece of metal is attracted to the magnet. Thus we've finally answered the question of how the magnet sticks to the refrigerator.

Sample Reading Quiz or Discussion Questions

1. What is the SI unit for the strength of the magnetic field?
2. What is the shape of the trajectory that a charged particle follows in a magnetic field?

3. The magnetic field of a small current element is given by
   a. Biot-Savart's law.   d. Faraday's law.
   b. Gauss's law.   e. Ohm's law.
   c. Ampère's law.   f. Einstein's law.

4. The magnetic field of a straight, current-carrying wire is
   a. parallel to the wire.   d. inside the wire.
   b. perpendicular to the wire. e. zero.
   c. around the wire.

Sample Exam Questions

1. Two current loops in a uniform magnetic field are shown in the figure. One of the loops is in a stable position, so it will return to this position if it is rotated slightly. The other loop is unstable, like an upside-down pendulum. It will flip over if it is rotated slightly away from this position. Which loop is which? Give a careful explanation, including force diagrams.

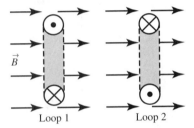

2. A uniform magnetic field points upward, in the plane of the paper. A wire is perpendicular to the paper. When the wire carries a current, the net magnetic field at point 2 is zero.

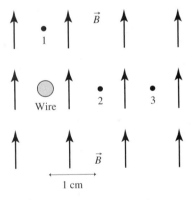

   a. What is the direction of the current in the wire—into or out of the paper? Explain.
   b. Point 1 is the same distance from the wire as point 2. Use a vector diagram to determine the net magnetic field at point 1.
   c. Point 3 is twice as far from the wire as point 2. Use a vector diagram to determine the net magnetic field at point 3.
   d. If the net magnetic field at point 3 is 1.0×10^{-4} T, how much current is flowing in the wire?

3. A positive helium ion He⁺ is released from rest from the surface of a +1000 V electrode. It crosses (in vacuum) the 1 mm gap, passes through a small hole in a 0 V electrode and into a magnetic field, then begins to curve as shown. The magnetic field is uniform in the region to the right of the 0 V electrode and is zero to the left of the 0 V electrode.

   a. What is the direction of the magnetic field? Explain your reasoning, then *show* the field on the diagram.

   b. The magnetic field strength is 1.0 T. Does the ion collide with the 0 V electrode? Explain.

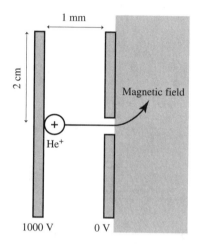

4. a. Two parallel wires carrying current in the same direction are found to attract each other. Our basic model of magnetism is that it is an interaction between moving charges. Give a step-by-step explanation, making explicit reference to moving charges, of *how* the wires attract each other. Your explanation should consist of sentences and pictures, but no equations.

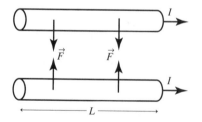

   b. Suppose the wires have a linear mass density of 2 g/m and carry equal, parallel currents. The upper wire is fixed in position. What value of the current will allow the lower wire to "float" 1 cm below the upper wire?

5. Three long straight wires of linear mass density 50 g/m are directed into the page. They each carry equal currents in the directions shown. The lower two wires are 4 cm apart and attached to a table. What current *I* will allow the upper wire to "float" so as to form an equilateral triangle with the lower wires?

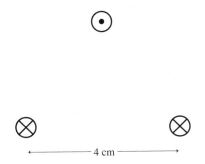

6. You have a horizontal cathode ray tube (CRT) for which the controls have been adjusted such that the electron beam *should* make a single spot of light exactly in the center of the screen. You observe, however, that the spot is deflected to the right. It is possible that the CRT is broken. But as a clever scientist, you realize that your laboratory might be in either an electric or magnetic field. Assuming that you do not have a compass, any magnets, or any charged rods, how can you use the CRT itself to determine whether the CRT is broken, is in an electric field, or is in a magnetic field? You cannot remove the CRT from the room.

Electromagnetic Induction

Background Information

Faraday's law and electromagnetic induction complete and provide closure for the subject of electricity and magnetism. There are two important ideas. First, the discovery of induced electric fields that are independent of charges is important evidence for the reality of fields. The electric field was introduced in a rather *ad hoc* fashion, and it is important for students to understand why, in retrospect, we attach such importance to the field concept. Second, electromagnetic induction allows a qualitative description of electromagnetic waves. While the details require Maxwell's equations, students can at least be aware that an electromagnetic wave is a self-sustaining field configuration in which each field is induced by the changing of the other field.

There is little research that bears directly on students' understanding of Faraday's law or Lenz's law. As most instructors are aware, many students think that the induced field opposes the applied field itself rather than opposes the *change* in the applied field. It is likely that this error arises simply from lack of sufficient opportunities for qualitative reasoning, with appropriate feedback, rather than from any fundamental misconceptions about the nature of magnetic induction.

To the extent that Faraday's law depends on the idea of flux, the limited research on Gauss's law at the University of Washington may be relevant. They found that less than 15% of students could use Gauss's law to reason about electric fields. Although the specific difficulties have not been explored, at least one seems to be a difficulty with the idea of flux. Faraday's law is easier in the sense that we consider flux only through planar surfaces, rather than through the spherical and cylindrical surfaces of Gauss's law. Nonetheless, instructors should anticipate needing a significant number of demonstrations and exercises to convey the idea of flux.

Student Learning Objectives

■ To observe the experimental evidence for electromagnetic induction.
■ To understand and use Lenz's law for induced currents.
■ To learn of Faraday's law as a new law of nature.
■ To understand basic applications of electromagnetic induction to technology.

Pedagogical Approach

It turns out there are three different ways to create an electric field:

■ A Coulomb field created by individual charges.
■ A field created by charge separation due to magnetic forces (motional emf).
■ A non-Coulomb field created by dB/dt.

It is surprising, but nonetheless true, that the second and third situations—which are very different physically—are *both* described by Faraday's law. The link between them, of course, is that both involve a change in magnetic flux. However, only the third is *new* physics that cannot be explained by the previous theory of electric and magnetic fields and forces. Many textbooks do not make a clear distinction between motional emf and the non-Coulomb emf due to a time-dependent magnetic field.

Motional emf is a direct and obvious extension of the magnetism chapters. However, many students get stuck here and think that *all* of electromagnetic induction is due to motional effects. An important goal is to convince students that an induced emf can be caused by two very different means—but Faraday's law characterizes both.

I much prefer to introduce Lenz's law *before* Faraday's law, and I've never understood the traditional approach of treating Lenz's law as an adjunct to Faraday's law. In fact, Lenz discovered his "law" for finding the direction of an induced current before Faraday and others determined the quantitative relationship that we today call Faraday's law. Starting with Lenz's law has the pedagogical advantage of getting students to *reason* about induced currents before having to worry about the numerical value of induced emfs. After all, most of the demonstrations of induction are concerned only with the *direction* of the induced current.

An early introduction of Lenz's law also allows a cleaner statement of Faraday's law:

An emf $\mathcal{E}$ is induced in a conducting loop if the magnetic flux through the loop changes. The magnitude of the emf is

$$|\mathcal{E}_{\text{loop}}| = \left|\frac{d\Phi}{dt}\right|,$$

and the direction of the emf is such as to drive an induced current in the direction given by Lenz's law.

This statement avoids the troublesome minus sign. Although one can define a sign convention for the area vector $\vec{A}$ and thus determine the direction of an induced current, physicists and engineers invariably use Lenz's law in all practical situations.

The minus sign is of real significance only when Faraday's law is incorporated into Maxwell's equations.

Many of the applications of magnetic induction, such as eddy currents or demonstrations that shoot aluminum rings into the air, are consequences of magnetic *forces* on induced currents. These applications can be understood either in terms of attractive/repulsive forces between parallel/antiparallel currents or between opposite/like magnetic poles.

Using Class Time

Most instructors have many favorite demonstrations of electromagnetic induction. Although I strongly encourage the use of demonstrations, Faraday's law is one point where there's a danger of going overboard and becoming more "gee whiz" than instructive. From a pedagogical perspective, it is best to keep the demonstrations limited to ones that

- You have time to go over and explain thoroughly.
- Students can *understand* by straightforward application of Lenz's law, Faraday's law, and their recently acquired knowledge of magnetic fields and forces.

Initial demonstrations should be simple things, such as pushing bar magnets in and out of coils, opening and closing switches, and so on. It's nice to use a large galvanometer or current meter that students can see deflect back and forth. The initial points to make are that

- The meter deflects only when something is *changing*. Holding a magnet inside a coil does nothing.
- Reversing the motion reverses the meter deflection.
- Pushing a north pole into a coil has the opposite effect of pushing a south pole into the same coil. The situation is more complicated than simply "push a magnet in" or "pull a magnet out."
- The effect occurs both for moving the coil toward the magnet and for moving the magnet toward the coil.

Students who understood the basic information of the magnetism chapters should have no major difficulties understanding the idea of motional emf, although you'll probably want to review what is meant by "emf." Comment that a battery is a *chemical emf* because it separates charge via chemical reactions, thus causing a potential difference. The new effect is a *motional emf* because charge separation, and thus a potential difference, occurs via motion of the charge carriers in a magnetic field. (Note that the potential difference between the ends of a moving slide wire is *not* necessarily the emf $\mathcal{E} = vLB$. The potential difference is $\Delta V = \mathcal{E} - IR$, where I is the induced current and R is the resistance of the slide wire.)

Some books introduce Faraday's law right away and use it to calculate the emf of a slide wire. Although the calculation is valid, this approach obscures the significance of Faraday's law. Motional emf should first be treated, as it truly is, as a straightforward extension of the basic ideas of the magnetism chapters.

Motional emf and the magnetic force on induced currents is sufficient to understand eddy currents. Most departments have, or can easily build, an eddy-current pendulum where a copper disk on a wooden rod swings through the pole tips of a laboratory magnet. Essentially no students have ever seen such a demonstration, and it is highly effective. In a small class, you can have students try pulling a copper sheet out of a magnet. All are very surprised by the unexpected forces that make this hard to do.

Magnetic flux can be introduced in different ways. If you have a sink at your lecture table, hold a small loop in the flowing water and ask how much water flows through the loop as you rotate it through different angles. Alternatively, hold a loop in front of a fan and ask how much air flows through the loop. Have students, in discussions with their neighbors, recognize that the three parameters that determine the answer are

- The size of the loop.
- The rate of flow of the water or air.
- The angle between the loop and the water/air flow.

It is then quite convincing to define the *flux* of water or air as $A \times$ rate of flow $\times \cos \theta$, where you define θ as the angle between the flow and the *axis* of the loop. Point out that $A \cos \theta$ is the *effective* area of the loop as seen by the water or air.

This visual image of a flow through an area is very important to help students understand the concept of flux. It is then straightforward to use an image of "magnetic field arrows flowing through a loop" to define the magnetic flux, introducing the notion $\vec{A} \cdot \vec{B}$. It's worth giving the students several simple practice examples—square and circular loops at different angles—before proceeding. Other good exercises are:

By what factor does the flux through a loop change if the magnetic field is halved and the sides of the loop are doubled?

At what angle is the flux through a loop half of the maximum possible flux?

Once you've introduced flux, you can return to demonstrations where you showed that moving either a coil *or* a magnet produced the same effect. While motional emf can explain what happens when you move the coil in the field, it doesn't explain why there's an effect when you move the magnet but the charge carriers are at rest. The similarity between the two, you can note, is that both involve a *change in the flux* through the coil.

This observation leads to Lenz's law, which you can introduce as a new law of nature. Students will need quite a few practice examples before they catch on to the idea that the field of the induced current is opposing the *change* of flux rather than opposing the flux itself. Some students will still need practice using the right-hand rule to relate the field direction to the current direction through a coil or solenoid.

Several good exercises are as follows:

In Figure 1, does the loop of wire have a clockwise current, a counterclockwise current, or no current if: a) The magnetic field strength is increasing? b) The magnetic field strength is constant? c) The magnetic field strength is decreasing?

In Figure 2, does the lower loop of wire have a clockwise current, a counterclockwise current, or no current: a) Before the switch is closed? b) Immediately after the switch is closed? c) Long after the switch is closed? d) Immediately after the switch is opened?

In Figure 3, does the current through the meter flow right to left, left to right, or is there no current if: a) The magnet is held at rest? b) The magnet is pulled out of the coil?

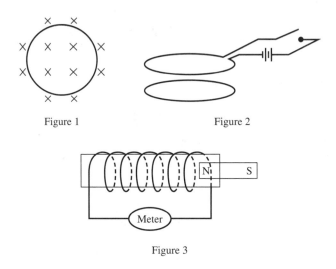

Figure 1   Figure 2

Figure 3

You can then introduce Faraday's law and use it for standard examples of calculating induced emfs and induced currents. You'll want to revisit the motional emf of the slide wire, pointing out that Faraday's law provides a new perspective but, in this case, no new physics. The new and unexpected physics arises in situations where the flux is changing while the conductor is stationary.

Applications of Faraday's law to multiply connected loops around regions of changing magnetic field are very subtle and can easily confuse experienced physicists. This is an introductory course, so instructors are urged to limit exercises and examples to single loops of simple geometry around simple, well-defined fields.

Practical applications of Faraday's law to generators, transformers, metal detectors, magnetic recording, and so on are important to convey the significance of Faraday's discovery. Although the discussion must be more qualitative than quantitative, you should certainly demonstrate and talk about as many applications of electromagnetic induction as you have time for.

If you're not going on to a full development of Maxwell's equations, a mini-lecture on electromagnetic waves is an appropriate way to end electricity and magnetism. You want students to recognize that

- Electromagnetic waves are a self-sustaining field configuration, independent of charges and currents. This is a consequence of electromagnetic induction.
- The basic wave consists of oscillating $\vec{E}$ and $\vec{B}$ fields that are mutually perpendicular to each other and to the direction of propagation.
- Maxwell's theory *predicts* that electromagnetic waves of *all* frequencies travel with the same speed $c = 1/\sqrt{\varepsilon_0 \mu_0}$. Thus *static* measurements of ε_0 and μ_0, which are part of the basic definitions of the electric and magnetic fields of a point charge, end up predicting the speed of light.

Sample Reading Quiz or Discussion Questions

1. Cite two practical examples of electromagnetic induction.

2. Currents circulate in a piece of metal that is pulled through a magnetic field. What are these currents called?

3. Electromagnetic induction was discovered by
   a. Faraday.
   b. Henry.
   c. Maxwell.
   d. Both Faraday and Henry.
   e. Both Faraday and Maxwell.
   f. All three.

4. The direction that an induced current flows in a circuit is given by
   a. Faraday's law.
   b. Henry's law.
   c. Maxwell's law.
   d. Lenz's law.
   e. Hertz's law.
   f. Edison's law.

Sample Exam Questions

1. Two metal loops face each other. The upper loop is suspended by plastic springs and can move up or down. The lower loop is fixed in place and is attached to a battery and a switch. *Immediately* after the switch is closed,
   a. Is there a force on the upper loop? If so, in which direction will it move? Explain your reasoning.
   b. Is there a torque on the upper loop? If so, which way will it rotate? Explain your reasoning.

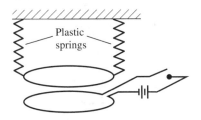

2. The figure below shows a loop 10 cm in diameter in three different magnetic fields. The loop's resistance is 0.1 Ω. For each situation, determine the induced emf, the induced current, and the direction of current flow.

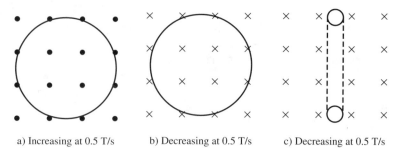

a) Increasing at 0.5 T/s   b) Decreasing at 0.5 T/s   c) Decreasing at 0.5 T/s

3. The outer coil of wire is 10 cm long, 2 cm in diameter, wrapped tightly with one layer of 0.5-mm-diameter wire, and has a total resistance of 1.0 Ω. It is attached to a battery, as shown, that steadily decreases in voltage from 12 V to 0 V in 0.5 s, then remains at 0 V for $t > 0.5$ s. The inner coil is 1 cm long, 1 cm in diameter, has 10 turns of wire, and has a total resistance of 0.01 Ω. It is connected to a current meter.

   a. As the voltage to the outer coil begins to decrease, what direction (left-to-right or right-to-left) does current flow through the meter? Explain.

   b. Draw a graph showing the current in the inner coil as a function of time for $0 \leq t \leq 1$ s. Include a numerical scale on the vertical axis.

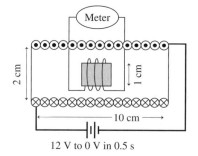

12 V to 0 V in 0.5 s

4. The figure below shows a vertically oriented loop. The loop has dimensions 20 cm × 20 cm, a mass of 10 g, and a resistance of 0.01 Ω. A uniform 1 T magnetic field passes through the upper half of the loop and is perpendicular to the loop. The loop is released from rest and allowed to fall.

   a. Will the loop fall with constant speed or constant acceleration? Explain.

   b. How long will it take the loop to leave the field? How does this compare to the time it would take the loop to fall the same distance in the absence of a field?

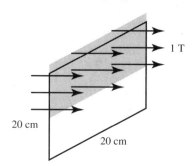

21

Geometrical Optics

Background Information

Students approach optics with a naive conception of light and its properties. All students know that "light travels in a straight line," but few have a clear understanding of what this actually means. Many students don't recognize that light is a physical entity, with an existence apart from its sources and effects. They don't make a clear distinction between light and vision.

As an example of their naive views, interviews find that many students from middle school through college think that light has a finite range before "running down." After all, they don't see their headlight beams during the day, so apparently the light from the headlights only travels a few feet during daylight hours! And while most students know that "white light consists of all colors," few can interpret this statement. Many students think that a color filter *adds* color to white light.

In short, students have scattered ideas about light, but not a coherent model that can be used for reasoning. We want them to learn *two* models, the ray model and the wave model, in quick succession. Students who go on to modern physics will be presented with a third model, the photon model. The research evidence is that students emerge from optics with a confused, hybrid mix of the ray and wave models. After instruction, the majority of students give incorrect answers to simple qualitative questions about image formation and about interference. This chapter will focus on difficulties with geometrical optics; physical optics is considered in the next chapter.

Student understanding of geometric optics was studied by Goldberg and McDermott (1987) and by Wosilait et al. (1998). Much of this work was based on post-instruction interviews with students in the calculus-based physics course. They found that students had difficulty applying the most basic ideas of the ray model of light. When presented with a wide aperture (1 cm) and a point source of light, the majority of students thought that the pattern of light on a screen would be dominated by diffraction effects. Even after being led to recognize that the image would be a bright rectangle, *none* of the interviewed students could determine its size. They also had great difficulty predicting the pattern of light on the screen if an aperture was illuminated by two point sources of light. Few students had developed a working model of light propagation that would allow them to reason about simple situations.

Goldberg and McDermott were interested in how students understand image formation by a lens. Subjects were shown a luminous object, a lens, and a focused inverted image on a screen. They were asked a series of "what if" questions, but none of the answers were demonstrated.

- Asked what would be seen on the screen if the lens were removed, the majority replied that there would still be an image on the screen, but it would be upright rather than inverted.
- Asked what would be seen on the screen if a cardboard mask covered the top half of the lens, 75% of post-instruction students said that half of the image would vanish.
- Asked what would happen if the screen were moved forward or backward, 65% of students indicated there would always be an image, but it might change size or be "fuzzy."

Many students explained their answers with a drawing similar to that shown in the figure. They showed only two rays from the object, both parallel to the axis, and they indicated that an image would be formed wherever these rays strike the screen. It's not hard to see that this incorrect model of light rays and image formation can account for the three responses given. For example, a mask over the top half of the lens blocks one of the two rays, so only half the image "makes it" to the screen.

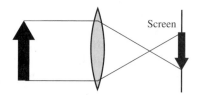

Arons (1997) comments on how students respond when asked to describe how light emanates from an object, passes through a lens, and ends up forming an image on a screen:

> A substantial number tell a story that effectively boils down to something like "The image travels from the object to the lens; in the lens it is turned upside down; then it travels to the screen."

In effect, many students think there is a "potential image" that moves through space from the object to the screen, where it is given realization. This is what they are attempting to show with drawings such as the one in the figure, and this incorrect model of light propagation explains their responses to the interview questions. Many of these students also predicted that a mask over a lens would prevent *any* image from appearing on the screen if a hole in the mask was smaller than the object. After all, they explained, "the image doesn't fit through the hole." These interviews *followed* instruction in geometrical and physical optics.

Even students who had developed a better understanding of light propagation were not immune to major conceptual difficulties with image formation. Many of these students drew a textbook ray diagram for image formation, as shown, but then went on to note that a mask would block half the rays and thus cause half the image to vanish. Many of these same students continued to hold a belief that the screen somehow "captures" the image, and they had difficulty conceiving that an image could exist in empty space.

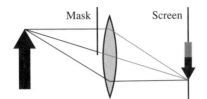

Altogether, research leads to the following conclusions:

- Most students do not understand the fundamental ideas of the ray model of light.
- Students think of rays as physical entities, not abstract representations.
- Students don't understand the role of the three principal rays of a lens.
- Students don't understand the function of a converging lens in image formation.
- Students don't understand the role of the screen in image formation.
- Students don't understand the role of the eye in "seeing images."

Successful use of the lens equation does not indicate an understanding of optics. Interviews with students who could successfully solve for an image position found that many were unaware that the image position is unique.

Student Learning Objectives

- To understand the ray model and its domain of applicability.
- To apply ray tracing from situations ranging from apertures to image formation.
- To understand images and image formation.
- To understand how, when, and why images are seen.

Pedagogical Approach

There's nothing difficult about the mathematics of lenses and mirrors. The technical demands of this chapter are minimal when compared to many previous chapters. But successful use of the lens equation does not indicate an understanding of optics. The teaching focus needs to be on developing the ray model of light as a useful working tool.

Light rays are an abstract concept, a representation of light valid in a restricted domain of applicability. The standard textbook approach assumes that rays and the basic features of the ray model are obvious to students, but the research shows that this is clearly not the case. It's important to be explicit in developing the ray model, and to then apply it in increasingly complex situations.

I start by defining a light ray as "a line in the direction along which light energy is flowing." A laser beam, which you'll use for many demonstrations, is really a bundle of many parallel light rays. The *ray model* then holds that

- Light travels through a transparent medium in straight lines, called light rays, at speed $v = c/n$, where n is the index of refraction of the medium.

■ Light rays do not interact with each other.

■ A light ray continues forever unless it has an interaction with matter that causes the ray to change directions or to be absorbed.

■ Light has four different ways in which it can interact with matter.

At an interface between two media, light can be *reflected* or *refracted.*

Within a medium, light can be *scattered* or *absorbed.*

These are illustrated in the figure below.

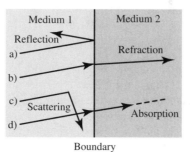

Light interacts with matter by a) reflection, b) refraction, c) scattering, or d) absorption.

■ An *object* is a source of light rays. We make no distinction between self-luminous objects and reflective objects. Rays originate from *every* point on the object, and each point sends rays in *all* directions. This is shown in the figure below, which is pretty messy. To simplify the picture, we use a *ray diagram* that shows only a few important rays. A ray diagram does *not* imply that these are the only rays.

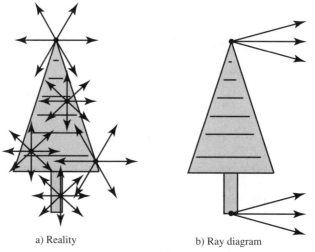

a) An object emits light rays in all directions from all points.
b) A ray digram is a simplified view.

■ The eye "sees" an object when bundles of *diverging* rays from each point on the object enter the pupil and are focused to an image on the retina. Based on how the eye's lens has to move to focus the image, your brain "computes" the distance d at which the rays originated. You then perceive the object as being at distance d. The details of the image will be presented later.

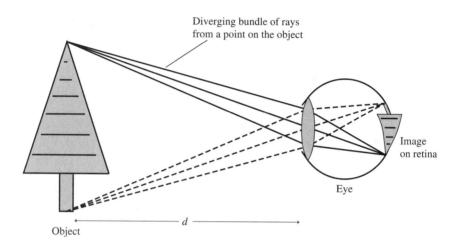

Many of these items—such as an object emitting rays from all points in all directions—seem so obvious to the physicist that they're never explicitly stated. But the research cited above shows that many students are unaware of these aspects of light and light rays, so they proceed through the chapter without ever grasping the most basic ideas.

I've found that it's very important to give early attention to "seeing," even if the details of image formation are deferred until later. Students need to recognize that you see something when a bundle of diverging rays enters the pupil and is then focused by the eye. Many of the student difficulties with understanding images, especially virtual images, stems from a confusion between "the image" versus "seeing the image." A simple model of vision, such as this, makes it much easier to talk about images.

An explicit ray model leads naturally into ray tracing diagrams. I prefer to develop a ray tracing analysis of image formation *before* introducing the lens equation. This puts the emphasis on understanding light propagation, with the equation only providing more mathematical accuracy than can be gained by ray tracing alone.

I omit image formation by curved mirrors and the discussion of optical instruments (other than the eye and the magnifying glass) as part of my effort to make the course less encyclopedic. Students will have no difficulty learning about these at a later time if they leave introductory physics with a working knowledge of the ray model and its application to simple lenses.

Using Class Time

Laser beams travel in nice straight lines. The sharp shadows of sunlight suggest that light travels in straight lines. All students know the phrase "light travels in a straight line," but few have given any thought as to what light *is* or to how we know that it travels in straight lines. Few have ever made a distinction between self-luminous objects, such as a light bulb, versus reflective objects, such as a tree. And few are aware of what it means to "see" an object. Failure to grasp some of these most basic ideas is what prevents many students from developing a conceptual understanding of geometrical optics. Instruction needs to start with these fundamental issues.

I begin with self-luminous objects, such as the sun or a light bulb. Shadows suggest that light travels in a straight line. This leads to the definition of a light ray, but also to the realization that a light source emits many rays in all directions. Students "see" the light bulb because a small bundle of *diverging* rays enters their eye from each point on the object. A picture of the eye "seeing an object," such as the figure on the previous page, is useful at this point.

Now turn to a nonluminous object, such as the wall, the desk, or yourself, and ask how students see those. This leads to a more generalized idea of an *object*. You can then present the ray model, noting that—as with any model—its usefulness will depend on whether it provides a satisfactory explanation of phenomena.

You can first apply the ray model to an analysis of light passing through an aperture. Textbooks skip over this as being too trivial, but the research cited above found that students had a difficult time predicting how light would appear on a screen after passing through an aperture. Have them consider a point source of light and a noncircular aperture, such as a triangle. Ask what they would see on the screen and how it would change if the light-aperture or aperture-screen distances were changed. Can they calculate the image size? How would the screen look if there were two point sources separated by a small distance? In discussing these, you want to convey the idea that rays from the object *fill* the aperture. This will be important later for understanding image formation by a lens.

I like to discuss the camera obscura and the pinhole camera. Consider an *extended* object and an aperture that has shrunk to "a point." You want students to recognize that only one ray from each point on the object can pass through a point aperture, and this creates an inverted image. Of course, the intensity has also shrunk to zero! What happens to the image if you have a small but finite aperture? (Don't worry about diffraction, although you might want to briefly mention that there are "practical difficulties" with a small aperture that they'll learn about in the next chapter.) It's important that students be able to explain, in both words and picture, why a camera obscura shows an image of the *object* whereas your first example, with a point source, showed an image of the *aperture*. A successful explanation shows that they're beginning to assimilate the basic ideas of the ray model.

You can next apply the ray model to an analysis of the plane mirror. Start by using a laser beam to demonstrate the law of reflection. But there's still a conceptual gap from knowing the law of reflection to understanding how this leads to image formation. Two steps are needed. First, as a class exercise, draw several rays from an object point P to the mirror, as shown on the left in the figure. Have students use a straight-edge to draw the rays and their reflections, based on the law of reflection. Most students can get the reflection angles fairly close without needing a protractor. Then have them extend the reflected rays back to the virtual image point P′ as shown below. A figure like this is undoubtedly included in the textbook, but actually carrying out the operations makes an impression that is missed from simply reading the book.

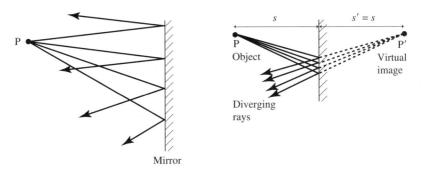

But locating the image point is only half the battle. The most obvious thing about a plane mirror is that we *see* an image in the mirror. Here's where the simple model of vision starts to pay off. A drawing similar to the one shown here, but probably not in your textbook, shows the students how the diverging rays from the virtual image allow them to *see* the image. Now you have not only a mirror

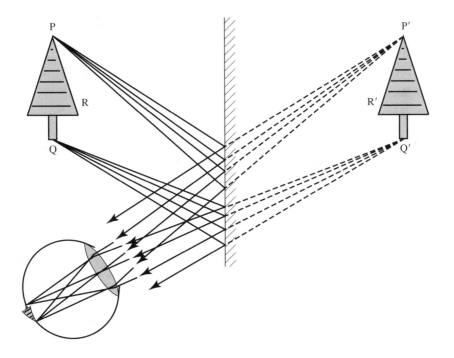

equation, $s' = s$, but you've used the ray model to provide a satisfactory explanation of how we see images in mirrors. As a practical matter, you'll probably want to prepare a picture like this in advance, as a transparency, rather than trying to draw it on the board.

Specular reflection is analyzed because we have a "law" to guide us. But diffuse reflection is far more important in the real world, and it deserves more discussion than it receives in most textbooks. How is it that we see the walls, our hands, this page? Students need help making the connection between specular and diffuse reflection. This is important because one of the research findings is that most students don't have a clear understanding about the role of the screen.

Demonstrations are very important for refraction, since this seems to violate the rule that light travels in straight lines. I like the smoky-colored plastic that allows students to easily see the propagation of a laser beam, but a tank of water with a pinch of corn starch or artificial sweetener also works well. You want students to recognize that the trajectory has a "kink," but it remains a straight line in each medium. You can either use a wave front diagram to explain this, using the different propagation speeds, or simply accept it as an empirical finding.

Be careful with overly simplified textbook drawings, like the one on the left. It can convey the wrong impression. These drawings need to be preceded by diagrams showing the parallel rays in a laser beam or the diverging rays from a point. Only then does the single-ray picture make sense.

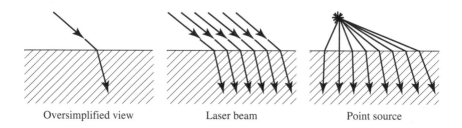

Oversimplified view   Laser beam   Point source

The mathematical aspects of Snell's law are straightforward, with the single exception of identifying the correct angles. Students easily mistake the angle between the ray and the surface for the angle of incidence. The mathematics of Snell's law is easy compared with what you've been doing recently, and it poses no major difficulties. The difficulty in more complex situations, such as analyzing a prism, lies in the geometry.

Refraction at a plane surface leads to the formation of an image. Finding the image is often relegated to a homework problem, such as "A coin is at the bottom of a 10-foot-deep swimming pool; how far does it appear below the surface?" Not only is this much too complex for nearly all students to solve as a homework problem, it passes up an important teaching opportunity. An analysis of image formation by a plane surface introduces the paraxial ray approximation, virtual images, and the role of the eye in seeing a virtual image. A discussion now, where all you need is a simple application of Snell's law, will make these issues easier when you soon encounter them with lenses. I like to use the example of seeing a fish in an aquarium as the basis for an analysis.

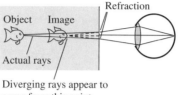

Diverging rays appear to come from this point.

I introduce image formation by lenses with ray tracing rather than by deriving the lens equation. This is easily done empirically, both with photographs like the ones below and with class demonstrations of laser beams passing through lenses. These establish the idea of a focal point and provide the basis for two of the three principal rays.

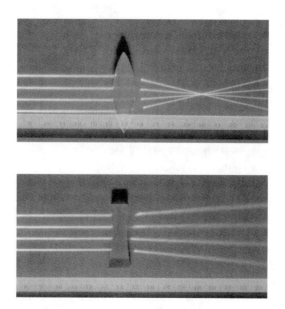

The third principal ray, which passes undeviated through the center, is easily introduced by analogy with light passing through a window pane after noting that the sides of the lens at this point are parallel. This is all you need to teach students how to study image formation by ray tracing.

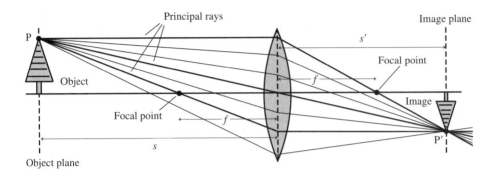

One caution is needed, however. The research cited earlier showed that many students have a hard time interpreting ray diagrams that show only the principal rays. Your first ray tracing needs to show *many* rays emanating from the object, with the principal rays highlighted as shown above. This is one of the fundamental ideas of the ray model. If you previously noted that rays from a point source *fill* an aperture, then students should be able to extend that idea to a lens. This is the critical reasoning step that allows them to realize that any portion of a lens creates a full image on the screen. Note that the figure below also shows the rays continuing past the image point. This is important for discussing the role of the screen and for allowing students to understand that an image can exist in empty space. Figures that always show the rays terminating at an image on a screen are misleading.

The point of introducing ray tracing prior to the lens equation is, of course, to keep students focused on reasoning rather than calculating. You can now begin to ask questions about what happens if you move the lens, move the screen, cover the lens, and so on. These ideas are easily extended to virtual images, diverging lenses, and even simple lens combinations. The only situation to avoid in introductory physics is a combination that involves a virtual object.

Ray tracing is a good place for Jeopardy questions. Present a lens and an image, then ask where the object was. Or present an object and an image and ask where the lens was. You can ask similar questions later with the lens equation, but it's nice to first have students think about them simply in terms of the rays.

Deriving and using the lens equation is now rather anticlimactic, but that's OK. The lens equation adds numerical accuracy, but students have gained a clear understanding of the ray model and of image formation.

I like to end geometrical optics with a short qualitative discussion of chromatic and spherical aberration. Ray tracing in the paraxial, thin-lens approximation suggests that all optical tasks can be accomplished by a single converging lens. Students often ask where you would use a diverging lens or a lens combination. A simple ray-diagram discussion of aberrations is interesting to many students, and they can see qualitatively how a positive-negative lens combination can be chosen so as to (ideally) cancel the aberrations.

Sample Reading Quiz or Discussion Questions

1. What is a light ray?

2. What is specular reflection?

3. A paraxial ray
   a. moves in a parabolic path.
   b. is a ray that has been reflected from parabolic mirror.
   c. is a ray that moves nearly parallel to the optical axis.
   d. is a ray that moves exactly parallel to the optical axis.

4. A virtual image is
   a. the cause of optical illusions.
   b. a point from which rays appear to diverge.
   c. an image that only seems to exist.
   d. the image that is left in space after you remove a viewing screen.

5. The focal length of a converging lens is
   a. the distance at which an image is formed.
   b. the distance at which an object must be placed to form an image.
   c. the distance at which parallel light rays are focused.
   d. the distance from the front surface to the back surface.

Sample Exam Questions

1. A point source of light is 1 m from a large opaque screen. The screen has an L-shaped aperture with the dimensions shown in the figure. Light from the point source is viewed on a second screen, 2 m behind the aperture.

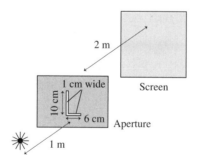

Draw a picture showing what you would see on the second screen. Your picture should a) indicate if boundaries are sharp or fuzzy, and b) show any relevant dimensions.

2. A student has built a 20-cm-long pinhole camera for a science fair project. She wants to photograph the Washington Monument, which is 550 feet tall, and have the image on the film be 2 inches high. How far should she stand from the Washington Monument?

3. A light bulb is suspended from the ceiling and hangs 1 m from a wall mirror, directly out from the mirror's left edge. The mirror is 2 m wide, and for convenience we'll assume

that it lies on the x-axis between $x = 0$ and $x = 2$ m. A student walks parallel to the mirror, 2 m away from the wall. Over what range of x can the student see an image of the light bulb in the mirror?

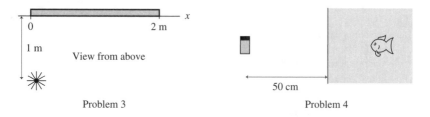

View from above

Problem 3   50 cm   Problem 4

4. A fish looks out the side of an aquarium at a can of fish food. The can is 50 cm from the edge of the aquarium. To the fish, does the can of food appear to be closer, farther, or equal to 50 cm from the edge? Support your answer with a ray diagram. (You can ignore the glass. Assume that the light passes directly from the air to the water.)

5. A 2-cm-high object is placed 60 cm from a 4-cm-diameter lens. A sharply focused 2-cm-high image is seen on a screen on the other side of the lens.

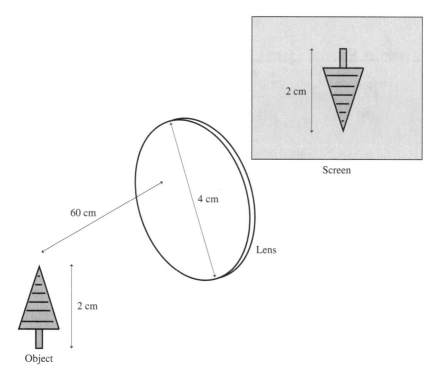

Screen

60 cm

4 cm

Lens

2 cm

Object

a. What is the focal length of the lens?
b. Suppose the screen is moved to be twice as far from the lens. Describe what you will see on the screen. Give a reason! (Your reason can consist of both words and pictures.)

   c. Restore the screen to its original position. Suppose a mask is placed over the lens. The mask has a 1-cm-diameter hole in the center, but everywhere else it is opaque. Describe what you will see on the screen. Give a reason!

6. Two converging lenses with focal lengths of 40 cm and 20 cm are placed 10 cm apart. A 2-cm-tall object is 15 cm from the 40 cm focal length lens, as shown in the figure.

   a. Use ray tracing to locate the image of the lens combination. Use a ruler to do this accurately. Make your best estimate of the image distance by making measurements on your diagram.

   b. Is the image real or virtual? Is it upright or inverted?

   c. Calculate the exact location and height of the image.

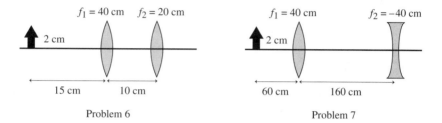

Problem 6   Problem 7

7. A converging lens of focal length 40 cm and a diverging lens of focal length -40 cm are placed 160 cm apart. A 2-cm-tall object is 60 cm from the converging lens, as shown in the figure.

   a. Use ray tracing to locate the image of the lens combination. Use a ruler to do this accurately. Make your best estimate of the image distance by making measurements on your diagram.

   b. Is the image real or virtual? Is it upright or inverted?

   c. Calculate the exact location and height of the image.

22

Physical Optics

Background Information

Interference and diffraction are important wave phenomena. Interference is widely used in science and engineering to measure small distances, vibrations, and optical surfaces. Holographic techniques are used from art to cockpit displays. Phased-array radar tracks aircraft with extremely high accuracy. Diffraction limits the density of optical data storage. And interference will later establish the wave-like nature of matter.

There is no clear boundary between *interference* and *diffraction*. The usage is primarily historical. In a general sense, diffraction represents the "spreading" of a wave, but the analysis is in terms of the interference of many sources along a wave front. And a "diffraction grating" is better thought of as multiple-slit interference.

Interference is a difficult topic for students. They must understand wave fronts, phase, and superposition, then they must be able to visualize what happens when two wave patterns overlap. Textbooks use pictures like the one shown here, but many students find this picture difficult to interpret. For example, many students think that

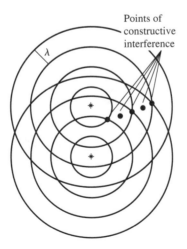

Points of constructive interference

an overlap of two crests causes an interference maximum while an overlap of two troughs causes an interference minimum. Students cannot visualize this pattern as moving outward from the sources, and they can't relate this picture to the standard photograph and demonstration of double-slit interference. Students need an opportunity to articulate what happens at different points in space as the wave patterns move outward. Then they need to think about and explain in their own words the intensity pattern of light as these waves impinge on a viewing screen.

The primary research on student understanding of interference and diffraction is from the McDermott group (Ambrose et al., 1999). They found that many students have serious difficulties understanding the basic features of the geometrical model and the wave model of light. Student explanations of interference and diffraction phenomena tended to be a confused and undifferentiated mixture of features from both models. Even the strongest students in the class had significant conceptual difficulties, and these were found to persist among physics majors in sophomore and junior-level courses.

In particular, their research has found that

- After studying physical optics, many students treated all apertures, regardless of width, as narrow slits. These students drew pictures of Huygen waves spreading out from 1-cm-wide apertures. This is a misapplication of the wave model in the domain of geometrical optics.
- On a post-instruction exam, only 20% of students correctly predicted *with correct reasoning* that the minima in a single-slit diffraction pattern would move further from the middle if the slit was narrowed. Some students misapplied geometrical optics reasoning to predict that the minima would move in. Others made a correct prediction, but their reasoning was incorrect and based on incorrect models of light.
- One group of students employed a hybrid model in which they interpreted the diffraction maximum as being the geometric image of the slit, and they attributed the fringes to "edge effects" of the slit. One student stated, "Light that strikes the edges will be diffracted off."
- Many students think that *no* light will pass through a slit if its width a is less than the wavelength λ. They state that the light will not "fit" through the slit in this situation. These students did, however, seem to recognize that wavelength is measured along the direction of propagation, perpendicular to the slit dimensions.
- Other students thought that diffraction occurs only if $a < \lambda$. One obtains a geometric image of the slit if $a > \lambda$, but the light "has to bend in order to fit through the slit" when the width is less than the wavelength, and this causes diffraction.
- Students were shown a two-slit interference pattern (with roughly equal intensity) and asked to predict what would happen if one slit were covered. Only 40% responded correctly. 25% predicted the pattern would be dimmer but unchanged, implying a belief that each slit alone produces the entire pattern.

20% predicted that the right or left half of the pattern would vanish, depending on which slit was covered. Both errors represent geometrical optics reasoning in an improper domain.

■ Many students think that the standard drawing of a wave represents an actual spatial extent of the wave. Student diagrams of light passing through a slit show that "part of the amplitude is cut off." This misconception is likely related to the misconception that electric field vectors extend through space. Students are interpreting diagrams literally, rather than as abstract representations of the situation.

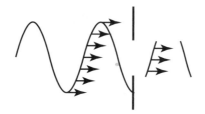

A practical source of trouble for students is the standard figure used to "explain" double-slit interference. This figure has two difficulties. First, it's rarely explained that we're seeing the slits "from above," along the direction of the slits. Without that information, students place a host of misinterpretations on this figure. Second, the figure overlays a graph on a picture of the experiment, with the graph rotated 90° such that intensity is graphed toward the left. This may be an efficient way to portray the situation, but students need a careful explanation of what is being shown here (and in similar figures later for diffraction), followed by exercises in which they must interpret the figure to answer questions.

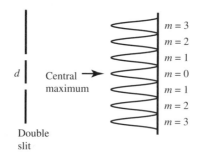

Student Learning Objectives

■ To see evidence for the wave nature of light.
■ To understand how and why interference occurs.

- To understand how minima and maxima are related to the path-length difference between two waves.
- To study the inevitable spreading of waves due to diffraction.
- To see how diffraction limits the resolution of optical instruments.

Pedagogical Approach

Most textbooks place physical optics after electricity and magnetism, long after the earlier chapters on traveling and standing waves. The treatment of superposition and interference in those earlier chapters is usually cursory, and long since forgotten by the time students reach optics. So it shouldn't be surprising, as McDermott's group (Ambrose et al., 1999) found, that students latch onto $d \sin \theta = m\lambda$ and apply it blindly with little understanding of what interference is all about.

Students need to acquire a new *model* of light in order to understand interference and diffraction. They need to visualize how waves move and how they interact. Instruction needs to be focused more on building conceptual foundations, less on routine calculations of fringe positions. My personal preference, as I've done it in my textbook *Physics: A Contemporary Perspective,* is to place all of wave physics, from strings through optics, in a single section of the book. This allows students to think about waves for an extended period of time and to develop the necessary understanding. Physical optics then flows smoothly out of analogous situations with string waves and sound waves. If you don't have this option, you'll need to take the time to develop the ideas of interference using more familiar waves before you start into physical optics.

The principle of superposition is hard for students to understand, as the research of Wittman et al. (1999) has shown. Many students focus only on the peaks and don't realize that the displacements add at *every* point on the wave. It's useful to start with copropagating and counterpropagating one-dimensional wave pulses. Students can draw the individual waves, then the net wave. Counterpropagating sinusoidal waves create a standing wave, and it's useful to make the connection between nodes/antinodes and destructive/constructive interference.

One of the student learning goals of the chapter on traveling waves was to understand the idea of phase. This will need a review if you haven't just recently finished traveling waves. Students need to recognize that a phase difference between two waves can arise because of

- a phase difference $\Delta \phi_0$ between the sources,
- different path lengths from the sources to the point where the waves combine, or
- a combination of these two.

Physical optics usually considers only path lengths, but interference in acoustics can easily arise from loudspeakers that are out of phase.

The two-dimensional wave pattern that was shown on the first page of this chapter is crucial for developing an understanding of interference. This picture can initially represent sound waves from two loudspeakers, then later be modified for double-slit interference. Although the figure is easily drawn, most students do not understand its implications unless you lead them through a series of questions and exercises. In particular, students need to

- Determine the path length difference Δr by explicitly counting the rings.
- Explain, at points of both constructive and destructive interference, what the individual waves are doing and what happens when they are superimposed.
- Map out the lines along which the interference is always constructive and lines along which the interference is always destructive (nodal and antinodal lines). This helps to connect two-dimensional interference with easier-to-understand one-dimensional standing waves.
- Explain what happens if the phase of one of the loudspeakers is changed.
- Explain what happens if the phase difference between the speakers is randomly fluctuating (incoherent sources).
- Imagine how the sound intensity will vary from point to point as these waves impinge on a wall. This is the acoustic version of the two-slit interference pattern.

These exercises establish the conceptual framework and provide a graphical representation that can be used to reason about interference.

Students readily believe that sound waves radiate out in all directions. After all, they know that a loudspeaker continues to be heard as they walk around it. Thus two nearby speakers create a region of space where two waves are overlapped. But it's *not* obvious that the conditions for interference—coherent sources and overlapped waves—are valid in the optical domain. Starting physical optics with double-slit interference, as most textbooks do, places students in a situation where they can't envision what is happening.

I find that it's better to start with single-slit diffraction. Not a quantitative analysis, but a qualitative display that light *spreads out* after passing through a narrow slit. Further, the spreading increases as the slit width decreases. You want students to recognize that

- This is not behavior that can be understood from the ray model of light.
- The light is acting rather like water waves passing through a hole in a jetty. This is an analogy that makes sense to many, perhaps most, students.

This allows you to make a *hypothesis* that light is a wave. This hypothesis will be justified only if it leads to successful predictions.

Once students accept that light spreads out after passing through a narrow slit, then you can introduce the double-slit experiment. Use a laser to illuminate both slits, rather than the older method of first using a single slit to generate a coherent wave. The points that you need to make clear are:

- Light spreads out after passing through each of the narrow slits. Consequently, there is a region where light from the two slits is overlapped. This is the critical point that is overlooked in many presentations.
- If light is a wave, the region of overlap is just like the overlapped circular waves you've been considering, except that the overlap has only a finite extent.

- If light is a wave, then the "sources" are coherent and in phase because both slits are illuminated by the same laser beam.
- Thus you have the conditions under which waves interfere. If you observe interference maxima and minima, then you have evidence that light is a wave.

This is reasoning that students can easily follow if you lead them through it, but very few would think of this on their own.

Now you're at a point where the analysis of the interference pattern makes sense. The extension of these ideas to interferometers, diffraction gratings, and single-slit diffraction is fairly straightforward.

I omit any quantitative analysis of the intensity patterns. I consider it a success if students learn enough of the wave model of light to *explain* interference and diffraction patterns and do basic calculations of the positions of maxima and minima. Details about the intensities can easily be deferred to upper division courses.

Using Class Time

Interference and diffraction need 5 or 6 days if you're starting from scratch. Four days can suffice if you've recently covered superposition and interference in an initial treatment of waves.

The ideas of interference can be introduced in one dimension with copropagating and counterpropagating waves. This starting point may seem far from reality, but thin-film antireflection coatings provide an example of interference between two copropagating light waves.

Present students with two counterpropagating pulses on a string, such as the ones shown below. Ask them to draw the individual pulses at intervals of 1 s, then the string's net displacement. Research has found that students have difficulty with this task because of an inadequate grasp of the principle of superposition. You want them to recognize that the waves add point by point along the entire length of the string, not just when the peaks overlap. As a follow up, flip one pulse over so that it's a negative displacement, then repeat the exercise.

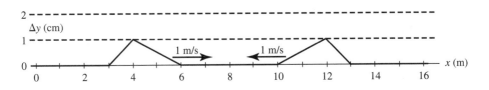

Sinusoidal waves can be introduced by asking students to consider sound waves emitted along a line by two loudspeakers. The speakers can be moved back and forth, and the relative phase between the speakers can be changed. The figure below shows situations of constructive and destructive interference. Note the wave front diagrams. Having students think about interference in terms of wave front diagrams will help when you move into two dimensions. You also want them to articulate that constructive interference occurs when both the crests *and the troughs* are aligned.

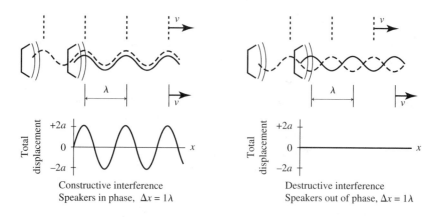

Constructive interference
Speakers in phase, $\Delta x = 1\lambda$

Destructive interference
Speakers out of phase, $\Delta x = 1\lambda$

If you make several transparencies of speakers and waves, you can slide them back and forth on an overhead projector to change the distance, and you can flip one over to change its phase by 180°.

The primary point of this initial exercise is that the phase difference between the two waves

$$\Delta\phi = 2\pi\frac{\Delta x}{\lambda} + \Delta\phi_0$$

can occur *either* due to a path length difference *or* to a phase difference between the sources. Regardless of how it happens, $\Delta\phi = m \cdot 2\pi$ is constructive interference and $\Delta\phi = \left(m + \frac{1}{2}\right) \cdot 2\pi$ is destructive. *If* the sources are in phase, constructive interference occurs when $\Delta x = m\lambda$.

An example that can be demonstrated graphically and worked numerically is to consider two loudspeakers emitting 343 Hz notes ($\lambda = 1$ m at room temperature) of amplitude a. What is the amplitude of the net wave if the distance between the two speakers is 0 cm, 25 cm, 50 cm, 75 cm, and 100 cm? You can have students consider the speakers themselves to be in phase, 180° out of phase, and 90° out of phase.

Counterpropagating sinusoidal waves are particularly interesting. Make two or three transparencies of each of the sinusoidal waves shown below, then cut them into narrow strips. (You'll also find these useful for double-slit interference.)

Slide the strips against each other in quarter-wavelength steps and ask students about the net displacement. They've previously learned that two counterpropagating waves create a standing wave, although they may have forgotten this if it's been a while since you covered standing waves. The point you want to make now is that standing wave antinodes are points of constructive interference and that nodes are points of destructive interference.

As a final important point of these one-dimensional exercises, have students draw a graph of *intensity* versus position for the counterpropagating waves. They may not have a rigorous definition of intensity, but they'll readily accept that the intensity is large where the string oscillates with a large amplitude, small where the amplitude is small. The exact shape isn't critical now, but you want them to recognize that

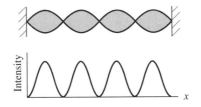

- The intensity graph looks more or less as shown in the figure.
- The intensity pattern is *stationary,* even though the waves are moving and the standing wave is oscillating. This is a difficult idea for many students.

This graph is a very important link for later understanding intensity in the double-slit interference experiment.

These one-dimensional examples will prepare students for a qualitative exploration of the overlapping wave front pattern of two circular waves, as was shown at the start of this chapter. You can introduce this picture as representing two loudspeakers that emit identical tones. It's useful to make two overhead transparencies of circular rings. You can place these on the projector with different distances between the sources. There are a number of points to make:

- Constructive interference occurs not only where two crests overlap but also where two troughs overlap. Destructive interference occurs where a crest overlaps a trough. As noted earlier, students have a tendency to think that destructive interference occurs at points of overlapping troughs.
- Constructive and destructive interference occur along *lines* through space, not just at the discrete points where crests or troughs overlap. These lines (antinodal and nodal lines, respectively) are the two-dimensional equivalents of standing wave nodes and antinodes.
- Each line of constructive interference is characterized by a fixed value of $\Delta r = m\lambda$ (assuming in-phase sources). You'll want to demonstrate this by having students count the rings to several different points along one of the antinodal lines.
- Ask the students what they would *hear* if they walk along a line parallel to the plane of the sources. You want them to recognize how the intensity is modulated

as they cross the nodal and antinodal lines. Connect this back to the graph they drew of intensity along a standing wave, and prepare to extend this idea to double-slit interference.

Acoustical interference is well worth demonstrating, although the demonstration has to be kept short to avoid becoming exceedingly annoying. Place two loudspeakers about 1 m apart and send identical 1500 Hz signals to each. Use a microphone as a detector, with the microphone connected to an oscilloscope for display. Move the microphone along a line 2 m in front of the plane of the loudspeakers. There should be a clear maximum at the center and another about 25 cm on either side. You'll want to stay as far away from walls or other reflecting surfaces as possible to make the maxima and minima distinct.

You should be able to find three or four maxima on either side of center, with the distance between maxima increasing as you move out. (Unlike the double-slit experiment, the maxima are *not* equally spaced because the small angle approximation is not valid.) If you have a small class, you can let students move back and forth in front of the speakers to *hear* the maxima and minima. Reversing the connections to one speaker, so that the speakers are out of phase, places a minimum at the center. You may want to use the waves on transparencies to help explain the observations. This demonstration is very useful as a prelude to the double-slit experiment.

A simple numerical exercise of acoustic interference is to consider two loudspeakers placed 1.5 m apart. Let both speakers emit identical 343 Hz waves and be in phase. Ask students whether the interference at point A is constructive, destructive, or in between. Then at points B (destructive, $\Delta r = 1.5\lambda$) and C (destructive, $\Delta r = 0.5\lambda$). Also inquire about the situation at points A and C if the speakers are out of phase.

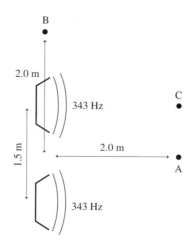

Ripple tanks and computer simulations, if you have them, are useful at this point in the presentation. It is important, however, not just to show them but to ask students to *explain* what is happening at different points in the pattern. Students should leave these introductory examples, in the familiar realm of string and sound, with a good understanding that interference patterns are explained by comparing the path length difference to the wavelength. It's all just geometry!

This is a roundabout way to get to physical optics, but students need time to develop a model of wave behavior that will allow them to think about and understand optical interference. Even so, making the leap from strings and acoustics to light isn't easy. Students accept that "spreads out" is a defining characteristic of waves, based on your demonstrations and their own experience. But few have ever seen evidence that light spreads out—and, in fact, you've likely just finished a chapter on geometric optics, where students learned that light travels in perfectly straight lines!

Single-slit diffraction—qualitative, not quantitative—is the appropriate transition. Set up a single-slit apparatus, but don't yet turn on the laser. Draw a picture on the board, and have students use geometrical optics reasoning to predict how the light intensity on the screen should look. Remind them of what they know about sunlight shining through windows. Turn on the laser only after all students are certain of the prediction of the ray model.

All you want to do initially is point out that the prediction of the ray model has failed. Instead, the light *spreads out*—just like sound waves or water waves passing through a small opening. There is no need at this time to call their attention to the minima in the intensity. Suppose you propose the hypothesis, "Light is some kind of wave." The questions you need to answer are then:

- What kind of evidence would be convincing that light is a wave?
- If light is a wave, what is its wavelength?
- If light is a wave, then why does the ray model work so well for lenses and mirrors?

You can promise that these questions will be answered before the chapter is over.

This introduction prepares students to understand the double-slit experiment. Point out the fact that you have two very narrow slits very close together. They just saw that a narrow slit causes light to spread. Consequently, light passing through two side-by-side slits will spread from each of them *and overlap*. (This is the critical point that isn't obvious if you haven't first demonstrated single-slit diffraction.) Now you can point out that the situation is exactly analogous to the acoustic interference they've just finished investigating.

You certainly want to spend a fair bit of time demonstrating and discussing the double-slit experiment. Many students don't understand the standard textbook drawing that shows the two slits from a bird's-eye perspective. Both the experimental arrangement and the drawing need a careful explanation to avoid unnecessary confusion. Point out explicitly that the interference pattern is strong evidence for your wave hypothesis.

With light interference, more than with acoustic interference, students tend to think that the interference exists *only* on the screen and not in the space between the slits and the screen. A useful demonstration is to orient the slits to make the interference pattern vertical, then use chalk dust to illuminate the "fans" of light heading toward the screen. This requires a very dark room. If your circumstances are less than ideal, you can use a diffraction grating to make the fingers of the fan brighter and more distinct. It's also possible to show the spatial pattern by sending the light through a tank of water to which a small pinch of corn starch or artificial sweetener has been added.

The sinusoidal transparency strips that you used for counterpropagating waves can again come in handy. Place them on top of a second transparency that shows two "slits" and a "screen." By pivoting the two transparencies about the slits, you can show—by explicitly counting the crests—that there are points on the screen where the path length difference Δr is 0λ, $\frac{1}{2}\lambda$, 1λ, $\frac{3}{2}\lambda$, 2λ, and so on. This simple demonstration is very helpful to many students. You can also use this demonstration to show why the interference pattern expands as the spacing between the slits decreases.

You'll also want to refer to the earlier graph of the wave intensity of a standing wave on a string and to the picture of two spreading circular waves. You want students to recognize that:

- The maxima and minima are where the antinodal and nodal lines intersect the screen. This will now be quite believable if you were able to demonstrate the fans of light heading toward the screen.
- The intensity pattern is stationary, even though the light waves are moving.

I like to conclude the double-slit demonstration by measuring the fringe spacing and the distance from the slits to the screen. I give students the slit spacing and then have them compute the wavelength of the laser (633 nm for a HeNe). Only modest care in the measurements is needed to get the wavelength within ± 30 nm, a 5% measurement. Now you have an experimental basis to talk about the wavelength of light, noting how small it is in comparison to sound and water waves.

Whether or not you discuss interferometers or thin-film reflection will depend on how much time you have. Students do need some time to think about what's going on, so it's preferable to omit interferometers rather than to rush through them. If you do have time, many departments have an acoustic interferometer demonstration in which a trombone slide is pulled out to lengthen one path. There is a very clear oscillation between minima and maxima at the detector. This demonstration makes the Michelson interferometer much easier to understand.

Most textbooks go from double-slit interference to single-slit diffraction and then to diffraction gratings. The reasoning behind this presentation has always escaped me, since the diffraction grating is really an extension of interference from two slits to many slits. Moving from the double-slit experiment to diffraction gratings is not only logical, it also prepares students for the analysis of a single slit by getting them to think about the interference of many waves.

Show a picture of three slits (see figure on next page), then "close" slit S_3. Consider a point on a viewing screen where there is a bright spot due to constructive interference of the light from slits S_1 and S_2. Will this still be a bright spot when S_3 is opened? Let students discuss this for a while without being in a rush to provide the answer. You want them to realize that the constructive interference condition

$\Delta r = m\lambda$ between S_1 and S_2 will also be true, at the same angle, for S_2 and S_3. So light from S_3 is in phase with the light from S_1 and S_2, and that point on the screen will continue to be bright. You can then ask about 4 slits, 5 slits, and N slits.

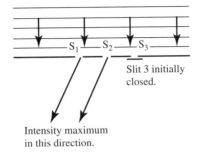

Slit 3 initially closed.

Intensity maximum in this direction.

If the amplitude of the wave from each slit is A, N in-phase waves give a total amplitude NA. Not only are the maxima of N slits in the same positions as the maxima of two slits (same slit spacing), but the intensity is much brighter. Going from two slits to many slits reinforces the maxima.

I don't attempt to make a rigorous justification of why the maxima become sharper with the diffraction grating. Two hand waving arguments are enough to satisfy students. First, the maxima are much brighter because all N waves are reinforcing each other there. By energy conservation, this would imply that there must be less light elsewhere on the screen. Second, the maxima are very bright because the light from each slit is in phase with all the others. This is a very unique condition. At other angles, the phases of light coming from the different slits are more or less random. They're going to tend to cancel when we add them up, so other points on the screen should be fairly dark.

Finally, you can return to the single slit. The 3, 4, . . . , N slit exercise with the diffraction grating will help students understand the single slit analysis, although you'll need to emphasize strongly that the analysis in this case locates the *minima*. This is hard for many students to remember, due in large part to the similarity of the single-slit-minimum equation to the double-slit-maximum equation. You'll want students to articulate the fact that the diffraction pattern is due to the entire slit. Some students believe that the central maximum is a geometric image of the slit while the fringes are due to light that "diffracts off" the edges of the slit.

A nice exercise is to calculate the slit spacing for which the diffraction central maximum is the same size as the geometric image of the slit. For $\lambda = 500$ nm and a screen distance $L = 1$ m, the slit width is $a = 1$ mm. This helps answer the final question that you posed with the wave hypothesis, "Why does geometric optics seem to work?" Interference and diffraction effects will just *start* to show up when aperture sizes are about 1 mm, and they won't be obvious until the apertures are quite a bit smaller than this. Now you have established a *criterion* for when the ray model and the wave model of light are valid. Lenses and mirrors are always much larger than 1 mm, so we can use ray optics and not worry about the spreading of the light.

Circular diffraction is somewhat difficult to demonstrate because of the alignment difficulties, but it is worth the effort to do so. Most textbooks have a photograph, but students don't get any sense of scale from a photo. You can supply, without derivation, $\theta_1 = 1.22\lambda/D$ for the angle of the first minimum.

Available time will determine whether or not you cover Rayleigh's criterion for image resolution. But it's definitely worth the time needed to discuss diffraction-limited focusing of a lens of focal length f and diameter D. The standard model of an ideal lens followed by a circular aperture of diameter D leads to a circular diffraction pattern in the focal plane of diameter $d_{\text{focal spot}} = 2.44\lambda f/D$. This has obvious application to the issue of the maximum density of optical storage, such as the density of pits on a compact disk. Students find this quite interesting.

Sample Reading Quiz or Discussion Questions

1. What was the first experiment to show that light is a wave?

2. Draw a picture of two waves that are out of phase.

3. What is a diffraction grating?

4. When laser light shines on a screen after passing through two closely spaced slits, you see
   a. a diffraction pattern.
   b. two dim, closely spaced points of light.
   c. interference fringes.
   d. constructive interference.

5. This chapter discussed the
   a. Michelson interferometer.
   b. Nicholson interferometer.
   c. both a and b.
   d. neither.

6. The spreading of waves behind an aperture is
   a. more for long wavelengths, less for short wavelengths.
   b. less for long wavelengths, more for short wavelengths.
   c. the same for long and short wavelengths.
   d. not discussed in this chapter.

7. Apertures for which diffraction is studied in this chapter are
   a. a single slit.
   b. a circle.
   c. a square.
   d. both a and b.
   e. both a and c.
   f. all of a, b, and c.

Sample Exam Questions

1. Two loudspeakers are placed 1.8 m apart. They play tones of equal frequency. If you stand 3 m in front of the speakers, and exactly between them, you hear a maximum of intensity. As you walk parallel to the speakers, staying 3 m from the plane of the speakers, the sound intensity decreases until reaching a minimum when you are directly in front of one of the speakers.
   a. What is the frequency of the sound?
   b. Draw, as accurately as you can, a wave front diagram. On your diagram, label the positions of the two speakers, the point at which the intensity is maximum, and the point at which the intensity is minimum.

    c. Use the wave front diagram to explain why the intensity is a minimum at a point 3 m directly in front of one of the speakers.

2. Sound of frequency f is broadcast into the enclosed tube, shown in the diagram, which is filled with room-temperature air. The upper section, of length L, can be extended like a trombone slide. A microphone at the far end of the tube detects maxima of the sound intensity when L is 24 cm, 32 cm, and 40 cm, and at no positions in between.

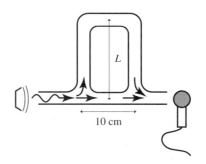

    a. What is the sound frequency?
    b. The slide is returned to the position $L = 24$ cm. Then the frequency of the sound is slowly decreased. What is the first frequency for which the sound intensity is a minimum?

3. The figure below shows light passing through two narrow, closely spaced slits. The graph shows the light intensity pattern seen on a screen behind the slits.

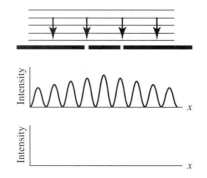

    a. Draw a graph on the axes provided to show the intensity pattern if the right slit is blocked, allowing light to go only through the left slit.
    b. Explain why the graph will look this way.

4. A screen is placed 200 cm behind two narrow slits that are separated by 0.480 mm. The figure below shows the light intensity on the screen.
    a. What is the wavelength of the light?
    b. Consider point A on the screen. What is Δr, the difference between the distance from A to one slit and the distance to the other slit?

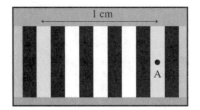

Problem 4

5. The figure below shows the intensity of light on a screen behind a single slit of width a. The light's wavelength is λ. Is $a > \lambda$, $a = \lambda$, or $a < \lambda$? Explain your reasoning.

Problem 5

6. A diffraction grating is placed 1.00 m from a viewing screen. Light from a hydrogen lamp goes through the grating. A hydrogen spectral line with a wavelength of 656 nm is seen 60.0 cm to one side of the center. Then the hydrogen lamp is replaced with a mercury lamp.
   a. Mercury has a spectral line with a wavelength of 546 nm. Where is this line seen on the screen?
   b. A spectral line is seen on the screen 36.4 cm away from the center. What is its wavelength?

7. A music CD is "read" by a laser beam that is focused onto the surface. The laser beam is reflected from small "pits" in the surface. Each pit stores one bit of digital data. A bright reflection is read as a digital 1, a dark reflection as a digital 0. The density of pits is limited by the ability of the laser to focus on them.

A CD has a maximum play time of 75 minutes. The surface of a CD has an inner diameter of 4 cm and an outer diameter of 12 cm. There are two stereo channels. Each channel is read at 44,000 "words" per second, and each word contains 16 bits. The lens used to focus the laser has a focal-length-to-diameter ratio $f/D = 0.8$, which is about the minimum ratio that is technically possible.

Assume that the pits are placed on the CD in a square pattern. What is the maximum possible wavelength of the laser? (A square pattern is not the most efficient, and actual CDs use a more complex pattern. Consequently, the actual wavelength used in CD players is $\approx 25\%$ larger than your calculation. Nonetheless, your calculation is a very good estimate.)

Quantum Physics

Background Information

Quantization is an essential part of our understanding of physics at the beginning of the twenty-first century. Quantum physics is increasingly important for students in engineering and other sciences, with applications such as quantum chemistry, materials engineering, nanostructures, scanning tunneling microscopes, quantum-well devices, and lasers. A significant fraction of scientists and engineers of the twenty-first century will need some knowledge of quantum physics.

Several of the curricula tested under the auspices of the Introductory University Physics Project included a significant component of quantum physics. Independent evaluation of the IUPP curricula found that students were receptive to and interested in this topic. Many students have an awareness of quantum ideas from other science courses as well as from popular literature and movies, and most are well motivated. Although the ideas are "abstract," the mathematical rigor of elementary quantum mechanics is no higher than that of wave physics.

The three keys to success identified by the IUPP class tests are

- Keep the presentation closely tied to phenomena and applications.
- Keep the mathematical level modest.
- Allow adequate time. These are strange new ideas, and they can't be rushed.

It is important to note that the modern physics or quantum physics chapters at the end of an introductory course are not intended to replace the usual sophomore-level "modern physics" course. The goals here, much more limited, are to introduce basic phenomena, concepts, and terminology to those students for whom this is the last course in physics. Those students who do go on to modern physics will be better prepared for the increased mathematical sophistication by this initial exposure, on a more physical level, to the major ideas.

All students are familiar with and use the terms "atom" and "electron," but most students have only vague conceptions as to what these terms mean. Most have a collection of half-remembered "facts" from chemistry, but no clear mental image of either a classical or a semi-classical atom. For example, a significant fraction of students have little or no idea what ions or isotopes are, and many think that protons can be added to or removed from atoms as easily as electrons. Virtually none of the students can cite any evidence by which we know about atoms or their structure, so their knowledge is primarily "book learning" not tied to any particular physical phenomena.

The primary difficulty with quantum and atomic physics, of course, is that the basic entities are not directly perceptible to our senses. Our knowledge of atoms, at least until quite recently, was *inferred* from a variety of macroscopic experiments. No one single experiment tells us all about atoms, but taken together they form a very secure "web of knowledge."

As Arons (1997) emphasizes, genuine understanding of quantum physics requires students to recognize *how* and *why* we know that atoms exist and that energies are quantized. Memorized facts and definitions that aren't connected to physical phenomena don't represent real knowledge. Hence primary functions of these closing chapters are to:

- Acquaint students with some of the experimental evidence by which we know about atoms and photons.
- Let students recognize that many of the phenomena associated with light and atoms cannot be predicted or explained by classical physics.

There is very little research on students' understanding of atoms. Earlier chapters in this guidebook have noted the many misconceptions students have about the nature of light. For example, many students think that a red filter *adds* red light to white light rather than subtracting non-red wavelengths. These misconceptions are relevant to their understanding of emission and absorption spectra. A large fraction of students think that radioactivity is somehow due to orbital electrons, and most students think that an object will be radioactive after being exposed to radioactivity.

The only aspect of quantum physics on which there has been a reasonable amount of research is the photoelectric effect. Steinberg, et al. (1996), in a study of sophomore students in a modern physics class, found that most students had significant conceptual difficulties with the photoelectric effect itself and with the photon model used to explain the results. Using informal surveys, I've found that only a small fraction of students in a modern physics class understand the significance or the implications of the photoelectric effect. When asked on an exam to explain how the photoelectric effect was inconsistent with classical physics, the majority of students wrote that the mere existence of the photoelectric effect violated classical physics. Only a very small minority could articulate how the photon model succeeds where the classical model fails.

Three issues that cause particular trouble are

- Not understanding the experiment itself—how it works or what is measured.
- Not understanding the basic ideas of the photon model of light.
- Not being able to use either the classical or the photon model to *reason* about how the experimental results would change if some parameter of the experiment is varied.

The usual presentation of the photoelectric effect experiment assumes that the students understand basic circuits and the meaning of potential difference. As earlier chapters in this guidebook have documented, most students who have completed conventional instruction in electricity and electric circuits *cannot* reason with or about potential difference. For example, most cannot correctly determine the potential difference across a gap in a circuit—which is what the photoelectric tube is. This has serious implications for their ability to understand the photoelectric effect.

The Steinberg et al. (1996) research found that less than 50% of students, on the final exam, could explain why the current never goes negative or why the current is not zero when $\Delta V = 0$. Most "explanations" were based on Ohm's law. When asked to draw a current-voltage graph for the phototube, one-third of the students drew the graph for a resistor—linear, passing through the origin, and with both positive and negative current. In addition, a significant fraction of students thought that electron ejection from the cathode was due to the battery. When asked to explain the source of the current, they referred to the battery but did not mention the light. These are difficulties with the basic experimental design. But without understanding how the experiment functions, students cannot understand the distinction between classical and quantum explanations.

In a post-instruction interview setting, students were shown a correct current-potential graph and were asked several "what if" questions. If asked to draw the graph for a different cathode whose work function exceeds the photon energy hf, less than 50% recognized that the current is zero for all ΔV. Many drew graphs with a positive value of the stopping voltage. If asked to draw the graph for a weaker light intensity but the same wavelength, more than 50% drew a graph in which the stopping potential decreased. If shown a situation in which no current was flowing because the light frequency was below threshold, 80% thought that current would begin to flow if the *numerical* value (in volts) of the tube's potential difference ΔV was increased to be larger than the *numerical* value (in eV) of the work function.

Altogether, the research suggests that most students

- Don't understand the experiment itself.
- Don't recognize the specific measurements where the classical model fails but the photon model succeeds.
- Don't distinguish between the role of ΔV, hf, and the work function E_0.
- Don't distinguish between photon flux and photon energy.
- Don't understand how the stopping potential is related to the photon energy.

Consequently, the significance of the photoelectric effect experiment is lost on most students.

On a more general issue, the University of Washington researchers found that a significant fraction of students have serious conceptual errors about how photons travel between two points. In particular, students tend to interpret the usual wave-packet diagram as the actual *trajectory* followed by a photon.

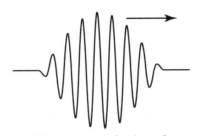

The *trajectory* of a photon?

I have confirmed this surprising discovery by having the instructor in a modern physics class give the following question to his class.

Light of wavelength 1 μm is emitted from point A. A photon is detected 5 μm away at point B. Draw on the figure the trajectory that a photon follows between points A and B.

Fully half the students in the class drew either a sinusoidal trajectory with five oscillations (since $\Delta x/\lambda = 5$) or drew a wave-packet trajectory.

This misinterpretation of a wave-packet diagram has other implications. Although not studied in detail, the researchers at the University of Washington have seen evidence that a significant number of students think that electrons also move along sinusoidal trajectories in experiments, such as the double slit, where electron interference is observed. Other researchers have found that many students interpret electron interference and diffraction as being due to collisions or repulsive interactions between the electrons. Most of these students predicted that the electron interference pattern would vanish if the electrons were fired one at a time.

Finally, on a more practical level, most instructors are aware that students have many difficulties with the energy unit *electron volt*. Part of the difficulty lies with the name itself, which suggests units of potential rather than energy, and part lies with the fact that many students never understood the concept of potential and cannot differentiate between potential and potential energy. Quite a number of class examples are needed to help students see how electron volts are used (and not used).

Another common difficulty, one relevant to atomic physics, is with negative energies for bound states. The fact that an orbiting electron has positive kinetic energy makes it especially difficult to see how the total energy is negative. You need to lead students through a series of questions where they first agree that an electron at rest infinitely far away has zero energy (easy to believe) and that an orbiting electron—because it is bound—has *less* energy (harder to believe).

Student Learning Objectives

■ To recognize phenomena that cannot be explained by classical physics, thus motivating the need for a new theory.

■ To establish experimental evidence by which we know about the existence of atoms and about their properties.

■ To understand the photoelectric effect experiment and its implications.

■ To understand the photon model and its application to the photoelectric effect.

■ To understand the evidence for matter waves and the de Broglie wavelength.

■ To recognize that the de Broglie standing wave of a confined particle requires energy quantization.

■ To understand Bohr's stationary-state model of the atom.

■ To use the Bohr model to explain discrete spectra and the observed differences between absorption and emission spectra.

■ To study Bohr's specific model of the hydrogen atom.

Pedagogical Approach

The basic entities of quantum and atomic physics are not perceptible to the senses. A major goal of these chapters is to introduce some of the basic phenomena that led to an understanding of atoms and that motivated the introduction of quantum theory. This is a topic for which the historical perspective is especially effective. However, some of the examples need to be explored in more depth than usually presented in order to examine how difficult and confusing the issues were at the time and how experimental evidence was actually *used* to decide between competing hypotheses.

Few textbooks give sufficient information about the experimental history of atoms and electrons. The story is quite interesting, and it provides a particularly good lesson on "What is science and how does science operate?" Chemical studies in the early nineteenth century began to establish an empirical basis for the existence of atoms, and by 1850 it was accepted that chemicals are composed of various combinations of a small number of basic elements and that the atoms of each element are all alike. But the electrical conductivity of solutions and gases—one of Faraday's major studies—was still mysterious and unexplained. His observations suggested that atoms are *not,* as the classical picture implied, the indivisible, fundamental entities of nature. Instead, atoms appeared to have some form of internal structure that determines their electrical properties.

An important technological breakthrough was the development, in the 1850s, of much improved vacuum pumps. In 1858, the German scientist Plücker began a study of Faraday's gas discharge tube, using lower gas pressures. He found that, at sufficiently low pressures, the cathode glow extends to the glass wall and, significantly, the glass itself emits a greenish glow at that point. Further, a solid object sealed in the tube casts a shadow on the glass wall. This discovery suggested that the cathode emits *cathode rays* of some form that travel in straight lines. But naming the rays did nothing to explain them. What were they?

The most systematic studies on the new cathode rays were carried out during the 1870s by Sir William Crookes. While earlier studies had been primarily qualitative observations, Crookes devised a set of tubes with which measurements could be made. These are, of course, what we today call Crookes tubes. His primary innovation was to elongate the tube, use yet lower pressures, and to introduce one or more holes to produce a collimated beam of cathode rays.

The work of Crookes and others demonstrated that

- All metal cathodes produce cathode rays, and the ray properties are independent of the cathode material.
- The rays are deflected by a magnetic field *as if* they are negative charges.
- The rays can exert forces on objects and can transfer energy to objects, causing a thin foil in the cathode beam to glow red hot.

Crookes's experiments led to more questions than they answered. Were the cathode rays some sort of particles? Or a wave? Note that magnetic deflection, although suggestive that the cathode rays are negative particles, is not by itself sufficient to demonstrate that they really are.

It is important for students to realize how difficult these questions were at the time and how experimental evidence was used to answer them. Crookes advanced a theory that molecules in the gas collided with the cathode, somehow acquired a negative charge, and then "rebounded" with great speed as they were repelled by the negative cathode. The "charged molecules" would travel away in a straight line, carry energy and momentum, be deflected by a magnetic field, and could cause the tube to glow, or *fluoresce,* where it was struck by the molecules. Crookes's theory predicted, of course, that the negative ions would also be deflected by an electric field. This would be a definitive demonstration that cathode rays are charged particles. Crookes attempted to demonstrate it by sealing electrodes into the tube and creating an electric field, but his efforts were inconclusive.

His theory was, however, immediately attacked. It was known that the cathode rays could travel the length of a 90-cm-long tube with no discernible deviation from a straight line. But the mean free path for molecules at this pressure, due to collisions with other molecules, is only about 6 mm. It was later discovered that the cathode rays could even penetrate very thin metal foils and, if these foils were at the end of the tube, then travel about 1 cm through the air. No atomic-size particle could penetrate these foils. Crookes's theory, seemingly adequate when it was proposed, was wildly inconsistent with subsequent experimental observations.

An alternative theory, more widely held than Crookes's, was that the cathode rays were electromagnetic waves. After all, waves travel in straight lines, cast shadows, carry energy and momentum, and can, under the right circumstances, cause materials to fluoresce. It was known that hot metals emit light—incandescence—so it seemed plausible that the cathode could be emitting waves. The major obstacle for the wave theory was the deflection of cathode rays by a magnetic field. But the theory of electromagnetic waves was still quite new at the time, and many characteristics of these waves were still unknown. Although visible light was not deflected by a magnetic field, it was easy to think that some other form of electromagnetic waves might be so influenced.

Shortly after Röntgen's 1895 discovery of X rays, J. J. Thompson began using them to study electrical conduction in gases. He found that X rays could discharge an electroscope and concluded that they must be ionizing the air molecules, thereby making the air conductive. This simple observation was of profound significance. Up until then, the only form of ionization known was the creation of positive and negative ions in solutions. Although not yet understood, the fact that two atoms could acquire charge as a molecule splits apart in solution did not jeopardize the idea that the atoms themselves were indivisible. But after seeing that even monatomic gases, such as helium, could be ionized by X rays, Thompson realized that the *atom itself* must have charged constituents. This was the first *direct* evidence that the atom is a complex structure made up, somehow, of charged pieces.

Thompson was also conducting experiments on cathode rays. He had previously leaned toward the charged-particle model of cathode rays, and this new insight about the atom reinforced his belief. Other scientists had placed an electrode in the cathode-ray beam and measured a current flow, as shown in the left figure. Although this seemed to demonstrate that the rays are charged particles, proponents of the wave model argued that the current might be a separate, independent event that just happened to be following the same straight line as the rays.

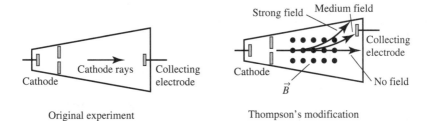

Original experiment   Thompson's modification

Thompson realized that he could use magnetic deflection of the cathode rays to settle the issue. He built a modified tube that had an electrode off to the side, as shown on the right, then placed the tube in a magnetic field which deflected the cathode rays to the side. He could determine their trajectory by the location of the green spot as it moved across the face of the tube. Just at the point when the field was strong enough to deflect the cathode rays onto the electrode, it began to collect a current! At an even stronger field, when the cathode rays were deflected completely to the other side of the electrode, the current ceased.

This was the first conclusive demonstration that cathode rays really are negatively charged particles. But why were they not deflected by an electric field? Thompson's first experiment to look for such a deflection met with the same inconclusive results that others had found. But it was Thompson's experience with the X-ray ionization of gases that led him to recognize the difficulty. He realized that the rapidly moving particles in the cathode-ray beam must be *ionizing* the molecules in the gas—which were still plentiful even at low pressures. The electric field created by the space charge effectively neutralized the field of the electrodes, hence there was no deflection.

Fortunately, vacuum technology was getting ever better. By using the most sophisticated techniques of his day, Thompson was eventually able to lower the pressure to a point where ionization of the residual gas was not a problem. And then, just as he had expected, the cathode rays *were* deflected by an electric field.

This was a decisive victory for the charged-particle model, but it still did not reveal the nature of the particles. It is at this point that Thompson devised the cross-field experiment described in all textbooks. But the point of that experiment is largely lost without knowing more about the context in which it occurred. Further, Thompson's reasoning regarding the outcome of his final experiment is usually not given. He had measured the charge-to-mass ratio of cathode rays and found it to be roughly 1000 times smaller than the charge-to-mass ratio of hydrogen, which was known from electrolysis. As he noted, this could imply either that cathode rays had a much smaller mass than hydrogen ions, or a much larger charge. He then referred back to the earlier experiments showing cathode rays travel in long straight-line trajectories, much longer than molecular mean free paths, and that they can penetrate thin foils. This evidence was earlier used to argue that cathode rays could not be particles. Now that it was certain that they were, Thompson turned the tables and used the same information to conclude that the particles must be vastly smaller than atoms and molecules. This was the important final step in the reasoning that concluded with the discovery of a *subatomic* particle.

It is important that students not see this historical material simply as entertainment. You need to have them *use* the experimental evidence to *reason* about what does—and does not—happen in different situations. For example, Thompson found that the cathode rays emitted by aluminum, iron, and lead cathodes all had the same charge-to-mass ratio. What is the significance of this discovery? Only after students can answer such questions do they have any genuine knowledge of atoms and atomic structure.

The analysis of the experiments of Thompson, Millikan, Rutherford, and others requires many ideas from classical physics. Some of these ideas will need a short review at this time, but their reintroduction gives students yet another opportunity to practice using basic concepts of classical physics. Not only is this information a springboard to quantum physics, it is also, in many respects, a synthesis of all the classical physics that has been covered.

From a pedagogical perspective, the photoelectric effect is the critical experiment that points the way toward quanta. Although Planck and the blackbody spectrum have historical precedence, this story cannot be presented in a meaningful way to introductory students. Not only are students unsure how to interpret a continuous spectrum, they don't have the knowledge or background to understand how this spectrum was a crisis for classical physics. The photoelectric effect experiment, by contrast, provides a straightforward distinction between classical and quantum ideas.

The standard presentation of the photoelectric effect presents students with "facts" about the experiment and moves quickly to the photon model as if the failure of classical physics is obvious. The research cited above has found that a majority of students in a sophomore modern physics class gain little understanding of the experiment or its implications. The researchers did find significant improvement if students carried out a series of exercises in which they were required to go through the chain of reasoning from the data to the implications. This is analogous to following the reasoning that Thompson used to "discover" the electron, and it is an important step toward Arons's goal that students recognize *how* and *why* we know that atoms exist and that energies are quantized.

The presentation of the photoelectric effect needs to focus more attention than is usual on

- Explaining the experiment.
- Stating clearly what was learned by Lenard in 1900 and known to Einstein. In particular, the existence of a threshold frequency, that the stopping voltage is independent of the light intensity, and that there is no perceptible delay before current starts to flow.
- Exploring which of the results could be understood classically and noting where the classical explanation fails.
- Giving a clear statement of the photon model.
- Exploring in detail how the photon model is applied to the experiment.

Note that many textbooks portray the experiment with a variable resistor whose slide wire is connected to the anode. Student difficulties surrounding the operation of the resistor circuit often obscures the physics ideas we're trying to convey. It is preferable to show the circuit simply with a variable battery, which can be reversed to get negative voltages.

Classroom exercises should emphasize *reasoning* about the experiment. Ask students how the current-voltage or current-frequency graphs would change if different parameters—intensity, wavelength, work function, and so on—are changed. They can explore which results can be explained by classical reasoning and which cannot. After the photon model is introduced, it is especially important for students to engage in qualitative reasoning prior to beginning numerical problems. There is good evidence that this approach is effective in allowing most students to reach a satisfactory understanding of the photoelectric effect.

The work function is usually symbolized as ϕ, for reasons I've never understood. This makes a somewhat difficult idea even more mysterious. The work function is the minimum energy needed to pull an electron out of a metal, so I prefer to give it the more obvious symbol E_0. This helps students to remember that the work function is an energy. Work functions are usually given in electron volts, so the confusion between volts and electron volts becomes an issue. Students will need multiple opportunities to practice using both volts and electron volts in the same problem.

The de Broglie wavelength is best introduced with theory and experimental evidence closely intertwined. De Broglie's initial supposition—if light waves can have particle-like properties, then perhaps particles can have wave-like properties—allows you to exploit analogies between electrons and photons as you develop the ideas of wave-particle duality. But trying to grasp what "electron wavelength" means is even more difficult for students than trying to understand photons. Electron interference patterns, neutron interference, and even recent experiments with atom interferometers all need to be mentioned and discussed to bring home the reality of this strange new aspect of matter.

The particle-in-a-box is an important application of the de Broglie wavelength. Students recall that a standing string wave or standing light wave occurs when the wave reflects back and forth in an enclosure. A confined particle in a box bounces back and forth between the ends. If a particle has wave-like properties, this should set up a standing de Broglie wave. The discrete frequencies of the standing wave lead directly to energy quantization. This is an important step on the road to understanding that atomic energies are quantized.

The semi-classical Bohr model of the atom is pedagogically very important, even if the specifics are later abandoned in favor of wave functions. Much of the language and terminology of quantum mechanics is introduced in the Bohr model, and it is an important stepping stone for students who are not ready to plunge straight into the subtleties of quantum physics. In addition, it's safe to say that the Bohr model is the basis of much of the day-to-day thinking about atoms by physicists themselves. Even though we recognize its limitations, there is a reasonably wide domain in which the Bohr model is an acceptable model of atomic behavior.

However, it is important to distinguish, which few textbooks do, between the *Bohr model,* which is a general model of atomic behavior, and the specific *Bohr atom,* which applies only to hydrogen and hydrogen-like ions. The Bohr model is based on the new concepts of stationary states, energy levels, and the emission and absorption of photons in transitions between stationary states. These are applicable to any atom, and they carry over into quantum mechanics. The Bohr atom, by contrast, is a specific model that establishes a procedure for finding the stationary states of one-electron atoms.

Thus our interest is more with the concepts of the Bohr model than with the specific details of the Bohr hydrogen atom. The Bohr model provides explanations for the stability of matter, for discrete spectra, for the origin of Einstein's light quanta, for why absorption spectra are simpler than emission spectra, for collisional excitation of atoms, and so on. These are very suggestive that a correct theory of atoms somehow will have to involve quanta.

From a pedagogical perspective, the Bohr hydrogen atom is best introduced not by angular momentum quantization but by requiring the electron's de Broglie wave to form a standing wave around the circumference of the classical orbit. Although not historically accurate, this approach has several strong advantages:

- It builds upon and extends the idea that a confined particle requires a de Broglie standing wave and that this leads to energy quantization.
- It makes explicit use of the idea of matter waves.
- It prepares students for thinking of bound-state wave functions as generalized standing waves that can exist only for certain discrete energies.

- It is more plausible than angular momentum quantization because students have already seen experimental evidence of matter waves.

Angular momentum quantization then follows as a *consequence* of requiring a standing wave.

As a historical note, Bohr derived the energy levels of the hydrogen atom from a rather round-about analysis using the correspondence principle, *not* by postulating the quantization of angular momentum. Angular momentum quantization was a *consequence* of his analysis, as it is of ours, not a postulate. Thus the standard textbook presentation that postulates angular momentum quantization is no more historically accurate than an approach based on the de Broglie wavelength.

There is a strong tendency in quantum physics for instructors who have been pursuing active learning to revert to lecture mode. It's true that there are fewer demonstrations and that the material gets more abstract. However, most students still need opportunities to practice basic reasoning with the concepts. In fact, the abstract nature of the concepts makes immediate feedback more important than ever. Instructors are urged to use a blend of focused mini-lectures, small-group exercises, and example problem solving (with student participation).

Using Class Time

Quantum physics is generally not a requirement of an introductory physics course, so there's a fair amount of flexibility in the time devoted to this material. I usually spend about 6 or 7 days. Demonstrations are often difficult, but there is still ample opportunity to have students actively involved in *reasoning* about the implications of historical experiments. This is a good opportunity to review and reinforce many topics from those covered earlier in the course.

The sequencing of topics varies from book to book. I like to start with spectroscopy, then follow a two-pronged approach. One path, through Balmer, Thompson, Millikan, and Rutherford, leads to atomic structure. But the Rutherford atom, for all its successes, is a dead end in the sense of still being a classical atom. The second path—Planck, Einstein, de Broglie—leads to quanta. These two routes converge on Bohr, who applied the principle of quantization to the Rutherford atom.

Although spectra seem quite obvious to experienced physicists, they are pretty mysterious to students. Students are not sure how they're created, what they represent, or what the significance is. Most departments have slide-mounted diffraction gratings that can be passed out to students to observe the line spectra of gas discharges. For comparison, be sure to also observe the continuous spectrum of a filament. Students need to recognize that atoms *create* the light in an emission spectrum. That each element has a unique spectrum must be telling us something about atoms—but what? The fact that the spectra are discrete is interesting, but until you get to a specific model of the atom, the Rutherford model, there's nothing to be gained by pointing out that classical atoms shouldn't have a discrete spectrum.

It is important to introduce the Balmer formula as an *empirical* result that very accurately describes the hydrogen spectrum. I find that most students don't really understand the idea of an "empirical result," so you need to point out very explicitly that the Balmer formula is not a "theory" of hydrogen and, in fact, could not be derived by any theory of the nineteenth century. So here's a puzzle, a mystery. It's a simple spectrum and a simple formula. Any decent theory of atoms ought to be able to *derive* the Balmer formula as a consequence of the theory. Can we meet the challenge?

The magnetic deflection of the electron beam in a Crookes tube should be demonstrated, even if students have already seen this during the chapter on magnetism. It's also nice to have a tube where students can see the green fluorescence on the face and see the "shadows" of objects. However, many of these require close viewing in a very dark room and are not easily shown in a large classroom. Later viewing during a recitation or laboratory may be possible and is recommended. A full crossed-field set up, if available and easily seen, is an especially effective demonstration.

Class exercises should ask students to predict how an experiment will turn out if cathode rays are waves, atomic-sized charged particles, or subatomic charged particles. It is important to recognize how hypotheses that seemed reasonable at the time, such as Crookes's hypothesis that cathode rays were negative ions, were ruled out by later experimental evidence.

Other important questions suitable for class discussion include

- What was the significance of the discovery that helium could be ionized by X rays?
- Why is gravitational deflection of the cathode rays not observed or considered? (Students who answer "Because the mass is too small" are forgetting that the acceleration due to gravity is independent of mass.)
- When a field is applied, the green spot deflects but remains well defined and does not smear out. What can you infer about the cathode ray particles from this observation?
- What is the significance of the fact that cathode rays from aluminum, iron, and lead cathodes all have the same charge-to-mass ratio?
- Given the discovery that the charge-to-mass ratio of cathode rays is about 1000 times that of hydrogen ions, why did Thompson assume that the two had equal charges rather than equal masses? What *experimental evidence* led to this choice?

It's surprisingly difficult for many students to employ deductive reasoning or to see how conclusions are inferred from evidence. These critical thinking exercises are good opportunities to practice basic scientific reasoning.

The Millikan oil drop experiment is historically interesting, but pedagogically less important. It is primarily a step on the road to the Rutherford model of the atom. Issues that give many students trouble, but that can be illustrated with the Rutherford model, are

- Positive and negative ions.
- Isotopes.
- The fact that ordinary chemical and atomic events involve the electrons but not the protons.

It's interesting to note that isotopes were discovered before the neutron was discovered, leading to an early model of the nucleus that contained $Z + N$ protons and N electrons.

Most students consider the Rutherford model to be the "obvious" model of an atom, and they wonder why Thompson and others didn't think of it. This may be due to the prevalence of the solar-system model in popular-science presentations of atoms. Thus it's interesting to have students focus on why the Rutherford model is *not* obvious and was initially resisted by many physicists. Although the full story is complex, the primary two objections were that

■ Physicists knew no mechanism or explanation of how the positive charges could be held in a nucleus against the overwhelming electrostatic repulsion.
■ Nuclear densities were inconceivable at the time.

The story thus far provides strong evidence that charged, subatomic particles are assembled into a nuclear atom. But the Rutherford model provided no explanation of atomic spectra, and this is the time to point that out. The missing piece, of course, is quantization.

I've already indicated that I consider the photoelectric effect to be the central topic of this introduction to quantum physics. I usually spend two full days on the photoelectric effect, which seems excessive, but I've found that it takes this long for most students to understand where the puzzle is and how the photon model is able to resolve the puzzle. This is time well spent if students can apply their understanding of quantization and photons to the Bohr model and discrete spectra.

You need to establish how the circuit of the photoelectric effect works before getting into the photoelectric effect itself. This is a good review of circuits and, especially, of the concept of potential difference. Begin with a simple battery and resistor, asking students for the potential difference $\Delta V_{ac} = V_a - V_c$ between points a and c. Then remove the resistor and again ask for ΔV_{ac}. Many students will have difficulty, as they did for similar exercises in the chapter on DC circuits, and you'll need to review the loop law reasoning that leads to $\Delta V_{ac} = \Delta V_{battery}$.

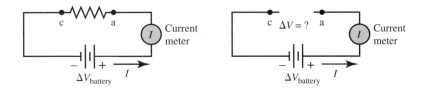

Put the resistor back in and ask students to draw a current-voltage graph. Before doing so, define a positive current for this circuit as counterclockwise and also note that "negative voltage" corresponds merely to reversing the battery. Have students articulate that an "ohmic response" is linear for both positive and negative voltages and passes through the origin. This graph shows how a *resistor* responds to a potential difference.

Then replace the resistor with a vacuum tube consisting of an anode and a thermionic cathode. You needn't explain thermal emission in detail, but you will need to introduce the idea that there's a minimum energy E_0—the work function—an electron must acquire to escape from the metal.

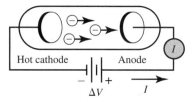

I use a swimming pool analogy. Water molecules are "bound" in the pool under ordinary circumstances, just as electrons are bound in a metal. To get a molecule out of the pool, you would need to do work on it to lift it up to the edge of the pool. The minimum energy is $E_0 = mgh$, where h is the distance from the water surface to the pool edge. Molecules near the surface would need only E_0 to escape, but deeper molecules would need more than this. So E_0 really is a *minimum* energy. On a windy day, when the water sloshes around, a few molecules may acquire enough energy to spontaneously escape from the pool.

Returning to the metal cathode, note that

- Electrons are bound in a metal at normal temperatures.
- There's a minimum energy E_0—the work function—an electron needs to escape.
- Heating the metal raises the kinetic energy of the electrons.
- At a sufficiently high temperature, some electrons have thermal energy $E_{th} > E_0$.
- These electrons escape and leave the cathode surface with kinetic energy $K \leq E_{th} - E_0$. They are emitted at all different angles.
- Because the electrons have a range of values of E_{th}, emitted electrons will have a range of values of kinetic energy K.
- This phenomena is entirely classical!

Before drawing the current-voltage diagram, have students consider electron trajectories for different values of ΔV. You want them to recognize (and articulate) that

- Positive ΔV accelerates electrons toward the anode. Essentially all electrons are collected by the anode for $\Delta V > 0$. Further increase of ΔV cannot increase the current.
- Negative ΔV pushes electrons away from the anode. However, electrons that started with enough energy can "get through." A good analogy is tossing balls toward the ceiling; all are slowing, but those that started with enough energy get there.
- There is no electric field if $\Delta V = 0$, so electrons move in straight lines. Those that were emitted toward the anode will reach it. Thus there *is* a current for $\Delta V = 0$.
- The potential difference $\Delta V = -V_{stop}$ that just stops the fastest electrons must, by energy conservation, be such that $eV_{stop} = K_{max}$. (I prefer to define the stopping voltage such that V_{stop} is a positive number, and the potential difference is then $\Delta V = -V_{stop}$.)

Although these ideas are stated in most textbooks, students must go through the reasoning themselves and articulate the ideas in their own words if it is to have meaning for them.

Now ask them to draw a current-voltage diagram. This will be difficult for most, even after the above discussion and exercises. To be specific, tell them the cathode emits 10^{16} electrons per second (a current of 1.6 mA) with a maximum kinetic energy of 1.0 eV. You may want to focus their attention on one piece of the graph at a time, dividing it into $\Delta V \geq 0$, $-V_{stop} \leq \Delta V < 0$, and $\Delta V < -V_{stop}$. Some students can draw the graph correctly for $\Delta V \geq -V_{stop}$, but they make errors if asked to extend their graph to the left of the stopping voltage. The I-V curve shown here is, of course, much oversimplified, but it conveys all the important points.

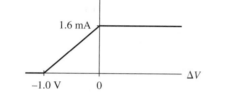

You'll need to inquire carefully to make sure everyone really understands the graph. Then congratulate them on having "discovered" a diode and note explicitly that it is a *non-ohmic* device. The voltage is still that provided by the battery, but the diode *responds* differently than a resistor. Note explicitly that the electrons are moving left-to-right, which corresponds to a positive current flowing right-to-left. Also note explicitly that current *never* flows left-to-right (negative current) for *any* value of ΔV.

To complete this exercise, ask them to redraw the graph after the temperature of the cathode is increased significantly. This may take a bit of discussion, but you want them to realize that both the number and the kinetic energy of the electrons increases. Thus the saturation portion of the graph goes up *and* the stopping voltage moves to the left.

These exercises may take half the class, but they are essential preparation for understanding the "surprises" of the photoelectric effect. You can demonstrate the photoelectric effect by clipping a recently sanded (to remove oxides) zinc plate to the knob on an electroscope, charging the electroscope negative, then shining an ultraviolet light on it. This quickly discharges the electroscope. A piece of glass (which absorbs UV) between the light and the electroscope prevents the discharge, so students can readily accept that short-wavelength radiation is somehow knocking electrons off the electroscope. The discharge doesn't occur if the zinc is removed, so the effect is dependent on the type of metal used.

An interesting question that will stump most students is, "If the UV light knocks electrons out of the metal, can I use the UV light to charge positively an electroscope that started out neutral?" After most agree that you can, show that *nothing happens* when you hold the UV light near an uncharged electroscope! So why not? A few students may realize, after some discussion, that as the electroscope becomes slightly positive it pulls back any electrons that are removed by the photoelectric effect. (A good review exercise is to calculate how much negative charge must be removed to give a 2 cm metal sphere a potential $V = +1$ V, enough to recapture electrons. The result, ≈ 0.001 nC, is ≈ 100 times less than needed to deflect the electroscope leaf.)

If you have a laboratory photocell you can show that the current is roughly constant for positive voltages and quickly drops to zero for negative voltages. Be sure to draw a picture to indicate that the light is incident only on the cathode, not the anode. Students should now recognize that the cathode is a source of electrons, just as in thermionic emission, and that the basic current-voltage graph will look the same.

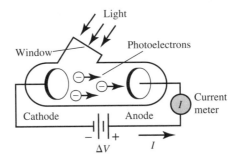

Because they've already given the basic circuit considerable thought, you can go fairly quickly through the main results of the experiment. In particular,

- Increasing the light intensity increases the current but *not* the stopping potential. This is in contrast to thermionic emission.
- There's a threshold frequency or wavelength that depends on the cathode material. As an exercise, tell them that 288 nm light shining on an aluminum cathode gives a 1 mA current for $\Delta V > 0$. What current do they *predict* if the wavelength is shifted slightly to 292 nm, but the intensity remains the same? You want them to recognize and articulate the classical expectation that this small change should make little or no difference in the current. But in fact, because the threshold for aluminum is 290 nm, the current abruptly drops to zero!
- Regardless of the light intensity, the current appears instantly with no delay.

The classical explanation of the photoelectric effect would be that light is absorbed, the electrons are gradually heated, and thermal emission occurs. Students need to articulate for themselves how the three major experimental results are inconsistent with this hypothesis.

You can expand this experimental discussion of the photoelectric effect by introducing the photon model. The three primary concepts of this model, not stated clearly in some textbooks, are

- Light of frequency f consists of individual, discrete quanta, each of energy $E = hf$. These are called photons.

- Photons are emitted or absorbed on an all-or-nothing basis.
- A photon, when absorbed by a metal, delivers its entire energy to a *single* electron. The light's energy is transformed into the electron's kinetic energy.

Note that this is a *hypothesis,* subject to experimental verification.

Most students find the "raindrop analogy" to be useful for thinking about photons. Each drop (photon) has a definite size (energy); the intensity of the rain (light) is determined by the number of drops (photons) per second; and for high intensities the sound (brightness) of the drops (photons) striking a surface seems continuous rather than discrete. You especially want them to recognize that the energy of a photon is independent of the intensity of the light. You'll also want to note explicitly that photons travel in *straight lines* at the speed of light!

The calculations associated with photons and the photoelectric effect are not difficult. Thus numerical exercises are of less importance than using class time to work out the *qualitative* implications of the photon model of the photoelectric effect. Continuing with the graphical exercises, draw a current-voltage graph and then ask students to show how it would look if

- The wavelength is changed.
- The intensity is changed.
- The cathode material is changed.

Have students use the photon model to articulate the *reasons* for the graph shapes.

Current-frequency graphs, such as the one shown below, are also useful. Have students redraw the graph if

- The intensity is increased.
- The work function is increased.
- The potential difference ΔV is increased.

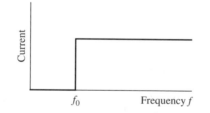

These exercises are time consuming, but the photoelectric effect is so important to the development of quantum physics that the time is well spent. At the conclusion of these exercises, most students do recognize how classical physics failed to explain the photoelectric effect and how the photon model succeeded. In other words, they now have significant *evidence* for why we believe that photons exist.

The photoelectric effect leads nicely into the de Broglie wavelength and the issue of wave-particle duality. De Broglie, after all, was simply asking the reverse question: If a wave can act like a particle, can a particle act like a wave? Unfortunately, few schools are equipped to make any kind of demonstration of this outrageous hypothesis. You'll probably have to rely on photographs of electron and neutron interference. If you have time, it's worth mentioning recent atom interferometer experiments in which atoms are the "waves" and a standing-wave laser beam is the "grating"—a complete reversal from the classical interferometer.

In addition to the usual calculations of electron wavelengths, an important application of de Broglie's ideas is to the one-dimensional particle-in-a-box. Classically, a particle bounces back and forth in the box. If the particle is wavelike, we need to think of this as two counterpropagating waves—exactly the conditions for a standing wave.

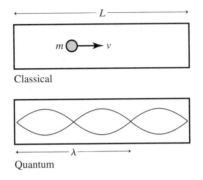

The standing wave requirement is

$$\lambda = \frac{h}{p} = \frac{2L}{n} \qquad n = 1, 2, 3, \ldots .$$

Combine this with $E = p^2/2m$, and you end up with

$$E_n = \left(\frac{h^2}{8mL^2}\right)n^2,$$

the standard particle-in-a-box result usually found from the one-dimensional Schrödinger equation. Not only does light have quantized energy, so does matter!

The important generalization you want students to recognize is that any *confined* particle has to have quantized energy. Confinement requires a standing wave, and students know that standing waves exist only for certain discrete wavelengths. So de Broglie's hypothesis is much more profound than simply allowing particles to exhibit interference. It provides a basis for recognizing that atomic energies have to be quantized.

Bohr united the atomic ideas and quantum ideas into the first successful quantum theory. Despite its limitations, it pointed the way to quantum mechanics a decade later. The important point, and one lost in most textbooks, is that Bohr proposed a *model* of a quantum system that is much more general than the specifics of the Bohr hydrogen atom. Although his quantization method failed, the *Bohr model* provides the conceptual foundation for quantum mechanics. The Bohr model, rather than the Bohr atom, should be the primary focus of teaching.

The Bohr model consists of six basic assumptions:

1. An atom consists of negative electrons orbiting a very small positive nucleus, as in the Rutherford model.

2. Atoms can exist only in certain *stationary states*. Each stationary state corresponds to a particular set of electron orbits around the nucleus. These states are discrete and can be numbered $n = 1, 2, 3, 4, \ldots$ Each stationary state has a unique, well-defined energy E_n, this being the energy of the orbiting electrons. That is, atomic energies are *quantized*.
3. The stationary states of an atom are numbered in order of increasing energy: $E_1 < E_2 < E_3 < E_4 < \ldots$. The lowest energy state of the atom, with energy E_1, is *stable* and can persist indefinitely. It is called the *ground state* of the atom.
4. An atom can "jump" from one stationary state to another, undergoing an energy change $\Delta E_{\text{atom}} = |E_f - E_i|$, by emitting or absorbing a photon of frequency

$$f_{\text{photon}} = \frac{\Delta E_{\text{atom}}}{h},$$

where h is Planck's constant. E_i and E_f are the energies of the initial and final states. Such a jump is called a *transition* or, sometimes, a *quantum jump*.
5. An atom can absorb energy in a collision with an electron or another atom, causing the electron to move from a lower energy state to a higher energy state. The energy absorbed has to be exactly $\Delta E = E_f - E_i$.
6. Atoms will seek the lowest energy state—the ground state. An atom in an excited state, if left alone, will jump to lower and lower energy states—emitting photons—until it reaches the ground state.

The implications are profound. The Bohr model explains the stability of matter, discrete spectra, the origin of photons, and why each element has a unique spectrum. It provides a mechanism for collisional excitation of energy levels in a discharge tube. With thermodynamics added, it can explain why absorption spectra differ from emission spectra.

The one thing the Bohr model does *not* do is provide a procedure for determining the energy levels of the stationary states. The Bohr *atom*, by contrast, is a specific quantization procedure. Its success with hydrogen made it clear that Bohr was on the right track, although it remained for quantum mechanics to provide a more general quantization procedure.

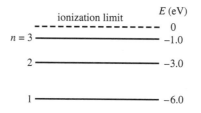

The three-level atom shown above provides a simple example where students can test drive many of the basic ideas of quanta and of atomic structure. This is the first time for most students to see an energy level diagram, so you'll need to explain what it represents. Many think that horizontal distances have a meaning, so you need to present the energy level diagram as yet another *pictorial model* of our knowledge of a situation. A series of questions and exercises based on this atom is

- Why are the energies negative?
- Which is the ground state? Which are excited states?
- What is meant by the *ionization limit?* What is the *ionization energy* of this atom?
- Can an atom in $n = 1$ emit a photon? If so, what is the wavelength? If not, why not?
- What wavelengths are seen in the emission spectrum of this atom? Are these infrared, visible, or ultraviolet wavelengths?
- What wavelengths are seen in the absorption spectrum of this atom?
- Why are the emission and the absorption spectra different?
- Why is light of wavelength 300 nm not absorbed by this atom?
- An electron with speed 1.2×10^6 m/s collides with the atom. What wavelength or wavelengths of light, if any, might be emitted by the atom after this collision? What is the electron's speed after the collision?

After students are comfortable with the idea of stationary states, you can pose the question of how we might determine the stationary states of an atom—specifically, the hydrogen atom. Classical physics—the particle view—requires the electrostatic force to be a centripetal force. Thus

$$\frac{e^2}{4\pi\varepsilon_0 r^2} = \frac{mv^2}{r}.$$

The new piece of information is that the electron also acts like a wave. The classical orbit is a one-dimensional motion "bent" into a circle, so this is much like the particle-in-a-box. A standing de Broglie wave around the circumference gives

$$\text{circumference} = 2\pi r = n\lambda = n\frac{h}{mv}.$$

Requiring the electron to act like both a particle and a wave means that it must satisfy these two simultaneous conditions. A little algebra (you might leave the details for a homework problem) leads to the standard results for the energies and radii of the stationary-state orbits. You'll need to spend more time *interpreting* the results than deriving them.

You'll want to go over the energy level diagram, the idea of an ionization limit, and the Lyman and Balmer series. This is all very basic, but it's new to students and they need to go through the reasoning for themselves. You'll probably want to touch briefly on the quantization of angular momentum, but it isn't a major result when the Bohr atom is approached this way. The few students who go on to take quantum mechanics can learn of the importance of angular momentum at that time.

Deriving the Balmer formula is the crowning result, showing that these crazy ideas about photons and matter waves succeed where the "obvious" ideas of classical physics failed. You'll want to mention the limitations of the Bohr atom, and give an advertisement for your quantum mechanics course for students who want to know the next step. But all in all, this brings introductory physics to a most satisfying close.

Sample Reading Quiz or Discussion Questions

1. What kind of drops did Millikan use to measure the fundamental unit of charge?

2. Cite one experimental discovery of the late nineteenth or early twentieth century that could *not* be understood on the basis of the laws of physics as they were known at the time.

3. Who first postulated the idea of light quanta?

4. What is a *quantum jump?*

5. What is the name of the diagram used to represent the stationary states of an atom?

6. The *electron volt* is a unit of
   a. charge.
   b. energy.
   c. potential.
   d. electric field.
   e. atomic power.
   f. atomic size.

7. The discoverer of the electron was
   a. Edison.
   b. Einstein.
   c. Faraday.
   d. Millikan.
   e. Rutherford.
   f. Thompson.

8. In the photoelectric effect experiment, current flows when the light frequency is
   a. less than the threshold frequency.
   b. equal to the threshold frequency.
   c. greater than the threshold frequency.
   d. less than the cathode's work function.
   e. equal to the cathode's work function.
   f. greater than the cathode's work function.

9. The minimum amount of energy needed to free an electron from a piece of metal is called
   a. Gibb's free energy.
   b. the liberation potential.
   c. the work function.
   d. quantum energy.
   e. threshold energy.

10. Which of these quantities are quantized in the Bohr hydrogen atom?
    a. Energy.
    b. The electron's speed.
    c. The orbital radius.
    d. Angular momentum.
    e. Both a and d.
    f. All of a–d.

11. Bohr successfully explained the spectrum of
    a. hydrogen.
    b. helium.
    c. hydrogen and helium.
    d. all the naturally occurring elements.
    e. all the elements in the periodic table.
    f. No spectra were successfully explained.

Sample Exam Questions

1. Cite two pieces of experimental evidence by which we know that atoms contain charged particles. Be specific! Don't just say "X's experiment," but say *how* X's experiment allows us to conclude something about the structure of atoms.

2. Draw a picture of a hydrogen-like ^{11}B ion, showing all the electrons, protons, and neutrons. Boron has $Z = 5$.

3. A 20 MeV alpha particle is fired straight toward a lead nucleus ($Z = 82$, $A = 207$). What is the distance between their centers at the instant of closest approach? You can assume that the lead nucleus remains stationary.

4. It is said that the photoelectric effect demonstrates the particle-like nature of light. Explain how this conclusion is reached. That is, what *experimental evidence* is consistent with particle-like behavior for light but not with wave-like behavior? Cite at least two pieces of evidence.

5. The graph below shows the current in a photoelectric effect experiment as a function of the anode-cathode potential difference ΔV. The light shining on the cathode has a wavelength of 178 nm. Some possibly relevant data about different metals:

Metal	Resistivity ρ (Ωm)	Work function E_0 (eV)
aluminum	2.8×10^{-8}	4.3
copper	1.7×10^{-8}	4.7
gold	2.4×10^{-8}	5.1
sodium	2.0×10^{-8}	2.7

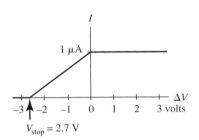

a. Of what metal is the cathode made?
b. Suppose that each photon striking the cathode has a 60% probability of ejecting a photoelectron. How many photons are incident on the cathode each second?
c. How many watts of 178 nm light power are incident on the cathode?

6. An atom has the following wavelengths in its emission and absorption spectra:
       Emission (nm):   207, 249, 355, 497, 829, and 1243.
       Absorption (nm):   207, 249, and 355.
Separate experiments show that the atom has an ionization energy of 8.0 eV.
a. Draw an energy-level diagram for the atom. Identify each energy level by its quantum number and its energy.
b. Identify the transition (such as $3 \rightarrow 2$) of each wavelength in the emission spectrum.

7. The first three energy levels of the fictitious element X are shown in the figure.

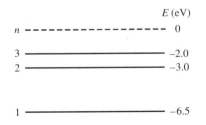

    a. What is the ionization energy of element X?

    b. What wavelengths are observed in the absorption spectrum of element X? Give your answers in nm.

    c. State whether each of your wavelengths in part b corresponds to ultraviolet, visible, or infrared light.

    d. An electron with a speed of 1.4×10^6 m/s collides with an atom of element X. Shortly afterward, the atom emits a 1240 nm photon. What was the electron's speed after the collision? Assume, because the atom is so much more massive than the electron, that the recoil of the atom is negligible.

8. Electrons with a speed of 2.0×10^6 m/s pass through a double-slit apparatus. A detector measures interference fringes with a fringe spacing of 1.5 mm.

    a. What will the fringe spacing be if the electrons are replaced by neutrons having the same speed?

    b. At what speed will neutrons produce interference fringes with a spacing of 1.5 mm?

9. A particle in a rigid one-dimensional box of width 10 fm has an energy level $E_n = 32.9$ MeV and an adjacent energy level $E_{n+1} = 51.4$ MeV.

    a. Determine the values of n and $n + 1$.

    b. What is the wavelength of a photon emitted in the $n + 1 \rightarrow n$ transition? Compare this to a typical visible-light wavelength.

    c. What is the mass of the particle? Can you identify it?

References
and Resources

References

This bibliography is a guide to the literature on physics education research and active learning. It is not intended to be exhaustive, but rather a starting point to locate some of the books and papers that have influenced this guidebook. References marked with an * are those I consider to be good entry points into the literature. Journal citations are limited primarily to *The American Journal of Physics, Physics Today,* and *The Physics Teacher* because most instructors will have ready access to these.

For those inclined to get deeper into the subject, the McDermott and Redish (1999) resource letter provides an extensive bibliography with short annotations for each citation. The references cited in Hammer (1996) and Redish (1994) are also a good place to start.

Arons's *Teaching Introductory Physics* (or his earlier *A Guide to Introductory Physics Teaching*) is a resource that no physics instructor should be without. It is an excellent introduction to the issues of physics education research. A more general book on cognitive science, which may be of interest to some instructors, is Howard Gardner's *The Mind's New Science: A History of the Cognitive Revolution* (1987). Sheila Tobias provides an insightful look at science education in *They're Not Dumb, They're Different* (1990).

B. S. Ambrose, P. S. Shaffer, R. N. Steinberg, and L. C. McDermott, "An investigation of student understanding of single-slit diffraction and double-slit interference," *Am. J. Phys.* **67,** 146–155 (1999).

A. B. Arons, "Cognitive level of college physics students," *Am. J. Phys.* **47,** 650–651 (1979).

A. B. Arons, "Phenomenology and logical reasoning in introductory physics courses," *Am. J. Phys.* **50,** 13–20 (1982).

A. B. Arons, "Student patterns of thinking and reasoning," Part 1: *The Phys. Teach.* **21,** 576–581 (1983); Parts 2 and 3: *The Phys. Teach.* **22,** 21–26 and 88–93 (1984).

* A. B. Arons, *A Guide to Introductory Physics Teaching,* John Wiley & Sons, New York, 1990.

A. B. Arons, *Teaching Introductory Physics,* John Wiley & Sons, New York, 1997.

* E. Bagno and B.-S. Eylon, "From problem solving to a knowledge structure: An example from the domain of electromagnetism," *Am. J. Phys.* **65,** 726–736 (1997).

J. Clement, "Students' preconceptions in introductory mechanics," *Am. J. Phys.* **50,** 66–71 (1982).

R. Cohen, B. Eylon, and U. Ganiel, "Potential difference and current in simple electric circuits: A study of students' concepts," *Am. J. Phys.* **51,** 407–412 (1983).

D. F. Dykstra, C. F. Boyle, and I. A. Monarch, "Studying conceptual change in learning physics," *Science Education* **76,** 615–652 (1992).

J. Evans, "Teaching electricity with batteries and bulbs," *The Phys. Teach.* **15,** 15–22 (1978).

H. Gardner, *The Mind's New Science: A History of the Cognitive Revolution,* Basic Books, New York, 1987.

* F. M. Goldberg and L. C. McDermott, "An investigation of student understanding of the real image formed by a converging lens or a concave mirror," *Am. J. Phys.* **55,** 108–119 (1987).

R. R. Hake, "Promoting student crossover to the Newtonian world," *Am. J. Phys.* **55,** 878–884 (1987).

R. R. Hake, "Socratic pedagogy in the introductory physics laboratory," *The Phys. Teach.* **30,** 546–552 (1992).

R. R. Hake, "Interactive-engagement vs traditional methods: A six-thousand student survey of mechanics test data for introductory physics courses," *Am. J. Phys.* **66,** 64–74 (1998).

I. A. Halloun and D. Hestenes, "The initial knowledge state of college physics students," *Am. J. Phys.* **53,** 1043–1055 (1985a).

* I. A. Halloun and D. Hestenes, "Common-sense concepts about motion," *Am. J. Phys.* **53,** 1056–1065 (1985b).

D. Hammer, "More than misconceptions: Multiple perspectives on student knowledge and reasoning, and an appropriate role for education research," *Am. J. Phys.* **64,** 1316–1325 (1996).

P. Heller, R. Keith, and S. Anderson, "Teaching problem solving through cooperative grouping. Part 1: Group versus individual problem solving," *Am. J. Phys.* **60,** 627–636 (1992).

P. Heller and M. Hollabaugh, "Teaching problem solving through cooperative grouping. Part 2: Designing problems and structuring groups," *Am. J. Phys.* **60,** 637–644 (1992).

* D. Hestenes, "Toward a modeling theory of physics instruction," *Am. J. Phys.* **55,** 440–454 (1987).

* D. Hestenes, M. Wells, and G. Swackhamer, "Force Concept Inventory," *The Phys. Teach.* **30,** 141–158 (1992a).

D. Hestenes and M. Wells, "A mechanics baseline test," *The Phys. Teach.* **30,** 159–166 (1992b).

R. Knight, "The Vector Knowledge of Beginning Physics Students," *The Phys. Teach.* **33,** 74–80 (1995).

* P. Laws, "Calculus-based physics without lectures," *Physics Today* **44**(12), 23–31 (December 1991).

* P. Laws, "Promoting active learning based on physics education research in introductory physics courses," *Am. J. Phys.* **65,** 13–21 (1997a).

P. Laws, *Workshop Physics Activity Guide,* John Wiley & Sons, New York, 1997b.

R. Lawson and L. C. McDermott, "Student understanding of the work-energy and impulse-momentum theorems," *Am. J. Phys.* **55,** 811–817 (1987).

A. J. Mallinckrodt and H. Leff, "All about work," *Am. J. Phys.* **60,** 356–365 (1992).

E. Mazur, *Peer Instruction: A User's Manual,* Prentice-Hall, Upper Saddle River, NJ, 1997.

* L. C. McDermott, "Research on conceptual understanding in mechanics," *Physics Today* **37**(7), 24–32 (July 1984).

L. C. McDermott, M. L. Rosenquist, and E. H. van Zee, "Student difficulties in connecting graphs and physics: Examples from kinematics," *Am. J. Phys.* **55,** 503–513 (1987). This article is paired with Rosenquist and McDermott (1987).

* L. C. McDermott, "What we teach and what is learned—closing the gap," *Am. J. Phys.* **59,** 301–315 (1991).

L. C. McDermott and P. S. Shaffer, "Research as a guide for curriculum development: An example from introductory electricity. Part I: Investigation of student understanding," *Am. J. Phys.* **60,** 994–1003 (1992). See Shaffer and McDermott (1992) for Part II.

L. C. McDermott, P. S. Shaffer, and M. D. Somers, "Research as a guide to teaching introductory mechanics: An illustration in the context of the Atwood's machine," *Am. J. Phys.* **62,** 46–55 (1994).

L. C. McDermott, P. S. Shaffer, and the Physics Education Group, *Tutorials in Introductory Physics,* Prentice-Hall, Upper Saddle River, NJ, 1998.

* L. C. McDermott and E. F. Redish, "Resource Letter: PER–1: Physics Education Research," *Am. J. Phys.* **67,** 755–767 (1999).

D. E. Meltzer and K. Manivannan, "Promoting Interactivity in Physics Lecture Classes," *The Phys. Teach.* **34,** 72–76 (1996).

J. Mestre and J. Touger, "Cognitive research—what's in it for physics teachers?" *The Phys. Teach.* **27,** 447–456 (1989).

T. O'Kuma, D. P. Maloney, and C. J. Hieggelke, *Ranking Task Exercises in Physics,* Prentice-Hall, Upper Saddle River, NJ, 2000.

* E. F. Redish, "Implications of cognitive studies for teaching physics," *Am. J. Phys.* **62,** 796–803 (1994).

F. Reif, "Understanding and teaching important scientific thought processes," *Am. J. Phys.* **63,** 17–32 (1995).

F. Reif and J. I. Heller, "Knowledge structure and problem solving in physics," *Educational Psychologist* **17,** 102–127 (1982).

M. L. Rosenquist and L. C. McDermott, "A conceptual approach to teaching kinematics," *Am. J. Phys.* **55,** 407–415 (1987). This article is paired with McDermott et al, (1987).

P. S. Shaffer and L. C. McDermott, "Research as a guide for curriculum development: An example from introductory electricity. Part II: Design of instructional strategies," *Am. J. Phys.* **60,** 1003–1013 (1992). See McDermott and Shaffer (1992) for Part I.

D. R. Sokoloff and R. K. Thornton, "Using interactive lecture demonstrations to create an active learning environment," *The Phys. Teach.* **35,** 340–347 (1997).

D. R. Sokoloff, R. K. Thornton, and P. W. Laws, *RealTime Physics,* John Wiley & Sons, New York, 1999.

R. N. Steinberg, G. E. Oberem, and L. C. McDermott, "Development of a computer-based tutorial on the photoelectric effect," *Am. J. Phys.* **64,** 1370–1379 (1996).

* R. K. Thornton and D. R. Sokoloff, "Learning motion concepts using real-time microcomputer-based tools," *Am. J. Phys.* **58**, 858–867 (1990).

R. K. Thornton and D. R. Sokoloff, "Assessing student learning of Newton's laws: The force and motion conceptual evaluation and the evaluation of active learning laboratory and lecture curricula," *Am. J. Phys.* **66**, 338–346 (1998).

S. Tobias, *They're Not Dumb, They're Different,* Research Corporation, Tucson, AZ, 1990.

D. E. Trowbridge and L. C. McDermott, "Investigation of student understanding of the concept of velocity in one dimension," *Am. J. Phys.* **48**, 1020–1028 (1980); "Investigation of student understanding of the concept of acceleration in one dimension," *Am. J. Phys.* **49**, 242–253 (1981).

* A. Van Heuvelen, "Learning to think like a physicist: A review of research-based instructional strategies," *Am. J. Phys.* **59**, 891–897 (1991a).

A. Van Heuvelen, "Overview, Case Study Physics," *Am. J. Phys.* **59**, 898–907 (1991b).

A. Van Heuvelen, *ALPS: Mechanics (Vol. 1) and Electricity and Magnetism (Vol. 2),* Hayden-McNeil Publishing, Plymouth, MI, 1994.

A. Van Heuvelen and D. P. Maloney, "Playing Physics Jeopardy," *Am. J. Phys.* **67**, 252–256 (1999).

A. Van Heuvelen and Xueli Zou, "Multiple representations of work-energy processes," *Am. J. Phys.* **69**, 184–194 (2001).

J. Wilson, "The CUPLE Physics Studio," *The Phys. Teach.* **32**, 518 (1994).

* M. C. Wittman, R. N. Steinberg, and E. F. Redish, "Making sense of how students make sense of mechanical waves," *The Phys. Teach.* **37**, 15–21 (1999).

K. Wosilait, P. R. L. Heron, P. S. Shaffer, and L. C. McDermott, "Development and assessment of a research-based tutorial on light and shadow," *Am. J. Phys.* **66**, 906–913 (1998).

Resources

Research-Based Physics Textbooks

Ruth Chabay and Bruce Sherwood, *Matter and Interactions* and *Electric and Magnetic Interactions,* John Wiley & Sons (2002), www.wiley.com/college/.

Randall Knight, *Physics: A Contemporary Perspective,* Preliminary Edition, and *Physics: A Contemporary Perspective Student Workbook,* Addison Wesley (1997), www.awlonline.com/physics/. (Forthcoming in 2004 as *Physics for Scientists and Engineers: A Strategic Approach,* First Edition)

Thomas Moore, *Six Ideas That Shaped Physics,* McGraw-Hill (2003), www.mhhe.com/catalogs/sem/physics/.

Frederick Reif, *Understanding Basic Mechanics,* John Wiley & Sons (1995) www.wiley.com/college/.

Lillian McDermott and the Physics Education Group at the University of Washington, *Physics by Inquiry,* John Wiley & Sons (1996), www.wiley.com/college/.

Hardware and Software

PASCO Scientific
10101 Foothills Blvd.
Roseville, CA 95678-9011
www.pascoscientific.com

Physics Academic Software
Box 8202
North Carolina State University
Raleigh, NC 27695-8202
www.aip.org/pas

Vernier Software
8565 S.W. Beaverton-Hillsdale Highway
Portland, OR 97225-2429
www.vernier.com

AAPT Physical Science Resource Center

www.psrc-online.org

Notes

Notes

Notes

Notes

Notes

Notes

Notes

Notes